Mathematica® for Theoretical Physics

T0156126

Mathematica®

for Theoretical Physics

Classical Mechanics
and Nonlinear Dynamics

Second Edition

Gerd Baumann

Gerd Baumann
Department of Mathematics
German University in Cairo GUC
New Cairo City
Main Entrance of Al Tagamoa Al Khames
Egypt
Gerd.Baumann@GUC.edu.eg

This is a translated, expanded, and updated version of the original German version of the work "*Mathematica®* in der Theoretischen Physik," published by Springer-Verlag Heidelberg, 1993 ©.

Library of Congress Cataloging-in-Publication Data
Baumann, Gerd.
    [Mathematica in der theoretischen Physik. English]
    Mathematica for theoretical physics / by Gerd Baumann.—2nd ed.
        p. cm.
    Includes bibliographical references and index.
    Contents: 1. Classical mechanics and nonlinear dynamics — 2. Electrodynamics, quantum mechanics, general relativity, and fractals.
    ISBN 0-387-01674-0
        1. Mathematical physics—Data processing.   2. Mathematica (Computer file)   I. Title.

QC20.7.E4B3813   2004
530'.285'53—dc22   2004046861

ISBN   1-4899-9323-1   ISBN 0-387-25113-8 (eBook)   Printed on acid-free paper.
ISBN   978-1-4899-9323-6

Additional material to this book can be downloaded from http://extras.springer.com

9 8 7 6 5 4 3 2 1

To Carin,
for her love, support, and encuragement.

Preface

As physicists, mathematicians or engineers, we are all involved with mathematical calculations in our everyday work. Most of the laborious, complicated, and time-consuming calculations have to be done over and over again if we want to check the validity of our assumptions and derive new phenomena from changing models. Even in the age of computers, we often use paper and pencil to do our calculations. However, computer programs like *Mathematica* have revolutionized our working methods. *Mathematica* not only supports popular numerical calculations but also enables us to do exact analytical calculations by computer. Once we know the analytical representations of physical phenomena, we are able to use *Mathematica* to create graphical representations of these relations. Days of calculations by hand have shrunk to minutes by using *Mathematica*. Results can be verified within a few seconds, a task that took hours if not days in the past.

The present text uses *Mathematica* as a tool to discuss and to solve examples from physics. The intention of this book is to demonstrate the usefulness of *Mathematica* in everyday applications. We will not give a complete description of its syntax but demonstrate by examples the use of its language. In particular, we show how this modern tool is used to solve classical problems.

This second edition of *Mathematica in Theoretical Physics* seeks to prevent the objectives and emphasis of the previous edition. It is extended to include a full course in classical mechanics, new examples in quantum mechanics, and measurement methods for fractals. In addition, there is an extension of the fractal's chapter by a fractional calculus. The additional material and examples enlarged the text so much that we decided to divide the book in two volumes. The first volume covers classical mechanics and nonlinear dynamics. The second volume starts with electrodynamics, adds quantum mechanics and general relativity, and ends with fractals. Because of the inclusion of new materials, it was necessary to restructure the text. The main differences are concerned with the chapter on nonlinear dynamics. This chapter discusses mainly classical field theory and, thus, it was appropriate to locate it in line with the classical mechanics chapter.

The text contains a large number of examples that are solvable using *Mathematica*. The defined functions and packages are available on CD accompanying each of the two volumes. The names of the files on the CD carry the names of their respective chapters. Chapter 1 comments on the basic properties of *Mathematica* using examples from different fields of physics. Chapter 2 demonstrates the use of *Mathematica* in a step-by-step procedure applied to mechanical problems. Chapter 2 contains a one-term lecture in mechanics. It starts with the basic definitions, goes on with Newton's mechanics, discusses the Lagrange and Hamilton representation of mechanics, and ends with the rigid body motion. We show how *Mathematica* is used to simplify our work and to support and derive solutions for specific problems. In Chapter 3, we examine nonlinear phenomena of the Korteweg–de Vries equation. We demonstrate that *Mathematica* is an appropriate tool to derive numerical and analytical solutions even for nonlinear equations of motion. The second volume starts with Chapter 4, discussing problems of electrostatics and the motion of ions in an electromagnetic field. We further introduce *Mathematica* functions that are closely related to the theoretical considerations of the selected problems. In Chapter 5, we discuss problems of quantum mechanics. We examine the dynamics of a free particle by the example of the time-dependent Schrödinger equation and study one-dimensional eigenvalue problems using the analytic and numeric capabilities of *Mathematica*. Problems of general relativity are discussed in Chapter 6. Most standard books on Einstein's theory discuss the phenomena of general relativity by using approximations. With *Mathematica*, general relativity effects like the shift of the perihelion can be tracked with precision. Finally, the last chapter, Chapter 7, uses computer algebra to represent fractals and gives an introduction to the spatial renormalization theory. In addition, we present the basics of fractional calculus approaching fractals from the analytic side. This approach is supported by a package, FractionalCalculus, which is not included in this project. The package is available by request from the author. Exercises with which *Mathematica* can be used for modified applications. Chapters 2–7 include at the end some exercises allowing the reader to carry out his own experiments with the book.

Acknowledgments Since the first printing of this text, many people made valuable contributions and gave excellent input. Because the number of responses are so numerous, I give my thanks to all who contributed by remarks and enhancements to the text. Concerning the historical pictures used in the text, I acknowledge the support of the http://www-gapdcs.st-and.ac.uk/~history/ webserver of the University of St Andrews, Scotland. My special thanks go to Norbert Südland, who made the package FractionalCalculus available for this text. I'm also indebted to Hans Kölsch and Virginia Lipscy, Springer-Verlag New York Physics editorial. Finally, the author deeply appreciates the understanding and support of his wife, Carin, and daughter, Andrea, during the preparation of the book.

Ulm, Winter 2004

Gerd Baumann

Contents

Volume II

1
Introduction

This first chapter introduces some basic information on the computer algebra system *Mathematica*. We will discuss the capabilities and the scope of *Mathematica*. Some simple examples demonstrate how *Mathematica* is used to solve problems by using a computer.

All of the following sections contain theoretical background information on the problem and a *Mathematica* realization. The combination of both the classical and the computer algebra approach are given to allow a comparison between the traditional solution of problems with pencil and paper and the new approach by a computer algebra system.

1.1 Basics

Mathematica is a computer algebra system which allows the following calculations:

- symbolic

- numeric

- graphical

- acoustic.

Mathematica was developed by Stephen Wolfram in the 1980s and is now available for more than 15 years on a large number of computers for different operating systems (PC, HP, SGI, SUN, NeXT, VAX, etc.).

The real strength of *Mathematica* is the capability of creating customized applications by using its interactive definitions in a notebook. This capability allows us to solve physical and engineering problems directly on the computer. Before discussing the solution steps for several problems of theoretical physics, we will present a short overview of the organization of *Mathematica*.

1.1.1 Structure of *Mathematica*

Mathematica and its parts consist of five main components (see figure 1.1.1):

- the *kernel*

- the *frontend*

- the standard *Mathematica* packages

- the *MathSource* library

- the programs written by the user.

The *kernel* is the main engine of the system containing all of the functions defined in *Mathematica*. The *frontend* is the part of the *Mathematica* system serving as the channel on which a user communicates with the kernel. All components interact in a certain way with the kernel of *Mathematica*.

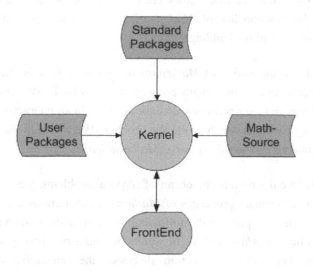

Figure 1.1.1.   *Mathematica* system

The *kernel* itself consists of more than 1800 functions available after the initialization of *Mathematica*. The kernel manages calculations such as symbolic differentiations, symbolic integrations, graphical representations, evaluations of series and sums, and so forth.

The standard packages delivered with *Mathematica* contain a mathematical collection of special topics in mathematics. The contents of the packages range from vector analysis, statistics, algebra, to graphics and so forth. A detailed description is contained in the technical report Guide to Standard *Mathematica* Packages [1.4] published by Wolfram Research Inc.

MathSource is another source of *Mathematica* packages. *MathSource* consists of a collection of packages and notebooks created by *Mathematica* users for special purposes. For example, there are calculations of Feynman diagrams in high-energy physics and Lie symmetries in the solution theory of partial differential equations. *MathSource* is available on the Internet via http://library.wolfram.com/infocenter/MathSource/.

The last part of the *Mathematica* environment is created by each individual user. *Mathematica* allows each user to define new functions extending the functionality of *Mathematica* itself. The present book belongs to this part of the building blocks.

The goal of our application of *Mathematica* is to show how problems of physics, mathematics, and engineering can be solved. We use this computer program to support our calculations either in an interactive form or by creating packages which tackle the problem. We also show how non standard problems can be solved using *Mathematica*.

However, before diving into the ocean of physical problems, we will first discuss some elementary properties of *Mathematica* that are useful for the solutions of our examples. In the following, we give a short overview of the capabilities of *Mathematica* in symbolic, numeric, and graphical calculations. The following, section discusses the interactive use of *Mathematica*.

1.1.2 Interactive Use of *Mathematica*

Mathematica employs a very simple and logical syntax. All functions are accessible by their full names describing the mathematical purpose of the function. The first letter of each name is capitalized. For example, if we wish to terminate our calculations and exit the *Mathematica* environment, we type the termination function **Quit[]**. This function disconnects the kernel from the frontend and deletes all information about our calculations.

Any function under *Mathematica* can be accessed by its name followed by a pair of square brackets which contain the arguments of the respective function. An example would be **Plot[Sin[x],{x,0,π}]**. The termination function **Quit[]** is the one of the few functions that lacks an argument.

After activating *Mathematica* on the computer by typing *math* for the interactive version or *mathematica* for the notebook version, or using just a double click on the *Mathematica* icon, we can immediately go to work. Let us assume that we need to calculate the ratio of two integer numbers. To get the result, we simply type in the expression and press Return in the interactive or Shift plus Return in the notebook version. The result is a simplified expression of the rational number.

69 / 15
$\dfrac{23}{5}$

The input and output lines of *Mathematica* carry labels counting the number of inputs and outputs in a session. The input label is **In[no]:=** and the related output label is **Out[no]=**. Another example is the exponentiation of a number. Type in and you will get

2^10
1024

The two-dimensional representation of this input can be created by using *Mathematica* palettes or by keyboard shortcuts. For example, an exponent is generated by *CTRL+6* on your keyboard

2^{10}
1024

Multiplication of two numbers can be done in two ways. In this book, the multiplication sign is replaced by a blank:

2 5
10

You can also use a star to denote multiplication:

```
2 * 5
```

```
10
```

In addition to basic operations such as addition (+), multiplication (*), division (/), subtraction (-), and exponentiation (^), *Mathematica* knows a large number of mathematical functions, including the trigonometric functions **Sin[]** and **Cos[]**, the hyperbolic functions **Cosh[]** and **Sinh[]**, and many others. All available *Mathematica* functions are listed in the handbook by Stephen Wolfram [1.1]. Almost all functions listed in the work by Abramowitz and Stegun [1.2] are also available in *Mathematica*.

1.1.3 Symbolic Calculations

By symbolic calculations we mean the manipulation of expressions using the rules of algebra and calculus. The following examples give a quick idea of how to use *Mathematica*. We will use some of the following functions in the remainder of this book.

A function consists of a name and several arguments enclosed in square brackets. The arguments are separated by commas. One function frequently used in the solution process is the function **Solve[]**. **Solve[]** needs two arguments: the equation to be solved and the variable for which the equation is solved. For each *Mathematica* function, you will find a short description of its functionality and its purpose if you type the name of the function preceded by a question mark. For example, the description of **Solve[]** is

```
? Solve
```

```
Solve[eqns, vars] attempts to solve an equation or set
    of equations for the variables vars. Solve[eqns,
    vars, elims] attempts to solve the equations
    for vars, eliminating the variables elims. More…
```

A hyperlink to the *Mathematica* help browser is available via the link on More…. If you click on the hyperlink, the help browser of *Mathematica* pops up and delivers a detailed description of the function. Each help page contains additional examples demonstrating the application of the function.

The help facility of *Mathematica* **?** or **??** always gives us a short description of any function contained in the *kernel*. For a detailed description of the functionality, the reader should consult the book by Wolfram [1.1].

Let us start with an example using **Solve[]** applied to a quadratic equation in *t*:

```
Solve[t² - t + a == 0, t]
```

$$\left\{\left\{t \to \frac{1}{2}\left(1 - \sqrt{1 - 4a}\right)\right\}, \left\{t \to \frac{1}{2}\left(1 + \sqrt{1 - 4a}\right)\right\}\right\}$$

It is obvious that the result is identical with the well-known solutions following from the standard solution procedure of algebra.

Next, let us differentiate a function with one independent variable. The differential is calculated by using the derivative symbol $\partial_\square$, which is equivalent to the derivative function **D[]**. Both functions are used for ordinary and partial differentiation:

```
∂t Sin[t]
```

$\cos(t)$

The inverse operation to a differentiation is integration. Integration of a function is executed by

---

```
Integrate[tᵃ, t]
```

$$\frac{t^{a+1}}{a+1}$$

The same calculation is carried out by the symbolic notation in the StandardForm:

$$\int t^a \, dt$$

$$\frac{t^{a+1}}{a+1}$$

Mathematica allows different kinds of input style. The first or input notation is given by the spelled out mathematical name. The second standard form is a two-dimensional symbolic representation. The third way to input expressions is traditional mathematical forms. The integral from above then looks like

$$\int t^a \, dt$$

$$\frac{t^{a+1}}{a+1}$$

Each input form has its pro and con. The spelled out input form is always compatible with the upgrading of *Mathematica*. The traditional form has some features which prevents the compatibility but increases the readability of a mathematical text. In the following, we will mix the different input forms and choose that one which is appropriate for the representation. For interactive calculations, we use the standard or traditional form; for programming, we switch to input notations. The different representations are also available in the output expressions. They can be controlled by the Cell button in the command menu of *Mathematica*.

Next, let us examine some operations from calculus. The calculation of a limit is given by

$$\text{Limit}\left[\frac{\text{Sin}[t]}{t}, \ t \rightarrow 0\right]$$

1

The expansion of a function $f(t)$ in a Taylor series around $t = 0$ up to third order is given by

$$\text{Series}[f[t], \{t, 0, 3\}]$$

$$f(0) + f'(0)\, t + \frac{1}{2}\, f''(0)\, t^2 + \frac{1}{6}\, f^{(3)}(0)\, t^3 + O(t^4)$$

The calculation of a finite sum follows from

$$\sum_{n=1}^{10}\left(\frac{1}{2}\right)^n$$

$$\frac{1023}{1024}$$

The result of this calculation is represented by a rational number. *Mathematica* is designed in such a way that the calculation results are primarily given by rational numbers. This kind of number representation allows a high accuracy in the representation of results. For example, we encounter no rounding errors when using rational representations of numbers.

The Laplace transform of the function **Sin[t]** is calculated using the standard function **LaplaceTransform[]**:

$$\text{LaplaceTransform}[\text{Sin}[t], t, s]$$

$$\frac{1}{s^2 + 1}$$

Ordinary and some kind of partial differential equations can be solved using the function **DSolve[]**. A practical example is given by the relaxation equation $u' + \alpha u = 0$. The solution of this equation follows from

```
DSolve[∂ₜu[t] + αu[t] == 0, u, t]
```

$\{\{u \rightarrow \text{Function}[\{t\}, e^{-t\alpha} c_1]\}\}$

In addition to the standard functions, *Mathematica* allows one to incorporate standard packages dealing with special mathematical tasks (see Figure 1.1.1). To load such standard packages, we need to carry out the **Get[]** function abbreviated by << followed by the package name. Such a standard package is available for the purpose of vector analysis. Calculations of vector analysis can be supported using the standard package **VectorAnalysis**, which contains useful functions for cross-products of vectors as well as for calculating gradients of scalar functions. Some examples of this kind of calculation follow:

```
<< Calculus`VectorAnalysis`
```

```
CrossProduct[{a, b, c}, {d, e, f}]
```

$\{bf - ce, cd - af, ae - bd\}$

A more readable representation is gained by applying the function **MatrixForm[]** to the result:

```
CrossProduct[{a, b, c}, {d, e, f}] // MatrixForm
```

$$\begin{pmatrix} bf - ce \\ cd - af \\ ae - bd \end{pmatrix}$$

The suffix operator // allows us to append the function **MatrixForm[]** at the end of an input line. **MatrixForm[]** generates a column representation of a vector or a matrix. The disadvantage of this output form is that it is not usable in additional calculations. Another function available in the package **VectorAnalysis** is a gradient function for different coordinate systems (cartesian, cylindrical, spherical, elliptical, etc.). The following example applies the **Grad[]** in cartesian coordinates to a function depending on three cartesian coordinates x, y, and z:

```
Grad[f[x, y, z], Cartesian[x, y, z]] // MatrixForm
```

$$\begin{pmatrix} f^{(1,0,0)}(x, y, z) \\ f^{(0,1,0)}(x, y, z) \\ f^{(0,0,1)}(x, y, z) \end{pmatrix}$$

These examples give an idea of how the capabilities of *Mathematica* support symbolic calculations.

1.1.4 Numerical Calculations

In addition to symbolic calculations, we sometimes need the numerical evaluations of expressions. The numerical capabilities of *Mathematica* allow the following three essential operations for solving practical problems.

The solution of equations, for example the solution of a sixth-order polynomial $x^6 + x^2 - 1 = 0$, follows by

```
NSolve[x^6 + x^2 - 1 == 0, x]
```

$\{\{x \rightarrow -0.826031\}, \{x \rightarrow -0.659334 - 0.880844\,i\},$
$\quad \{x \rightarrow -0.659334 + 0.880844\,i\}, \{x \rightarrow 0.659334 - 0.880844\,i\},$
$\quad \{x \rightarrow 0.659334 + 0.880844\,i\}, \{x \rightarrow 0.826031\}\}$

To evaluate a definite integral in the range $x \in [0, \infty]$, you can use the numerical integration capabilities of **NIntegrate[]**. An example from statistical physics is

---

$$\texttt{NIntegrate}\left[\texttt{x}^3\,\texttt{e}^{-\texttt{x}^4},\,\{\texttt{x},\,0,\,\infty\}\right]$$

0.25

Sometimes, it is hard to find an analytical solution of an ordinary differential equation (ODE). The problem becomes much worse if you try to solve a nonlinear ODE. The function **NDSolve[]** may help you tackle such problems. An example of a second-order nonlinear ODE used in the examination of nonlinear oscillators demonstrates the solution of the initial value problem $y'' - y^2 + 2\,y = 0$, $y(0) = 0$, $y'(0) = \frac{1}{2}$. The initial value problem describes a nonlinear oscillator starting at $t = 0$ with a vanishing elongation and an initial velocity of $\frac{1}{2}$. The formulation in *Mathematica* reads

---

$$\texttt{NDSolve}\left[\left\{\texttt{y''[t]} - \texttt{y[t]}^2 + 2\,\texttt{y[t]} == 0,\right.\right.$$
$$\left.\left.\texttt{y[0]} == 0,\,\texttt{y'[0]} == \frac{1}{2}\right\},\,\texttt{y[t]},\,\{\texttt{t},\,0,\,10\}\right]$$

$\{\{y(t) \rightarrow \text{InterpolatingFunction}[(\,0.\quad 10.\,),\,<>][t]\}\}$

The result of the numerical integration is a representation of the solution by means of an interpolating function.

The above three examples serve to demonstrate that *Mathematica* is also capable of handling numerical evaluations. There are many other functions which support numerical calculations. As a rule, all functions which involve numerical calculations start with a capital **N** in the name.

1.1.5 Graphics

Mathematica supports the graphical representation of different mathematical expressions. *Mathematica* is able to create two- and three-dimensional plots. It allows the representation of experimental data given by lists of points, by parametric plots for functions in parametric form, or by contour plots for three-dimensional functions. It further allows the creation of short motion pictures by its function **Animation**. An overview of these capabilities is given next.

As a first example of the graphical capabilities of *Mathematica*, let us show how simple functions are plotted. The first argument of the plot function **Plot[]** specifies the function; the second argument denotes the plot range. All other arguments are options which alter the form of the plot in some way. A standard example in harmonic analysis is

```
Plot[Sin[x], {x, -π, π}, AxesLabel → {"x", "Sin[x]"}];
```

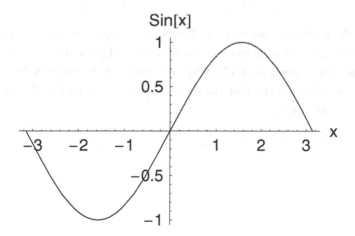

This plot can be improved in several directions: Sometimes you need a grid or other fonts for labeling or you prefer a frame around the plot. These properties are accessible by specifying the appropriate options of the following function:

```
Plot[Sin[x], {x, -π, π},
  AxesLabel → {StyleForm["x", FontWeight → "Bold",
    FontFamily → "Tekton"], StyleForm["Sin[x]",
    FontWeight → "Bold", FontFamily → "Tekton"]},
  Frame -> True, GridLines → Automatic,
  AxesStyle → {RGBColor[1, 0, 0], Thickness[0.01]},
  TextStyle → {FontSlant → "Italic", FontSize → 12}];
```

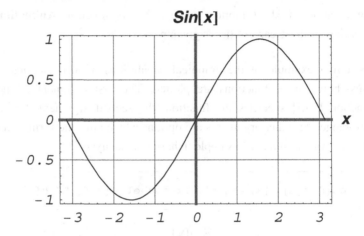

In three dimensions, we use **Plot3D[]** to represent the surface of a
function. A following example showing the surface in a rectangular water
tank. The arguments of **Plot3D[]** are similar to the function **Plot[]**. The
first specifies the function; the second and third specify the plot range; all
others are optional.

```
Plot3D[Sin[x] Cos[y], {x, -π, π}, {y, -2π, 2π},
  AxesLabel -> {"x", "y", "z"}, PlotPoints -> 35,
  TextStyle → {FontSlant → "Italic", FontSize → 12}];
```

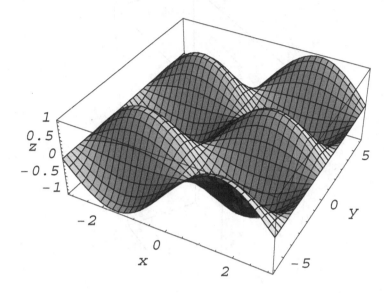

Sometimes you may know a solution of a problem only in a parametric representation. Consider, for example, the motion of an electron in a constant magnetic field. For such a situation, the track of the electron is described by a three-dimensional vector depending parametrically on time *t*. To represent such a parametric path, you can use the function **ParametricPlot3D[]**. The first argument of this function contains a list which describes the three coordinates of the curve. A fourth element of this list, which is optional, allows you to set a color for the track. We used in the following example the color function **Hue[]**. The second argument of the function **ParametricPlot3D[]** specifies the plot range of the parameter. All other arguments given to **ParametricPlot3D[]** are options changing the appearance of the plot.

```
ParametricPlot3D[{2 Sin[t], 5 Cos[t], t, Hue[0.4]},
  {t, 0, 4 π}, Axes -> False];
```

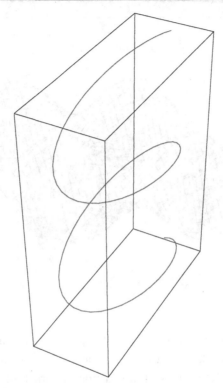

Another example is the movement of a planet around the Sun, for which the solution of the problem is in implicit form. According to Kepler`s theory (see Chapter 2, Section 2.5), a planet moves on an elliptical track around the Sun. The path of the planet is described in principal by a formula like $x^2 + 2 y^2 = 3$. To graphically represent such a path, we can use a function known as **ImplicitPlot[]** in *Mathematica*. This function becomes available if we load the standard package **Graphics`ImplicitPlot`**. A representation of the hypothetical planet track in x and y follows for the range $x \in [-2, 2]$ by

```
<< Graphics`ImplicitPlot`
```

```
pl1 = ImplicitPlot[x² + 2 y² == 3,
   {x, -2, 2}, PlotStyle -> RGBColor[1, 0, 0]];
```

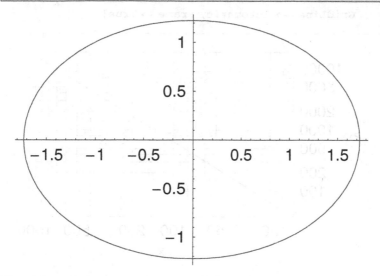

The color of the curve is changed from black to red by the option **PlotStyle→RGBColor[1,0,0]**.

If you have a function which is defined over a large range in x and y, such as in dynamical relaxation experiments, it is sometimes useful to represent the function in a log-log plot. For example, to show the graph of a scaling function like $f(x) = x^{1.4}$ in the range $x \in [1, 10^3]$, we can use **LogLogPlot[]** from the standard package **Graphics`Graphics`** to show the scaling behavior of the function. We clearly observe in the double logarithmic representation a linear relation between y and x which is characteristic for scaling (see Chapter 7 for more details).

```
<< Graphics`Graphics`;
LogLogPlot[x^1.4,
   {x, 1, 1000}, FrameLabel -> {"x", "y"},
   GridLines -> Automatic, Frame -> True];
```

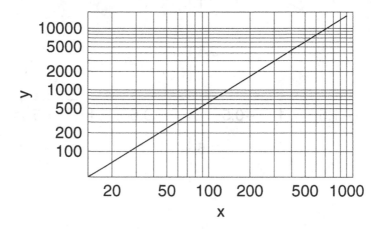

If you have to handle data from experiments, *Mathematica* can do much of the work for you. The graphical representation of a set of data can be done by the function **ListPlot[]**. This function allows you to plot a list of data. The input here is created by means of the function **Table[]**. The dataset, which we will represent by **ListPlot[]** consists of pairs $\{x, \sin(x)\,e^{-\frac{x}{4}}\}$ in the range $x \in [0, 6\pi]$. The data are located in the variable **tab1**. The graphical representation of these pairs of data is achieved by the function **ListPlot[]** using the dataset **tab1** as first argument. All other arguments are used to set temporary options for the function.

In[10]:=
```
tab1 = Table[{x, Sin[x] e^(-x/4)}, {x, 0, 6 π, 0.2}];
ListPlot[tab1, PlotStyle ->
    {RGBColor[0, 0, 0.500008], PointSize[0.015]},
   AxesLabel -> {"x", "y"}, PlotRange -> All];
```

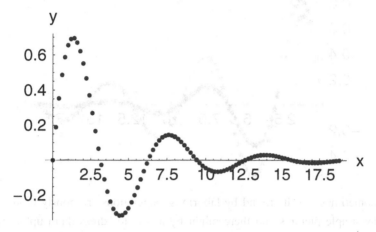

If you need to represent several sets of data in the same figure, you can use the function **MultipleListPlot[]** contained in the standard package **Graphics`MultipleListPlot`**. An example for two sets of data **tab1** and **tab2** is given below

```
<< Graphics`MultipleListPlot`
```

```
tab2 = Table[{x, Sin[x] e^(-x/8)}, {x, 0, 6 π, 0.2}];
```

```
MultipleListPlot[tab1, tab2,
  AxesLabel -> {"x", "y"}, PlotRange -> All];
```

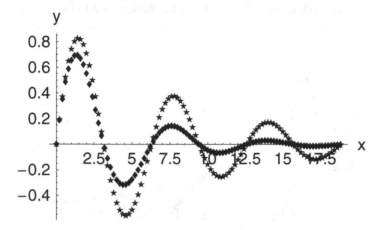

Sometimes, results found by laborious calculations are poorly represented by simple pictures and there might by a way to "dress them up" a bit. In many situations, you can vary a parameter or simply the time period to change the result in some way. The output of a small variation in parameters can be a great number of frames which all show different situations. To collect all of the different frames in a common picture, you can use the animation facilities of *Mathematica*. The needed functions are accessible if we load the standard package **Graphics`Animation`**. By using the function **Animate[]** contained in this package, you can create, for example, a flip chart movie for a planet moving around a star. The following animation combines two graphics objects, the first contained in the symbol **pl1** representing the track of the planet and the second consisting of a colored disk the planet.

```
<< Graphics`Animation`
```

```
pl2 = Animate[{pl1, Graphics[{RGBColor[0, 0, 1],
      Disk[{√3 Sin[x], √3 / 2 Cos[x]}, 0.1]}]}, {x, 0,
    2 π, 0.3}, PlotRange -> {{-1.9, 1.9}, {-1.5, 1.5}}]
```

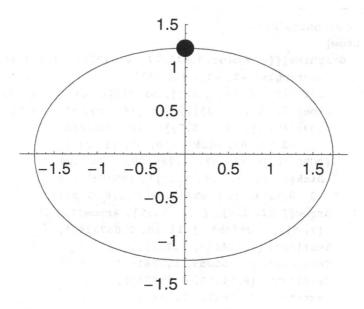

Note: In the printed version, we replace the animation by a single plot containing the different plots distinguished by different colors. We use this procedure to show the reader how an animation is generated and what kind of plots are generated.

If *Mathematica* does not provide you with the graphics you need, you are free to create your own graphics objects. By using graphics primitives like **Line[]**, **Disk[]**, **Circle[]**, and so forth, you can create any two- or three-dimensional objects you can imagine. A simple example to combine lines, disks, squares, and circles for depicting the scattering of α particles on a gold bar follows.

```
<< Graphics`Arrow`;
Show[
  Graphics[{{RGBColor[0.976577, 0.949233, 0.0195315],
     Rectangle[{-2, -2}, {2, 2}]},
    Line[{{0, 0}, {-12, 0}}], Line[{{0, 0}, {5, 6}}]],
    Line[{{0, 0}, {2, 0}}], Line[{{0, 0}, {3.4, 6.5}}],
    Line[{{0, 0}, {6.8, 5.7}}]], {RGBColor[0,
      0.500008, 0], Disk[{-10, 0}, {1, 2}]},
    {RGBColor[0.996109, 0.996109, 0.500008],
     Disk[{-10, 0}, {.6, 1.5}]}, {RGBColor[0,
      0, 0.996109], Disk[{5, 6}, {1.6, 1.3}]},
    Arrow[{-12, 1.5}, {-10, 1.5}], Arrow[{5, 6},
     {7, 8}], Text["b", {-11.88, 0.857724}],
    Text["J", {-11.4616, 2.392}],
    Text["Au", {-1.00059, 1.27616}],
    Text["dN", {6.74052, 6.85534}],
    Text["dΩ", {3.74171, 7.483}],
    Text["θ", {1.64952, 0.578765}],
    Text["db", {-8.88118, 0.997203}],
    Text["α", {-12.0892, 1.97356}]}],
  AspectRatio → Automatic];
```

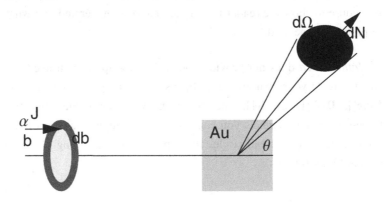

1.1.6 Programming

Mathematica not only is an interactive system but also allows one to generate programs supporting scientific calculations. By solving the following mathematical conjecture, we simultaneously demonstrate the creation of an interactive function in *Mathematica*. The iteration of the relation

$$f_{n+1} = f_{n-1} \int \left(\frac{f_n}{f_{n-1}}\right)^2 dx \tag{1.1.1}$$

under the initial conditions $f_0 = \cos(x)$ and $f_1 = \sin(x)$ results in a polynomial whose coefficients are given by trigonometric functions. The resulting polynomial can be represented in the form

$$f_\infty = \cos(x) \sum_{n=0}^{\infty} a_n x^n + \sin(x) \sum_{n=0}^{\infty} b_n x^n. \tag{1.1.2}$$

The related *Mathematica* representation is located in the variable **poly**. It reads

```
poly = Cos[x] ∑_{n=0}^{∞} a[n] x^n + Sin[x] ∑_{n=0}^{∞} b[n] x^n
```

```
Cos[x] ∑_{n-0}^{∞} a[n] x^n + Sin[x] ∑_{n=0}^{∞} b[n] x^n
```

The sums in the representation of the polynomial extend across the range $0 < n < \infty$. In the first step of the calculation, we introduce a list containing the initial conditions of the iteration. Lists in *Mathematica* are represented by a pair of braced brackets which contain the elements of the list separated by commas. To save the list for future use, we set the list equal to the variable **listf** by

```
listf = {Cos[x], Sin[x]}
```

```
{Cos[x], Sin[x]}
```

The first iteration step in Equation (1.1.1) is executed by the sequence

```
AppendTo[listf,
  listf[[1]] Integrate[( listf[[2]] / listf[[1]] )^2, x]] // Simplify
```

```
{Cos[x], Sin[x], -x Cos[x] + Sin[x]}
```

in which we append the result from an integration of the iteration formula to **listf** by means of the function **AppendTo[]**. The next step just changes the indices of the iteration and is given by

```
AppendTo[listf,
  listf[[2]] Integrate[( listf[[3]] / listf[[2]] )^2, x]] // Simplify
```

```
{Cos[x], Sin[x], -x Cos[x] + Sin[x],
  -1/3 x (-3 + x^2 + 3 x Cot[x]) Sin[x]}
```

Here, we increase the indices of the list elements in **listf** by one. The next interactive step results in

```
AppendTo[listf,
  listf[[3]] Integrate[( listf[[4]] / listf[[3]] )^2, x]] // Simplify
```

```
{Cos[x], Sin[x], -x Cos[x] + Sin[x],
  -1/3 x (-3 + x^2 + 3 x Cot[x]) Sin[x],
  1/45 x^3 (x (-15 + x^2) Cos[x] + 3 (5 - 2 x^2) Sin[x])}
```

Applying the function **Plus[]** to **listf** adds all elements of the list together, resulting in the representation of the polynomial in the form

```
poly = Apply[Plus, listf] // Simplify
```

$$\left(1 - x - x^2 - \frac{x^4}{3} + \frac{x^6}{45}\right) \text{Cos}[x] + \left(2 + x - \frac{2\,x^5}{15}\right) \text{Sin}[x]$$

The coefficients of the trigonometric functions **Cos[]** and **Sin[]** are accessed by

```
Coefficient[poly, Cos[x]]
```

$$1 - x - x^2 - \frac{x^4}{3} + \frac{x^6}{45}$$

and

```
Coefficient[poly, Sin[x]]
```

$$2 + x - \frac{2\,x^5}{15}$$

verifying the conjecture that the resulting function of the iteration is a polynomial with coefficients **Cos[x]** and **Sin[x]**. The disadvantage of this calculation is that we need to repeat the iteration. To avoid such repetition, we define a procedural function which performs the repetition automatically. Function **Iterate[]** derives the polynomial up to an iteration order *n*.

```
Iterate[initial_List,maxn_]:=Module[
(* --- local variables --- *)
   {df={},dfh,f=initial,fh},
(* --- iterate the formula and collect the results
--- *)
   Do[AppendTo[f,
      f[[n]] Integrate[(f[[n+1]]/f[[n]])^2,x]],
   {n,1,maxn}];
(* --- calculate the sum of all elements in f --- *)
   f = Expand[Apply[Plus,Simplify[f]]];
(* --- extract the coefficients from the polynom ---
*)
   fh = {Coefficient[f,initial[[1]]]};
   AppendTo[fh,Coefficient[f,initial[[2]]]];
(* --- return the result --- *)
   fh
   ]
```

The application of this sequential program **Iterate[]** to the starting functions **Cos[]** and **Sin[]** delivers

```
Iterate[{Cos[x],Sin[x]},4]
```

$$
\left\{ 1 - x - x^2 - \frac{x^4}{3} + \frac{x^6}{45} - \frac{x^7}{45} + \frac{2\,x^9}{945}, \right.
$$
$$
\left. 2 + x - \frac{2\,x^5}{15} + \frac{x^6}{45} - \frac{x^8}{105} + \frac{x^{10}}{4725} \right\}
$$

The result is a list containing the polynomial coefficients of the **Cos[]** and **Sin[]** functions, respectively. A more efficient realization of the iteration is given by the following functional program. The first part defines the iteration step:

```
Iterator[{expr1_, expr2_}] :=
  Expand[expr1 Integrate[(expr2 / expr1)^2, x]]
```

The second part extracts the last two elements from a list:

```
takeLastTwoElemets[l_List] := Take[l, -2]
```

The third part carries out the iteration:

```
Iterate1[input_, n_] :=
  Block[{F = input, t1}, t1 = Apply[Plus, Flatten[
     Simplify[Last[Table[Flatten[AppendTo[F, Expand[
        Apply[Iterator[takeLastTwoElemets[#]] &,
        {Flatten[F]}]]]], {n}]]]]];
  Map[Coefficient[t1, #] &, input]]
```

The results of the two functions can be compared by measuring the calculation time:

```
Iterate1[{Cos[x], Sin[x]}, 5] // Timing
```

$$\left\{15.87 \text{ Second}, \left\{1 - x - x^2 - \frac{x^4}{3} + \right.\right.$$
$$\frac{x^6}{45} - \frac{x^7}{45} + \frac{2 x^9}{945} - \frac{x^{11}}{4725} + \frac{x^{13}}{42525} - \frac{x^{15}}{4465125},$$
$$\left.\left. 2 + x - \frac{2 x^5}{15} + \frac{x^6}{45} - \frac{x^8}{105} + \frac{2 x^{10}}{4725} - \frac{4 x^{12}}{42525} + \frac{x^{14}}{297675}\right\}\right\}$$

```
Iterate[{Cos[x], Sin[x]}, 5] // Timing
```

$$\left\{5.82 \text{ Second}, \left\{1 - x - x^2 - \frac{x^4}{3} + \right.\right.$$
$$\frac{x^6}{45} - \frac{x^7}{45} + \frac{2 x^9}{945} - \frac{x^{11}}{4725} + \frac{x^{13}}{42525} - \frac{x^{15}}{4465125},$$
$$\left.\left. 2 + x - \frac{2 x^5}{15} + \frac{x^6}{45} - \frac{x^8}{105} + \frac{2 x^{10}}{4725} - \frac{4 x^{12}}{42525} + \frac{x^{14}}{297675}\right\}\right\}$$

The finding is that the procedural implementation is more efficient than the functional implementation. In addition to the efficiency, the two realizations of the programs demonstrate that a program in *Mathematica* can be generated in different ways. Other methods to implement algorithms are object-oriented programs, λ-calculus, rule-based programs,and so forth.

However, with **Iterate[]**, we can change the mathematical conjecture in the following way. Let us examine what happens if we use as initial conditions

hyperbolic functions instead of trigonometric functions. The result is easy
to derive if we use **Iterate[]** in the form of

```
Iterate[{Cosh[x],Sinh[x]},3]
```

$$\left\{ \frac{x^6}{45} + \frac{x^4}{3} - x^2 + x + 1,\ x - \frac{2x^5}{15} \right\}$$

Again, we obtain a polynomial whose coefficients are given by hyperbolic
functions. The interchange of initial conditions demonstrates that the
iteration

```
Iterate[{Sinh[x], Cosh[x]}, 3]
```

$$\left\{ \frac{x^6}{45} + \frac{x^4}{3} - x^2 + x + 1,\ x - \frac{2x^5}{15} \right\}$$

provides the same result. Meaning that the function is symmetric with
respect to the interchange of functions. However, the resulting polynomials
are different from the results gained from trigonometric functions:

```
Iterate[{Sin[x], Cos[x]}, 3]
```

$$\left\{ \frac{x^6}{45} - \frac{x^4}{3} + x^2 - x + 1,\ \frac{2x^5}{15} - \frac{2x^3}{3} + x \right\}$$

This small example demonstrates the capabilities of *Mathematica* for
finding solutions to a specific problem allowing us, at the same time, to
modify the initial question. However, the iterative solution of the
conjecture is not an exact proof. It only demonstrates the correctness of the
conjecture empirically. Yet, the empirical proof of the conjectured
behavior is the first step in proving the final result.

From the above example, we have seen that the use of *Mathematica*
facilitates our work insofar as special functions become immediately
available to us, not only analytically but also numerically and graphically.
This notwithstanding, we first need to be able to understand the physical

and mathematical relationships before we can effectively use *Mathematica* as a powerful tool.

In the following chapters, we will demonstrate how problems occurring in theoretical physics can be solved by the use of *Mathematica*. Note that we will not provide the reader with a detailed description of *Mathematica*. Instead, we will present a collection of mathematical steps gathered in a package. This package is useful for solving specific physical or mathematical problems by applying *Mathematica* as a tool. For a detailed description of the *Mathematica* functions, we refer the reader to the handbook by Wolfram [1.1] or the book by Blachman [1.3]. However, we hope that the reader will readily understand the solutions, because the code corresponds to notations in theoretical physics.

2
Classical Mechanics

2.1 Introduction

Classical mechanics denotes the theory of motion of particles and particle systems under conditions in which Heisenberg's uncertainty principle has essentially no effect on the motion and, therefore, may be neglected. It is the mechanics of Galilei, Newton, Lagrange, and Hamilton and it is now extended to include the mechanics of Einstein (Figure 2.1.1).

Figure 2.1.1. Galilei, Newton, Lagrange, Hamilton, and Einstein are the founding fathers of mechanics. These theoreticans remarkably defined the current understanding of mechanics.

This book is an attempt to present classical mechanics in a way that shows the underlying assumptions and that, as a consequence, indicates the boundaries beyond which its uncritical extension is dangerous. The presentation is designed to make the transition from classical mechanics to quantum mechanics and to relativistic mechanics smooth so that the reader will be able to sense the continuity in physical thought as the change is made.

The aim of classical mechanics and theoretical physics is to provide and develop a self-consistent mathematical structure which runs so closely parallel to the development of physical phenomena that, starting from a minimum number of hypotheses, it may be used to accurately describe and even predict the results of all carefully controlled experiments. The desire of accuracy, however, must be tempered by the need for reasonable simplicity, and the theoretical description of a physical situation is always simplified for convenience of analytical treatment. Such simplification may be thought of as arising both from physical approximations (i.e., the neglect of certain physical effects which are judged to be of negligible importance) and from mathematical approximations made during the development of the analysis. However, these two types of approximation are not really distinct, for usually each may be discussed in the language of the other. Representing as they do an economy rather than an ignorance, such approximation may be refined by a series of increasingly accurate calculations, performed either algebraically or numerically with a computer.

More subtle approximations appear in the laws of motion which are assumed as a starting point in any theoretical analysis of a problem. At present, the most refined form of theoretical physics is called quantum field theory, and the theory most accurately confirmed by experiment is a special case of quantum field theory called quantum electrodynamics. According to this discipline, the interactions among electrons, positrons, and electromagnetic radiation have been computed and shown to agree with the results of experiment with an over all accuracy of 1 part in 10^9. Unfortunately, analogous attempts to describe the interactions among mesons, hyperons, and nucleons are at present unsuccessful.

These recent developments are built on a solid structure which has been developed over the last three centuries and which is now called *classical mechanics*. Figure 2.1.2 illustrates how classical mechanics is related to other basic physical theories. The scheme is by no means complete. It represents a rough sketch of a discipline with great diversity.

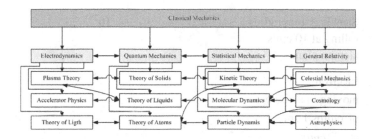

Figure 2.1.2.   Classical mechanics as proof for other disciplines in physics.

A theory that describes the motion of a particle at any level of approximation must eventually reduce to classical mechanics when conditions are such that relativistic, quantum, and radiative corrections can be neglected. This fact makes the subject basic to the student's understanding of the rest of the physics, in the same way that over the centuries it has been the foundation of human understanding of the behavior of physical phenomena.

Classical mechanics accurately describes the motion of a material system provided that the angular momentum of the system with respect to the nearest system which is influencing its motion is large compared with the quantum unit of angular momentum $\hbar = 1.054 \times 10^{-27} \ g \ cm^2 / s$. Examples of typical angular momentums are given in Table 2.1.1.

System	Approximate angular momentum in units of $\hbar$
Earth moving around the Sun	10^{64}
Steel ball 1cm radius rolling at 10 cm/s along a plane	10^{29}
Electron moving in a circle of radius 1 cm at 10^8 *cm/s*	10^8
Electron moving in an atom	$0,1,2,\ldots$

Table 2.1.1.   Comparison of fundamental scales.

Clearly, in all but the last case, the existence of a smallest unit of angular momentum is irrelevant, and the error introduced by using the approximation of classical mechanics will be small compared with both unavoidable experimental errors and other errors and approximations made in describing the actual physical situation theoretically. However, classical mechanics should not be studied only as an introduction to the more refined theories, for despite advances made during this century, it continues to be the mechanics used to describe the motion of directly observable macroscopic systems. Although an old subject, the mechanics of particles and rigid bodies is finding new applications in a number of areas, including the fields of vacuum and gaseous electronics, accelerator design, space technology, plasma physics, and magnetohydrodynamics. Indeed, more effort is being put into the development of the consequences of classical mechanics today than at any time since it was the only theory known. A recent development in classical mechanics is connected with chaotic behavior. Our aim is to provide a transition from traditional courses in classical mechanics to the rapidly growing areas of nonlinear dynamics and chaos and to present these old and new ideas in a broad and unified perspective.

2.2 Mathematical Tools

2.2.1 Introduction

This section introduces some of the mathematical tools necessary to efficiently describe mechanical systems. The basic tools discussed are coordinates, transformations, scalars, vectors, tensors, vector products, derivatives, and integral relations for scalars and vector fields.

Coordinates are the basic elements in mechanics used to describe the location of a particle in space at a certain time. These numbers are changed if we change the position in space. Thus, we need a procedure to describe the transition from the original position to the new position. The process of going from one location to another is carried out by a transformation. To describe the single elements of the coordinates, we need single figures, which are called scalars. If we arrange two or more of the scalars in a column or row, we get a vector. The arrangement of scalars in a two-dimensional or higher-dimensional array will lead us to tensors. Among the scalars, vectors, and tensors there exist algebraic and geometric relations which are defined in vector products, special derivatives, and integral relations.

2.2.2 Coordinates

In order to represent points in space, we first choose a fixed point O (the origin) and three directed lines through O that are perpendicular to each other, called the coordinate axes and labeled the x-axis, y-axis, and z-axis. Usually, we think of the x- and y-axes as being horizontal and the z-axis as being vertical and we draw the orientation of the axis as in Figure 2.2.1. Now, if P is any point in space, let a be the (directed) distance from the yz-plane to P, let b be the distance from the xz-plane to P, and let c be the distance from the xy-plane to P. We represent the point P by the ordered triple (a, b, c) of real numbers and we call a, b, and c the components of P; a is the x-component, b is the y-component, and c is the z-component. At the same time a, b, and c are the cartesian coordinates that describe the position of the point P relative to the coordinate system. Later, we will see that there exist other coordinates like angles and so forth. Thus, to locate the point (a, b, c) in space, we can start at the origin O and move a units along the x-axis, then b units along the y-axis, and then c units parallel to the z-axis. Coordinates are numbers in a system of reference. Usually, they define a point with respect to the origin in a coordinate system.

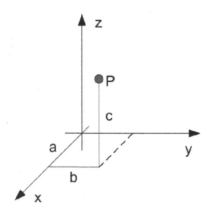

Figure 2.2.1.   Coordinate system with coordinates a, b, and c of a point P.

Very often, it is not convenient to describe the position of even a single particle in terms of rectangular cartesian coordinates referred to a particular set of coordinate axes. If for example, the particle moves in a plane under the influence of a force which is directed toward a fixed point in the plane and which is independent of the azimuthal angle θ, it is usually more convenient to use plane polar coordinates

$$q_1 = r = (x^2 + y^2)^{1/2} \tag{2.2.1}$$

and

$$q_2 = \phi = \tan^{-1}(\tfrac{y}{x}) \tag{2.2.2}$$

or if the force is spherically symmetric it is natural to use the spherical coordinates

$$q_1 = r = (x^2 + y^2 + z^2)^{1/2}$$
$$q_2 = \phi = \cot^{-1}\left(\tfrac{z}{(x^2+y^2)^{1/2}}\right) \tag{2.2.3}$$
$$q_3 = \theta = \tan^{-1}(\tfrac{y}{x}).$$

Here, $\tan^{-1}$ and $\cot^{-1}$ denote the inverse functions of tan and cot, respectively. The coordinates (2.2.3) are also used if the particle is constrained to move on a fixed circle or fixed sphere.

Sometimes, it is useful to look at the motion of the particle from the moving frame. In such a coordinate system, the coordinates q_1, q_2, q_3 is in uniform motion, for example, with respect to the x direction having velocity v relative to the system x, y, z

$$q_1 = x - vt,$$
$$q_2 = y, \qquad (v = \text{const}), \tag{2.2.4}$$
$$q_3 = z$$

or from a uniformly accelerated system

$$q_1 = x - \tfrac{1}{2} g t^2 \qquad (g = \text{const}),$$
$$q_2 = y, \tag{2.2.5}$$
$$q_3 = z.$$

In general, each transformation of the coordinate system x_i to a new set q_i may be expressed as a set of three equations of the form

$$x_i = x_i(q_1, q_2, q_3, t) \quad \text{with } i = 1, 2, 3. \tag{2.2.6}$$

For the stationary coordinate systems (2.2.2) and (2.2.3), the relations between x_i and q_i do not involve the time t.

If equations (2.2.6) are such that the three coordinates q_i can be expressed as functions of the x_i, we have

$$q_i = q_i(x_1, x_2, x_3, t) \quad \text{with } i = 1, 2, 3. \tag{2.2.7}$$

The q_i are as effective as the x_i in describing the position of the particle. The q_i are called *generalized coordinates* of the particle. The generalized coordinates may themselves be rectangular cartesian coordinates or they may be a set of any three variables, not necessarily with the dimension of length, which between them specify unambiguously the position of the particle relative to some set of axes.

2.2.3 Coordinate Transformations and Matrices

Let us consider a point P which has cartesian coordinates (x_1, x_2, x_3) with respect to a certain coordinate system. Next, consider a different coordinate system that can be generated from the original system by a single rotation; let the coordinates of the point P with respect to the new coordinate system be $(\tilde{x}_1, \tilde{x}_2, \tilde{x}_3)$. The transformation is illustrated for a two-dimensional case in Figure 2.2.2.

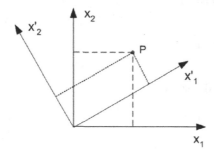

Figure 2.2.2.   Rotation of the original coordinate axis.

The new coordinate $\tilde{x}_1$ is the sum of the projection of x_1 onto the $\tilde{x}_1$ axis plus the projection of x_2 onto the $\tilde{x}_1$-axis. The $\tilde{x}_2$-coordinate is determined by similar projections of x_1 and x_2 onto the $\tilde{x}_2$-axis. The general relation for the coordinate transformation in three dimensions is given by

$$\tilde{x}_i = \sum_j \lambda_{ij} x_j \quad \text{with } i = 1, 2, 3, \tag{2.2.8}$$

where the λ_{ij} are the direction cosine of the $\tilde{x}_i$-axis relative to the x_j-axis. It is convenient to arrange the λ_{ij} into a square array called a matrix. The symbol λ will be used to denote the totality o the individual elements λ_{ij} when arranged in the following manner:

$$\lambda = \begin{pmatrix} \lambda_{11} & \lambda_{12} & \lambda_{13} \\ \lambda_{21} & \lambda_{22} & \lambda_{23} \\ \lambda_{31} & \lambda_{32} & \lambda_{33} \end{pmatrix}. \qquad (2.2.9)$$

Once the direction cosines which relate the two sets of coordinates are found, the general rules for specifying the coordinates of a point in either system. λ is called a transformation matrix.

The λ matrix has equal numbers of rows and columns and is therefore called a square matrix. It is not necessary that a matrix be square. In fact, the coordinates of a point may be written as a column matrix:

$$\tilde{x} = \begin{pmatrix} x_1 \\ x_2 \\ x_3 \end{pmatrix} \qquad (2.2.10)$$

or as a row matrix

$$\tilde{x} = (x_1 \quad x_2 \quad x_3). \qquad (2.2.11)$$

We must now establish rules whereby it is possible to multiply two matrices. Let us take a column matrix for the coordinates. Then, we have the following equivalent expressions:

$$\tilde{x}_i = \sum_j \lambda_{ij} x_j, \qquad (2.2.12)$$
$$\tilde{\vec{x}} = \lambda \vec{x} \qquad (2.2.13)$$

or in *Mathematica* notation

```
x̃ = ⎛ λ₁₁  λ₁₂  λ₁₃ ⎞ ⎛ x₁ ⎞
    ⎜ λ₂₁  λ₂₂  λ₂₃ ⎟ . ⎜ x₂ ⎟ ; x̃ // TableForm
    ⎝ λ₃₁  λ₃₂  λ₃₃ ⎠ ⎝ x₃ ⎠
```

```
x₁ λ₁₁ + x₂ λ₁₂ + x₃ λ₁₃
x₁ λ₂₁ + x₂ λ₂₂ + x₃ λ₂₃
x₁ λ₃₁ + x₂ λ₃₂ + x₃ λ₃₃
```

This relation completely specifies the operation of matrix multiplication for the case of a matrix of three rows and three columns operating on a matrix of three rows and one column. The next step is to generalize this result to matrices of $n \times n$ order.

The multiplication of a matrix A and a matrix B is defined only if the number of columns of A is equal to the number of rows of B. For such a case, the product $A.B$ is given by

$$C = A.B,$$
$$C_{ij} = \sum_k A_{ik} B_{kj}. \tag{2.2.14}$$

It is evident that matrix multiplication is not commutative. Thus, if A and B are both square matrices, then the sums

$$\sum_k A_{ik} B_{kj} \quad \text{and} \quad \sum_k B_{ik} A_{kj}$$

are both defined, but, in general, they will not be equal. This behavior is shown by the following example.

Example:

If A and B are the matrices

$$A = \begin{pmatrix} 2 & 4 \\ -5 & 1 \end{pmatrix}$$

$$\begin{pmatrix} 2 & 4 \\ -5 & 1 \end{pmatrix}$$

and

$$B = \begin{pmatrix} 5 & 2 \\ 9 & -4 \end{pmatrix}$$

$$\begin{pmatrix} 5 & 2 \\ 9 & -4 \end{pmatrix}$$

then

A.B

$$\begin{pmatrix} 46 & -12 \\ -16 & -14 \end{pmatrix}$$

but

B.A

$$\begin{pmatrix} 0 & 22 \\ 38 & 32 \end{pmatrix}$$

An important operation on a matrix is the transposition. A transposed matrix is a matrix derived from the original matrix by the interchange of rows and columns. The transposition of a matrix A is denoted by A^T. According to this rule, we have

$$\lambda_{ij}^T = \lambda_{ji} \tag{2.2.15}$$

If we define the λ matrix by

$$l = \begin{pmatrix} \lambda_{11} & \lambda_{12} & \lambda_{13} \\ \lambda_{21} & \lambda_{22} & \lambda_{23} \\ \lambda_{31} & \lambda_{32} & \lambda_{33} \end{pmatrix}$$

$$\begin{pmatrix} \lambda_{11} & \lambda_{12} & \lambda_{13} \\ \lambda_{21} & \lambda_{22} & \lambda_{23} \\ \lambda_{31} & \lambda_{32} & \lambda_{33} \end{pmatrix}$$

the transposed matrix is given by

I^T
$\begin{pmatrix} \lambda_{11} & \lambda_{21} & \lambda_{31} \\ \lambda_{12} & \lambda_{22} & \lambda_{32} \\ \lambda_{13} & \lambda_{23} & \lambda_{33} \end{pmatrix}$

Another property of matrices is that any matrix multiplied by the identity matrix is unaffected:

IdentityMatrix[3].I
$\begin{pmatrix} \lambda_{11} & \lambda_{12} & \lambda_{13} \\ \lambda_{21} & \lambda_{22} & \lambda_{23} \\ \lambda_{31} & \lambda_{32} & \lambda_{33} \end{pmatrix}$

or

I^T.**IdentityMatrix[3]**
$\begin{pmatrix} \lambda_{11} & \lambda_{21} & \lambda_{31} \\ \lambda_{12} & \lambda_{22} & \lambda_{32} \\ \lambda_{13} & \lambda_{23} & \lambda_{33} \end{pmatrix}$

Consider matrix λ to be known. The problem is to find the inverse matrix λ^{-1} such that

$$\lambda.\lambda^{-1} = \lambda^{-1}.\lambda = 1. \tag{2.2.16}$$

If C_{ij} is the cofactor of λ (i.e., the minor of λ with the sign $(-1)^{i+j}$), then the inverse is determined by

$$\lambda_{ij}^{-1} = \frac{C_{ji}}{\det(\lambda)} \tag{2.2.17}$$

if $\det(\lambda) \neq 0$. Note that in numerical work it sometimes happens that $\det(\lambda)$ is almost equal to 0. Then, there is trouble ahead. In *Mathematica*, the inverse of a matrix is calculated by the function **Inverse[]**. The application of this function to the matrix λ gives us

---

Inverse(l)

$\{\{(\lambda_{22}\,\lambda_{33} - \lambda_{23}\,\lambda_{32})\,/\,(-\lambda_{13}\,\lambda_{22}\,\lambda_{31} + \lambda_{12}\,\lambda_{23}\,\lambda_{31} +$
$\quad\lambda_{13}\,\lambda_{21}\,\lambda_{32} - \lambda_{11}\,\lambda_{23}\,\lambda_{32} - \lambda_{12}\,\lambda_{21}\,\lambda_{33} + \lambda_{11}\,\lambda_{22}\,\lambda_{33})\,,$
$\quad(\lambda_{13}\,\lambda_{32} - \lambda_{12}\,\lambda_{33})\,/\,(-\lambda_{13}\,\lambda_{22}\,\lambda_{31} + \lambda_{12}\,\lambda_{23}\,\lambda_{31} +$
$\quad\lambda_{13}\,\lambda_{21}\,\lambda_{32} - \lambda_{11}\,\lambda_{23}\,\lambda_{32} - \lambda_{12}\,\lambda_{21}\,\lambda_{33} + \lambda_{11}\,\lambda_{22}\,\lambda_{33})\,,$
$\quad(\lambda_{12}\,\lambda_{23} - \lambda_{13}\,\lambda_{22})\,/\,(-\lambda_{13}\,\lambda_{22}\,\lambda_{31} + \lambda_{12}\,\lambda_{23}\,\lambda_{31} +$
$\quad\lambda_{13}\,\lambda_{21}\,\lambda_{32} - \lambda_{11}\,\lambda_{23}\,\lambda_{32} - \lambda_{12}\,\lambda_{21}\,\lambda_{33} + \lambda_{11}\,\lambda_{22}\,\lambda_{33})\,\}\,,$
$\{(\lambda_{23}\,\lambda_{31} - \lambda_{21}\,\lambda_{33})\,/\,(-\lambda_{13}\,\lambda_{22}\,\lambda_{31} + \lambda_{12}\,\lambda_{23}\,\lambda_{31} +$
$\quad\lambda_{13}\,\lambda_{21}\,\lambda_{32} - \lambda_{11}\,\lambda_{23}\,\lambda_{32} - \lambda_{12}\,\lambda_{21}\,\lambda_{33} + \lambda_{11}\,\lambda_{22}\,\lambda_{33})\,,$
$\quad(\lambda_{11}\,\lambda_{33} - \lambda_{13}\,\lambda_{31})\,/\,(-\lambda_{13}\,\lambda_{22}\,\lambda_{31} + \lambda_{12}\,\lambda_{23}\,\lambda_{31} +$
$\quad\lambda_{13}\,\lambda_{21}\,\lambda_{32} - \lambda_{11}\,\lambda_{23}\,\lambda_{32} - \lambda_{12}\,\lambda_{21}\,\lambda_{33} + \lambda_{11}\,\lambda_{22}\,\lambda_{33})\,,$
$\quad(\lambda_{13}\,\lambda_{21} - \lambda_{11}\,\lambda_{23})\,/\,(-\lambda_{13}\,\lambda_{22}\,\lambda_{31} + \lambda_{12}\,\lambda_{23}\,\lambda_{31} +$
$\quad\lambda_{13}\,\lambda_{21}\,\lambda_{32} - \lambda_{11}\,\lambda_{23}\,\lambda_{32} - \lambda_{12}\,\lambda_{21}\,\lambda_{33} + \lambda_{11}\,\lambda_{22}\,\lambda_{33})\,\}\,,$
$\{(\lambda_{21}\,\lambda_{32} - \lambda_{22}\,\lambda_{31})\,/\,(-\lambda_{13}\,\lambda_{22}\,\lambda_{31} + \lambda_{12}\,\lambda_{23}\,\lambda_{31} +$
$\quad\lambda_{13}\,\lambda_{21}\,\lambda_{32} - \lambda_{11}\,\lambda_{23}\,\lambda_{32} - \lambda_{12}\,\lambda_{21}\,\lambda_{33} + \lambda_{11}\,\lambda_{22}\,\lambda_{33})\,,$
$\quad(\lambda_{12}\,\lambda_{31} - \lambda_{11}\,\lambda_{32})\,/\,(-\lambda_{13}\,\lambda_{22}\,\lambda_{31} + \lambda_{12}\,\lambda_{23}\,\lambda_{31} +$
$\quad\lambda_{13}\,\lambda_{21}\,\lambda_{32} - \lambda_{11}\,\lambda_{23}\,\lambda_{32} - \lambda_{12}\,\lambda_{21}\,\lambda_{33} + \lambda_{11}\,\lambda_{22}\,\lambda_{33})\,,$
$\quad(\lambda_{11}\,\lambda_{22} - \lambda_{12}\,\lambda_{21})\,/\,(-\lambda_{13}\,\lambda_{22}\,\lambda_{31} + \lambda_{12}\,\lambda_{23}\,\lambda_{31} +$
$\quad\lambda_{13}\,\lambda_{21}\,\lambda_{32} - \lambda_{11}\,\lambda_{23}\,\lambda_{32} - \lambda_{12}\,\lambda_{21}\,\lambda_{33} + \lambda_{11}\,\lambda_{22}\,\lambda_{33})\,\}\,\}$

Knowing the inverse of λ, we can check the definition (2.2.16)

Simplify[$l.l^{-1}$]

$$\begin{pmatrix} 1 & 0 & 0 \\ 0 & 1 & 0 \\ 0 & 0 & 1 \end{pmatrix}$$

---

which, in fact, reproduces the identity matrix.

For orthogonal matrices, there exist a connection between the inverse matrix and the transposed matrix. This connection is

$$\lambda^{-1} = \lambda^{T} \quad \text{only for orthogonal matrices!} \tag{2.2.18}$$

We demonstrate this relation for the 2×2 rotation matrices. A rotation by an angle ϕ in two dimensions is given by the matrix

$$R = \begin{pmatrix} \cos(\phi) & \sin(\phi) \\ -\sin(\phi) & \cos(\phi) \end{pmatrix}$$

$$\begin{pmatrix} \cos(\phi) & \sin(\phi) \\ -\sin(\phi) & \cos(\phi) \end{pmatrix}$$

The inverse of this matrix is

R^{-1} // **Simplify**

$$\begin{pmatrix} \cos(\phi) & -\sin(\phi) \\ \sin(\phi) & \cos(\phi) \end{pmatrix}$$

and the transpose is

R^T

$$\begin{pmatrix} \cos(\phi) & -\sin(\phi) \\ \sin(\phi) & \cos(\phi) \end{pmatrix}$$

obviously both matrices are equivalent. This result in two dimensions can be generalized to higher dimensions and to the general representation of λ. To demonstrate the relation (2.2.16) and the consequences from this definition for the transposes matrix, we write

$l.l^T$ == **IdentityMatrix[3]**

```
{{λ²₁₁ + λ²₁₂ + λ²₁₃ ,
    λ₁₁ λ₂₁ + λ₁₂ λ₂₂ + λ₁₃ λ₂₃ , λ₁₁ λ₃₁ + λ₁₂ λ₃₂ + λ₁₃ λ₃₃ },
  {λ₁₁ λ₂₁ + λ₁₂ λ₂₂ + λ₁₃ λ₂₃ , λ²₂₁ + λ²₂₂ + λ²₂₃ ,
    λ₂₁ λ₃₁ + λ₂₂ λ₃₂ + λ₂₃ λ₃₃ }, {λ₁₁ λ₃₁ + λ₁₂ λ₃₂ + λ₁₃ λ₃₃ ,
    λ₂₁ λ₃₁ + λ₂₂ λ₃₂ + λ₂₃ λ₃₃ , λ²₃₁ + λ²₃₂ + λ²₃₃ }} ==
  {{1, 0, 0}, {0, 1, 0}, {0, 0, 1}}
```

We observe that if the matrix λ is orthogonal, the off-diagonal elements have to vanish and the diagonal elements are identical to 1. This property can be verified if we replace the symbolic values λ_{ik} by their representations with directional cosines. If we can satisfy these conditions, the transpose and the inverse of the rotation matrix λ are identical. In fact, the transpose of any orthogonal matrix is equal to its inverse. The above relation allows an equivalent representation in components

$$\sum_j \lambda_{ij}\lambda_{kj} = \delta_{ik}, \tag{2.2.19}$$

where δ_{ik} is the Kronecker delta symbol

$$\delta_{ik} = \begin{cases} 0 & \text{if } i \neq k \\ 1 & \text{if } i = k. \end{cases} \tag{2.2.20}$$

This symbol was introduced by Leopold Kronecker (1823–1891). The validity of Equation (2.2.19) depends on the fact that the coordinate axes in each of the systems are mutually perpendicular. Such systems are said to be orthogonal and Equation (2.2.19) is the orthogonality condition.

The following examples demonstrate how rotations act on coordinate transformations. Let us first consider the case in which the coordinate axes are rotated counterclockwise through an angle of 90° about the x_3-axis. In such a rotation, $\tilde{x}_1 = x_2$, $\tilde{x}_2 = -x_1$, and $\tilde{x}_3 = x_3$. The only nonvanishing cosines are

$$\begin{aligned} \cos(\tilde{x}_1, x_2) &= 1 = \lambda_{12}, \\ \cos(\tilde{x}_2, x_1) &= -1 = \lambda_{21}, \\ \cos(\tilde{x}_3, x_3) &= 1 = \lambda_{33}. \end{aligned} \tag{2.2.21}$$

Thus the λ matrix for this case is

$$\lambda_{x_3} = \begin{pmatrix} 0 & 1 & 0 \\ -1 & 0 & 0 \\ 0 & 0 & 1 \end{pmatrix}$$

$$\begin{pmatrix} 0 & 1 & 0 \\ -1 & 0 & 0 \\ 0 & 0 & 1 \end{pmatrix}$$

The multiplication of this transformation matrix with vectors along the three coordinate axes shows us how the coordinate axes are changed. For the x_1-axis represented by

$$x1 = \begin{pmatrix} 1 \\ 0 \\ 0 \end{pmatrix};$$

we find

$$x1t = \lambda_{x_3}.x1$$

$$\begin{pmatrix} 0 \\ -1 \\ 0 \end{pmatrix}$$

which transforms the x_1-axis to the $-\tilde{x}_2$-axis. In case of the x_2-axis, we find

$$x2 = \begin{pmatrix} 0 \\ 1 \\ 0 \end{pmatrix};$$

$$x2t = \lambda_{x_3}.x2$$

$$\begin{pmatrix} 1 \\ 0 \\ 0 \end{pmatrix}$$

showing us that the x_2-axis is transformed to the $\tilde{x}_1$-axis. Finally, the x_3-axis remains unchanged:

$$x3 = \begin{pmatrix} 0 \\ 0 \\ 1 \end{pmatrix};$$

$$x3t = \lambda_{x_3} \cdot x3$$

$$\begin{pmatrix} 0 \\ 0 \\ 1 \end{pmatrix}$$

The following illustration demonstrates this kind of coordinate transformation:

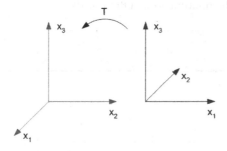

Another transformation about the x_1-axis is defined as follows:

$$\lambda_{x_1} = \begin{pmatrix} 1 & 0 & 0 \\ 0 & 0 & 1 \\ 0 & -1 & 0 \end{pmatrix}$$

$$\begin{pmatrix} 1 & 0 & 0 \\ 0 & 0 & 1 \\ 0 & -1 & 0 \end{pmatrix}$$

In the next step, let us apply the two rotations about the x_3- and the x_1-axis in such a way that we first carry out the rotation around the x_3-axis followed by a rotation about the x_1-axis. Defining the vector $\vec{x}$ by

$$\vec{x} = \begin{pmatrix} x_1 \\ x_2 \\ x_3 \end{pmatrix};$$

we first transform this vector to an intermediate vector xh:

$$\overrightarrow{xh} = \lambda_{x_3}.\vec{x}$$

$$\begin{pmatrix} x_2 \\ -x_1 \\ x_3 \end{pmatrix}$$

this vector is again used in the rotation around the x_1-axis:

$$\overrightarrow{xf} = \lambda_{x_1} . \overrightarrow{xh}$$

$$\begin{pmatrix} x_2 \\ x_3 \\ x_1 \end{pmatrix}$$

which results in a final vector with interchanged coordinates. This final state of the vector was generated by two transformations λ_{x_1} and λ_{x_3}, which can be verified by

$$\lambda_{x_1} . \lambda_{x_3} . \vec{x}$$

$$\begin{pmatrix} x_2 \\ x_3 \\ x_1 \end{pmatrix}$$

The result is the same as the sequential application of the rotations. Thus, the complete rotation can be represented by single transformation

$\lambda_{x_2} = \lambda_{x_1}.\lambda_{x_3}$

$$\begin{pmatrix} 0 & 1 & 0 \\ 0 & 0 & 1 \\ 1 & 0 & 0 \end{pmatrix}$$

which again delivers the same final state of the vector when applied to the original vector:

$\lambda_{x_2}.\vec{x}$

$$\begin{pmatrix} x_2 \\ x_3 \\ x_1 \end{pmatrix}$$

Note that the order in which the transformation matrices operate on $\vec{x}$ is important since the multiplication is not commutative. Changing the product order, we find

$\lambda_{x_3}.\lambda_{x_1}.\vec{x}$

$$\begin{pmatrix} x_3 \\ -x_1 \\ -x_2 \end{pmatrix}$$

which is different from the previous result because

$\lambda_{x_3}.\lambda_{x_1} \neq \lambda_{x_1}.\lambda_{x_3}$

True

Thus, an entirely different orientation results.

Next, consider a coordinate rotation around the x_3-axis which allows to continuously vary the transformation angle ϕ around the x_3-axis. Such a transformation is identical with a rotation in the x_1- x_2-plane. We denote this kind of rotation by

$$R_{x_3}(\phi\_) := \begin{pmatrix} \cos(\phi) & \sin(\phi) & 0 \\ -\sin(\phi) & \cos(\phi) & 0 \\ 0 & 0 & 1 \end{pmatrix}$$

The action of this transformation can be demonstrated by transforming an arbitrary vector $\vec{x}$

$$\vec{x} = \begin{pmatrix} x_1 \\ x_2 \\ x_3 \end{pmatrix};$$

by means of the transformation matrix R_{x_3}. The result of such a transformation is given by a vector containing the original coordinates x_1, x_2, and x_3:

$$rh = R_{x_3}(\phi).\vec{x}$$

$$\begin{pmatrix} \cos(\phi)\, x_1 + \sin(\phi)\, x_2 \\ \cos(\phi)\, x_2 - \sin(\phi)\, x_1 \\ x_3 \end{pmatrix}$$

If we change the angle ϕ continuously, the original vector undergoes a rotation around the x_3-axis. This behavior is demonstrated in the following illustration.

```
Map[ (Show[Graphics3D[{RGBColor[0, 0, 0.996109],
      Line[{{0, 0, 0}, rh // Flatten}] /.
       {x₁ → 1, x₂ → 1, x₃ → 1, ϕ → #}}]], PlotRange →
        {{-1.5, 1.5}, {-1.5, 1.5}, {0, 1.5}}]) &,
   Table[i, {i, 0, 2π, .3}]];
```

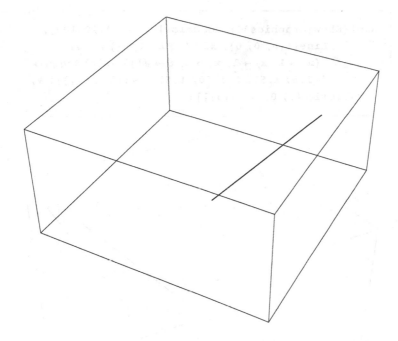

Another rotation frequently used in the theory of rigid bodies is a rotation around the x_2-axis.

$$R_{x_2}(\theta\_) := \begin{pmatrix} \cos(\theta) & 0 & -\sin(\theta) \\ 0 & 1 & 0 \\ \sin(\theta) & 0 & \cos(\theta) \end{pmatrix}$$

The application of this transformation matrix to the vector $\vec{x}$ gives us

$$\text{x2r} = R_{x_2}(\theta).\vec{x}$$

$$\begin{pmatrix} \cos(\theta)\, x_1 - \sin(\theta)\, x_3 \\ x_2 \\ \sin(\theta)\, x_1 + \cos(\theta)\, x_3 \end{pmatrix}$$

The graphical representation for specific coordinates looks like

```
Map[(Show[Graphics3D[{RGBColor[0, 0, 0.996109],
    Line[{{0, 0, 0}, x2r // Flatten}]] /.
      {x₁ → 1, x₂ → 1, x₃ → 1, θ → #}}]], PlotRange →
      {{-1.5, 1.5}, {-1.50, 1.5}, {-1.5, 1.5}}]) &,
   Table[i, {i, 0, 2 π, .3}]];
```

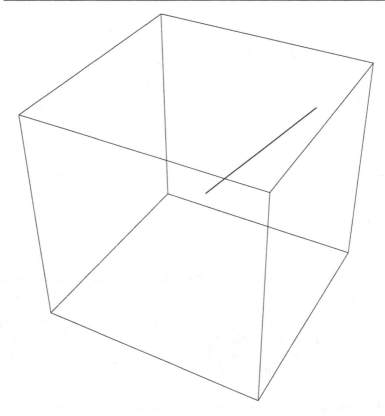

These two rotation matrices can be used to generate a general rotation in three dimensions. The three angles ϕ, θ, and ψ are known as Euler angles. Applications of this kind of transformation matrice will be discussed in Section 2.10 on rigid body motion.

---

genRot = $R_{x_3}(\psi).R_{x_2}(\theta).R_{x_3}(\phi)$

```
{{Cos[θ] Cos[φ] Cos[ψ] - Sin[φ] Sin[ψ],
  Cos[θ] Cos[ψ] Sin[φ] + Cos[φ] Sin[ψ], -Cos[ψ] Sin[θ]},
 {-Cos[ψ] Sin[φ] - Cos[θ] Cos[φ] Sin[ψ],
  Cos[φ] Cos[ψ] - Cos[θ] Sin[φ] Sin[ψ], Sin[θ] Sin[ψ]},
 {Cos[φ] Sin[θ], Sin[θ] Sin[φ], Cos[θ]}}
```

Our application here is just a general rotation in three dimensions:

```
Map[(Show[Graphics3D[{RGBColor[0, 0, 0.996109],
     Line[{{0, 0, 0}, genRot.x̄ // Flatten}] /.
     {x₁ → 1, x₂ → 1, x₃ → 1, φ → π/3, θ → π/4, ψ → #}}],
   PlotRange → {{-1.5, 1.5}, {-1.50, 1.5},
     {-1.5, 1.5}}]) &, Table[i, {i, 0, 2 π, .3}]];
```

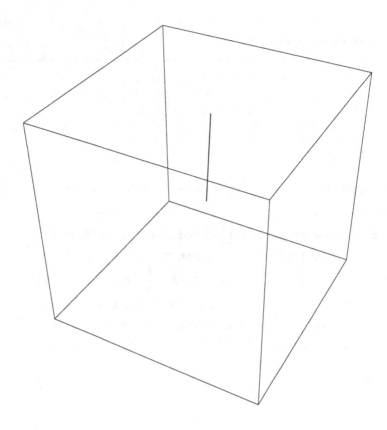

2.2.4 Scalars

In the mathematical description of physical processes, the values of a great many quantities can be specified by a single real number. For example, length, time, mass, and temperature are such quantities. The values of these quantities can be arranged on a single scale. They are called *scalars*.

The scale on which we measure the scalars is connected with a measuring unit. For the sake of consistency, we cannot always choose the units arbitrarily. What we can choose are a few so-called fundamental units. Other units are derived from this basic set and, thus, are uniquely determined. For example, if we choose to measure the length in centimeters (cm), meters (m), or kilometers (km), the units of area and volume are already given.

The smaller number of necessary units for physical quantities is bound to be small. There is an agreement that a number smaller than 3 is of no practical interest. Historically, there are different systems of measurement, the cgs, the mks, and the Giorgi system. The cgs system uses the fundamental units length, mass, and time measured in centimeter, gram and second. Even electrical and magnetic units are derived from this system. In the mks system, the units are meter, kilogram, and second. The ampere is taken to be a fundamental electric unit in the mks system. This additional unit turns the mks system into the mksa or the Giorgi system.

Scalars can be positive, as mass and volume, or both positive and negative such as the density of electric charge. Every physical quantity has what is called a given *dimension* as defined by the measuring units. However, the ratio between two quantities of the same kind is dimensionless or a pure number.

The calculus used for pure numbers is valid for scalars. However, in physics, only scalars of the same kind and of the same dimension can be added or subtracted. By multiplication and division, we get quantities of different dimensions expressed in other units.

Let us examine the real meaning of a scalar. For this, let us consider an array of particles with different masses. Each particle is labeled according to its mass (see Figure 2.2.3). The coordinate axes are shown so that t is possible to specify a particular particle by a pair of numbers (x, y).

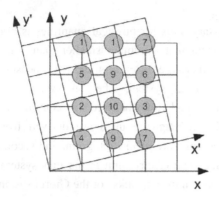

Figure 2.2.3.   Change of coordinates and action on the scalar quantities.

The mass m of the particle at (x, y) can be expressed as $m(x, y)$. Now, consider the axes rotated as shown in Figure 2.2.3. It is evident that each mass is now located at $(\tilde{x}, \tilde{y})$. However, because the masses itself did not change during the transformation, we can state

$$m(x, y) = m(\tilde{x}, \tilde{y}) \tag{2.2.22}$$

because the mass of any particle is not affected by a change in the coordinate axes.

Quantities which have the property that they are invariant under coordinate transformations are termed scalars.

Although it is possible to give the mass of a particle relative to any coordinate system by the same number, it is clear that there are some physical properties associated with the particle which cannot be specified in such a simple manner. For example, the direction of motion and the direction of force are such quantities. The description of these more complicated quantities require the use of vectors.

2.2.5 Vectors

Not all physical quantities can be characterized by a single number. There are a large number of quantities which need two or more numbers to provide an exact description of the quantity. Simply stated, the combination of two or more numbers in an array are called *vectors*. Vectors consist of components specifying a direction in space. The term *vector* is used to indicate a quantity that has both magnitude (a scalar) and direction. A vector is often represented by an arrow or a directed line segment. The length of the arrow represents the magnitude of the vector and the arrow points in the direction of the vector.

Physical quantities of the vector type are velocities, forces, torques, and so forth. Vectors can be two, three, or n dimensional. However, in this text, we consider vectors in three-dimensional Euclidian space. As an historical aside, it is interesting to note that the vector quantities listed are all taken from mechanics, but that vector analysis was not used in the development of mechanics and, indeed, had not been created. The need of vector analysis became apparent only with the development of Maxwell's electromagnetic theory and in appreciation of the inherent vector nature of quantities such as electric field and magnetic field (see Chapter 4).

Vectors are characterized by a magnitude and a direction in space. As we will see in a moment, vectors are defined by their transformation properties. Consider a coordinate transformation of the type

$$\tilde{x}_i = \sum_j \lambda_{ij} x_j \tag{2.2.23}$$

with

$$\sum_j \lambda_{ij} \lambda_{kj} = \delta_{ij}. \tag{2.2.24}$$

If under such a transformation a quantity $\phi = \phi(x_1, x_2, x_3)$ is unaffected, then ϕ is called a scalar.

If a set of quantities (A_1, A_2, A_3) is transformed from the x_i system to the $\tilde{x}_i$ system by means of a transformation matrix λ with the result

$$\tilde{A}_i = \sum_j \lambda_{ij} A_j, \tag{2.2.25}$$

then the quantities A_i transform as the coordinates of a point and the quantity $\vec{A} = (A_1, A_2, A_3)$ is termed a vector.

A vector can be conveniently represented by an arrow with length proportional to the magnitude. The direction of the arrow gives the direction of the vector, the positive sense of direction being indicated by the point. In this representation, vector addition, e.g.

$$\vec{C} = \vec{A} + \vec{B}$$

$$\vec{A} + \vec{B}$$

consists in placing the back end of vector $\vec{B}$ at the point of vector $\vec{A}$. Vector $\vec{C}$ is then represented by an arrow drawn from the back of $\vec{A}$ to the point of $\vec{B}$. This procedure, the triangle law of addition, assigns meaning to the Equation (2.2.25) and is illustrated in Figure 2.2.4.

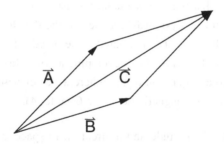

Figure 2.2.4.   Triangle law of vector addition.

By completing the parallelogram, we see that

$$\vec{C} == \vec{B} + \vec{A}$$

True

Note that the vectors are treated as geometrical objects that are independent of any coordinate system. Indeed, we have not yet introduced a coordinate system.

A direct physical example of this triangle addition law is provided by a weight suspended by two cords (Figure 2.2.5). If the junction point O is in equilibrium, the vector sum of the two forces $\vec{F}_1$ and $\vec{F}_2$ must just cancel the downward force of gravity, $\vec{F}_3$. Here, the triangle addition law is subject to immediate experimental verification.

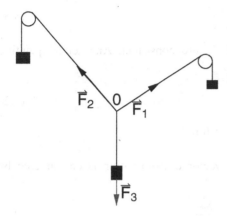

Figure 2.2.5. Equilibrium of forces. $\vec{F}_1 + \vec{F}_2 = \vec{F}_3$.

2.2.6 Tensors

Physical quantities can be of still higher complexity than scalars and vectors. For example, the inertia of a rigid body is described by a tensor. Tensors are distinguished by their rank. The combination of n vectors in an array generates, in general, an n-rank tensor. In this scheme, scalars are tensors of rank zero and vectors are first-rank tensors. A tensor of the second rank, for example, has $3^2 = 9$ components. A tensor can usually be said to define the dependence of a vector upon another vector.

In Section 2.2.4, a quantity that did not change under rotations of the coordinate system that is, an invariant quantity, was labeled a scalar. A quantity whose components transformed like those of the distance of a point from a chosen origin was called a vector (see Section 2.2.5). The transformation property was adopted as the defining characteristic of a vector. There is a possible ambiguity in definition (2.2.23)

$$\tilde{x}_i = \sum_j \lambda_{ij} x_j \tag{2.2.26}$$

in which λ_{ij} is the cosine of the angle between the $\tilde{x}_i$-axis and the x_j-axis.

If we start with our prototype vector $\vec{x}$, then

$$\tilde{x}_i = \sum_j \frac{\partial \tilde{x}_i}{\partial x_j} x_j \tag{2.2.27}$$

by partial differentiation. If we set

$$\lambda_{ij} = \frac{\partial \tilde{x}_i}{\partial x_j}, \tag{2.2.28}$$

Equations (2.2.26) and (2.2.27) are consistent. Any set of quantities x_j transforming according to

$$\tilde{x}_i = \sum_j \frac{\partial \tilde{x}_i}{\partial x_j} x_j \tag{2.2.29}$$

is defined as a contravariant vector.

A slightly different type of vector transformation is encountered by the gradient $\nabla \phi$, defined by

$$\nabla \phi = \vec{i} \frac{\partial \phi}{\partial x_1} + \vec{j} \frac{\partial \phi}{\partial x_2} + \vec{k} \frac{\partial \phi}{\partial x_3}, \tag{2.2.30}$$

where the vectors $\vec{i}$, $\vec{j}$, and $\vec{k}$ denote the unit vectors of the coordinate system. The gradient transforms as

$$\frac{\partial \tilde{\phi}}{\partial \tilde{x}_i} = \sum_j \frac{\partial \phi}{\partial x_j} \frac{\partial x_j}{\partial \tilde{x}_i}, \tag{2.2.31}$$

using $\phi = \phi(x_1, x_2, x_3)$ and $\tilde{\phi} = \phi(\tilde{x}_1, \tilde{x}_2, \tilde{x}_3)$ defined as a scalar quantity. Notice that this differs from Equation (2.2.29) in that we have $\partial x_j / \partial \tilde{x}_i$ instead of $\partial \tilde{x}_i / \partial x_j$. Equation (2.2.31) is taken as the definition of a *covariant vector* with the gradient as the prototype.

In cartesian coordinates,

$$\frac{\partial x_j}{\partial \tilde{x}_i} = \frac{\partial \tilde{x}_i}{\partial x_j} = \lambda_{ij}, \tag{2.2.32}$$

and there is no difference between contravariant and covariant transformations. In other systems, Equation (2.2.32), in general, does not apply, and the distinction between contravariant and covariant is real and must be observed. In the remainder of this section, the components of a contravariant vector are denoted by a superscript, x^i, whereas a subscript is used for the components of a covariant vector x_i.

To remove some of the fear and mystery from the term *tensor*, let us rechristen a scalar as a tensor of rank zero and relabel a vector as a tensor of first rank. Then, we proceed to define contravariant, mixed, and covariant tensors of second rank by the following equations:

$$\tilde{A}^{ij} = \sum_{kl} \frac{\partial \tilde{x}_i}{\partial x_k} \frac{\partial \tilde{x}_j}{\partial x_l} A^{kl}, \qquad (2.2.33)$$

$$\tilde{B}^i_j = \sum_{kl} \frac{\partial \tilde{x}_i}{\partial x_k} \frac{\partial x_l}{\partial \tilde{x}_j} B^k_l, \qquad (2.2.34)$$

$$\tilde{C}_{ij} = \sum_{kl} \frac{\partial x_k}{\partial \tilde{x}_i} \frac{\partial x_l}{\partial \tilde{x}_j} C_{kl}. \qquad (2.2.35)$$

We see that A^{kl} is contravariant with respect to both indices, C_{kl} is covariant with respect to both indices, and B^k_l transforms contravariantly with respect to the first index k but covariantly with respect to the second index l. Once again, if we are using cartesian coordinates, all three forms of the tensors of second rank, contravariant, mixed, and covariant, are the same.

The second-rank tensor A (components A^{ij}) can be conveniently represented by writing out its components in a square array (3×3 if we are in three-dimensional space):

$$A = \begin{pmatrix} A^{11} & A^{12} & A^{13} \\ A^{21} & A^{22} & A^{23} \\ A^{31} & A^{32} & A^{33} \end{pmatrix}. \qquad (2.2.36)$$

This does not mean that any square array of numbers or functions forms a tensor. The essential condition is that the components transform according Equations (2.2.33–2.2.35).

This transformation requirement can be illustrated by examining in detail the two-dimensional tensor:

$$T = \begin{pmatrix} -x\,y & -y^2 \\ x^2 & x\,y \end{pmatrix}$$

$$\begin{pmatrix} -x\,y & -y^2 \\ x^2 & x\,y \end{pmatrix}$$

In a rotated coordinate system the $\tilde{T}^{11}$ component must be $-\tilde{x}\,\tilde{y}$, as discussed for vectors. We check to see if this is consistent with the defining Equation (2.2.33):

$$\tilde{T}^{11} = -\tilde{x}\,\tilde{y} = \sum_{kl} \frac{\partial \tilde{x}_1}{\partial x_k} \frac{\partial \tilde{x}_1}{\partial x_l} T^{kl}$$
$$= \sum_{kl} \lambda_{1k} \lambda_{1l} T^{kl} \tag{2.2.37}$$

setting i and j equal to 1. Then, with the rotation matrix given by

$$\mathbf{lam} = \begin{pmatrix} \cos(\theta) & \sin(\theta) \\ -\sin(\theta) & \cos(\theta) \end{pmatrix}$$

$$\begin{pmatrix} \cos(\theta) & \sin(\theta) \\ -\sin(\theta) & \cos(\theta) \end{pmatrix}$$

we can represent the original vector $\vec{x} = (x,\,y)$ in the transformed system as

$$\vec{r} = \mathbf{Flatten}\left[\mathbf{lam}.\begin{pmatrix} x \\ y \end{pmatrix}\right]$$

$$\{x\cos(\theta) + y\sin(\theta),\ y\cos(\theta) - x\sin(\theta)\}$$

Combining the coordinates from the transformed vector and the right-hand side of Equation (2.2.37), we end up with an identity:

$$-\tilde{r}[\![1]\!]\ \tilde{r}[\![2]\!] \ == \ \sum_{k=1}^{2}\sum_{l=1}^{2} \mathbf{lam}[\![1,\,k]\!]\ \mathbf{lam}[\![1,\,l]\!]\ T[\![k,\,l]\!]\ //\ \mathbf{Simplify}$$

True

Repetition of the other three components verify that all transform in accordance of Equation (2.2.33) and that T is, therefore, a second-rank tensor.

This transformation property is not something to be taken for granted. For instance, if one algebraic sign were changed, if T^{22} were $-x\,y$ instead of $+x\,y$, then the array is

$$T = \begin{pmatrix} -x\,y & -y^2 \\ x^2 & -x\,y \end{pmatrix};$$

and condition (2.2.33) reduces to

$$-\vec{r}[\![1]\!]\ \vec{r}[\![2]\!] == \sum_{k=1}^{2}\sum_{l=1}^{2} \texttt{lam}[\![1,\ k]\!]\ \texttt{lam}[\![1,\ l]\!]\ \texttt{T}[\![k,\ l]\!]\ //\ \texttt{Simplify}$$

$$2\,x\,y\sin^2(\theta) == 0$$

which states that the equality is not satisfied and, thus, T is not a tensor because it does not require the transformation properties.

The addition and subtraction of tensors is defined in terms of the individual elements just as for vectors. To add or subtract two tensors, the corresponding elements are added or subtracted. If

$$A + B = C, \tag{2.2.38}$$

then

$$A^{ij} + B^{ij} = C^{ij}. \tag{2.2.39}$$

Of course, A and B must be tenors of the same rank and both expressed in a space of the same number of dimensions.

2.2.7 Vector Products

Having defined vectors, we now proceed to combine them. The laws for combining vectors must be mathematically consistent. From the possibilities that are consistent, we select two that are both mathematically and physically interesting.

The combination of $AB\cos(\theta)$, in which A and B are the magnitudes of two vectors and θ, the angle between them, occurs frequently in physics. For instance,

Work = Force $*$ Displacement $*$ $\cos(\theta)$

is usually interpreted as displacement times the projection of the force along the displacement. With such application in mind, we define

$$\vec{A}.\vec{B} == \sum_i A_i B_i$$

as the scalar product of $\vec{A}$ and $\vec{B}$. We note that for this definition $\vec{A}.\vec{B} = \vec{B}.\vec{A}$.

We have not yet shown that the word scalar is justified or that the scalar product is indeed a scalar quantity. First let us demonstrate that a vector $\vec{A}$ multiplied by itself is a scalar. For example

$$\vec{A}.\vec{A}$$

$$\vec{A}.\vec{A}$$

Now, let us define a vector $\vec{C}$ that is the sum of two other vectors $\vec{A}$ and $\vec{B}$:

$$\vec{C} = \vec{A} + \vec{B}$$

$$\vec{A} + \vec{B}$$

The scalar product of $\vec{C}$ with itself is thus

$\vec{C}.\vec{C}$ // **Expand**

$(\vec{A} + \vec{B}).(\vec{A} + \vec{B})$

Because the scalar product is commuting, we find with $\vec{C}.\vec{C} = C^2$

$\vec{A}.\vec{B} == \dfrac{1}{2}\,(-A^2 - B^2 + C^2)$

$\vec{A}.\vec{B} == \dfrac{1}{2}\,(-A^2 - B^2 + C^2)$

Because the right-hand side of this equation is invariant (i.e., a scalar quantity), the left-hand side, $\vec{A}.\vec{B}$, must also be invariant under rotation of the coordinate system. Hence, $\vec{A}.\vec{B}$ is a scalar.

Another property of the dot product is

$(\vec{A} - \vec{B}).(\vec{A} + \vec{B})$

$A^2 - B^2$

We next consider another method for the combination of two vectors, the so-called vector product or cross-product. For example, the angular momentum of a body is defined as

Angular momentum = Radius arm × Linear momentum

= Distance * Linear momentum* $\sin(\theta)$

First, we assert that this operation × does, in fact, produce a vector. The product considered here actually produces an axial vector, but the term *vector product* will be used in order to be consistent with popular usage. The vector product of $\vec{A}$ and $\vec{B}$ is denoted by a cross ×; older notation includes $[\vec{A}\,\vec{B}]$, $[\vec{A}.\vec{B}]$, and $[\vec{A} \wedge \vec{B}]$. For convenience in treating problems relating to quantities such as angular momentum, torque, and angular velocity, we define the cross-product as

$$\vec{C} = \vec{A} \times \vec{B}$$

$$\vec{A} \times \vec{B}$$

with

$$C = AB\sin(\theta) \qquad\qquad (2.2.40)$$

and where $\vec{C}$ is the vector that we assert results from this operation. The components of $\vec{C}$ are defined by the relation

$$C_i = \sum_{k,j} \epsilon_{ijk} A_j B_k \qquad\qquad (2.2.41)$$

where the symbol ϵ_{ijk} is the permutation symbol or Levi–Civita density and has the following properties:

$$\epsilon_{ijk} = \begin{cases} 0, & \text{if any index is equal to any other} \\ +1, & \text{if } i,\ j,\ k,\ \text{form an even permutation of 1, 2, 3} \\ -1, & \text{if } i,\ j,\ k\ \text{form an odd permutation of 1, 2, 3.} \end{cases}$$

We note that $\vec{A} \times \vec{B}$ is perpendicular to the plane defined by $\vec{A}$ and $\vec{B}$ because

$$\vec{A}.\left(\vec{A} \times \vec{B}\right)$$

0

and

$$\vec{B}.\left(\vec{A} \times \vec{B}\right)$$

0

Since a plane area can be represented by a vector normal to the plane and of magnitude equal to the area, evidently $\vec{C}$ is such a vector. The positive direction of $\vec{C}$ is chosen to be the direction of advance of a right-hand screw when rotated from $\vec{A}$ to $\vec{B}$.

We should note the following properties of the vector product which results from the definitions:

$$\vec{A} \times \vec{B} == -\vec{B} \times \vec{A}$$

but, in general,

$$\vec{C} =.$$

$$\vec{A} \times (\vec{B} \times \vec{C}) =!= (\vec{A} \times \vec{B}) \times \vec{C}$$

True

meaning that the cross-product is not associative. Another important result of the cross-product is

$$\vec{A} \times (\vec{B} \times \vec{C})$$

$$\vec{A}.\vec{C}\,\vec{B} - \vec{A}.\vec{B}\,\vec{C}$$

The scalar product of two cross-products is expressed by the difference of two dot products

$$(\vec{A} \times \vec{B}).(\vec{C} \times \vec{D})$$

$$\vec{A}.\vec{C}\,\vec{B}.\vec{D} - \vec{A}.\vec{D}\,\vec{B}.\vec{C}$$

The following identities useful in simplifying some expressions are stated without proof:

$\vec{A}.(\vec{B} \times \vec{C})$
$\vec{B} \times \vec{C}.\vec{A}$

$\vec{A} \times (\vec{B} \times \vec{C})$
$\vec{A}.\vec{C}\,\vec{B} - \vec{A}.\vec{B}\,\vec{C}$

$(\vec{A} \times \vec{B}) \times (\vec{C} \times \vec{D})$
$\vec{A} \times (\vec{B}.\vec{D})\,\vec{C} - \vec{A} \times (\vec{B}.\vec{C})\,\vec{D}$

$(\vec{A} \times \vec{B}).(\vec{A} \times \vec{B})$
$A^2\,B^2 - (\vec{A}.\vec{B})^2$

The sum of a cyclic permutation of a triple cross-product vanishes:

$\vec{A} \times (\vec{B} \times \vec{C}) + \vec{B} \times (\vec{C} \times \vec{A}) + \vec{C} \times (\vec{A} \times \vec{B})$
0

Applying the rules from above to the following example, we are able to simplify this expression to

$(\vec{A} - \vec{B}) \times (\vec{A} + \vec{B})$
$2\,\vec{A} \times \vec{B}$

2.2.8 Derivatives

If a scalar function $\phi = \phi(s)$ is differentiated with respect to the scalar variable s, then because neither part of the derivative can change under a coordinate transformation, the derivative itself cannot change and must therefore be a scalar; that is, in the x_i and $\tilde{x}_i$ coordinate systems, $\phi = \tilde{\phi}$ and $s = \tilde{s}$, so that $d\phi = d\tilde{\phi}$ and $ds = d\tilde{s}$. Hence,

$$\frac{d\phi}{ds} = \frac{d\tilde{\phi}}{d\tilde{s}} = \left(\frac{\tilde{d\phi}}{ds}\right). \tag{2.2.42}$$

In a similar manner, we can formally define the differentiation of a vector $\vec{A}$ with respect to a scalar s. The components of $\vec{A}$ transform according to

$$\tilde{A}_i = \sum_j \lambda_{ij} A_j. \tag{2.2.43}$$

Therefore, upon differentiation, we obtain, since λ is independent of $\tilde{s}$,

$$\frac{d\tilde{A}_i}{d\tilde{s}} = \frac{d}{d\tilde{s}} (\sum_j \lambda_{ij} A_j) = \sum_j \lambda_{ij} \frac{dA_j}{d\tilde{s}}. \tag{2.2.44}$$

Since s and $\tilde{s}$ are identical, we have

$$\frac{d\tilde{A}_i}{d\tilde{s}} = \left(\frac{\tilde{dA_i}}{ds}\right) = \sum_j \lambda_{ij} \frac{dA_j}{ds}. \tag{2.2.45}$$

Thus, the quantities dA_j/ds transform as do the components of a vector and, hence, are the components of a vector which we can write as $d\vec{A}/ds$.

The derivatives of vector sums and products obey the rules of ordinary vector calculus; for example,

$$\frac{\partial \left(\vec{A}(s) + \vec{B}(s)\right)}{\partial s}$$

$$\vec{A}'(s) + \vec{B}'(s)$$

The dot product differentiated gives

$\dfrac{\partial\left(\vec{A}(s).\vec{B}(s)\right)}{\partial s}$
$\vec{A}(s).\vec{B}'(s) + \vec{B}(s).\vec{A}'(s)$

Differentiation of a cross-product results in

$\dfrac{\partial\left(\vec{A}(s)\times\vec{B}(s)\right)}{\partial s}$
$\vec{A}(s)\times\vec{B}'(s) + \vec{A}'(s)\times\vec{B}(s)$

A product of a scalar and a vector yields

$\dfrac{\partial\left(\phi(s)\,\vec{A}(s)\right)}{\partial s}$
$\vec{A}(s)\,\phi'(s) + \phi(s)\,\vec{A}'(s)$

Knowing that a vector depending on a scalar can be differentiated without changing the nature, we now turn to the discussion of the most important member of a class called vector differential operators. The most important operator of this class is the gradient operator.

Consider a scalar ϕ which is an explicit function of the coordinates x_j and, moreover, is a continuous, single-valued function of these coordinates throughout a certain region of space. Then, under a coordinate transformation that carries the x_i into the $\tilde{x}_i$, $\tilde{\phi}(\tilde{x}_1, \tilde{x}_2, \tilde{x}_3) = \phi(x_1, x_2, x_3)$, and by the chain rule of differentiation, we can write

$$\frac{\partial\tilde{\phi}}{\partial\tilde{x}_1} = \sum_j \frac{\partial\phi}{\partial x_j}\frac{\partial x_j}{\partial\tilde{x}_1}. \tag{2.2.46}$$

Similarly, we obtain for $\partial\tilde{\phi}/\partial\tilde{x}_2$ and $\partial\tilde{\phi}/\partial\tilde{x}_3$, so that in general we have

$$\frac{\partial\tilde{\phi}}{\partial\tilde{x}_i} = \sum_j \frac{\partial\phi}{\partial x_j}\frac{\partial x_j}{\partial\tilde{x}_i}. \tag{2.2.47}$$

Now, the inverse coordinate transformation is

$$x_j = \sum_k \lambda_{kj} \, \tilde{x}_k. \tag{2.2.48}$$

Differentiating this relation, we find

$$\frac{\partial x_j}{\partial \tilde{x}_i} = \frac{\partial}{\partial \tilde{x}_i} \left(\sum_k \lambda_{kj} \, \tilde{x}_k \right) = \sum_k \lambda_{kj} \frac{\partial \tilde{x}_k}{\partial \tilde{x}_i}. \tag{2.2.49}$$

However, the term in the last expression is just δ_{ki}, so that

$$\frac{\partial x_j}{\partial \tilde{x}_i} = \sum_k \lambda_{kj} \delta_{ki} = \lambda_{ij}. \tag{2.2.50}$$

Substituting Equation (2.2.50) into Equation (2.2.47), we obtain

$$\frac{\partial \tilde{\phi}}{\partial \tilde{x}_i} = \sum_j \frac{\partial \phi}{\partial x_j} \lambda_{ij}. \tag{2.2.51}$$

Because it follows the correct transformation equation, the function $\partial \phi / \partial x_j$ is the jth component of a vector which is termed the gradient of the function ϕ. Note that even though ϕ is a scalar, the gradient of ϕ is a vector. The gradient of ϕ is written either as **grad** ϕ or as $\nabla \phi$.

Since the function ϕ is an arbitrary scalar function, it is convenient to define the differential operator described above in terms of the gradient operator

$$(\mathbf{grad})_i = \nabla_i = \frac{\partial}{\partial x_i} \tag{2.2.52}$$

We can express the complete vector operator as

$$\mathrm{grad} = \nabla = \sum_i \vec{e}_i \frac{\partial}{\partial x_i}. \tag{2.2.53}$$

The gradient of a scalar function is of extreme importance in physics expressing the relation between a force field and a potential field.

Force = $-\nabla$ potential

Thus, we can state that the gradient operator can operate directly on a scalar function as $\nabla \phi$, be used in a scalar product with a vector function in $\nabla . \vec{A}$ called the divergence of $\vec{A}$, or be used in a vector product with a vector function as in $\nabla \times \vec{A}$ which is known as the curl of $\vec{A}$.

The successive operation of the gradient operator produces

$$\nabla . \nabla = \sum_i \frac{\partial}{\partial x_i} \frac{\partial}{\partial x_i} = \sum_i \frac{\partial^2}{\partial x_i^2}. \tag{2.2.54}$$

This important product operator is called the Laplacian and is also written

$$\nabla^2 = \sum_i \frac{\partial^2}{\partial x_i^2}.$$

The following lines demonstrate some properties of the gradient. First, we check the product rule on scalar functions. Given two scalar functions $\phi = \phi(\vec{x})$ and $\psi = \psi(\vec{x})$, we are interested in the action of ∇ on the product:

```
Nabla[φ ψ ]
```

$\psi \nabla\phi + \phi \nabla\psi$

Another example is the calculation of the gradient for the function

$$S = (x^2 + y^2 + z^2)^{-3/2}$$

$$\frac{1}{(x^2 + y^2 + z^2)^{3/2}}$$

The gradient is a vector containing three components:

gr = Grad(S); gr // MatrixForm

$$\begin{pmatrix} -\dfrac{3x}{(x^2+y^2+z^2)^{5/2}} \\ -\dfrac{3y}{(x^2+y^2+z^2)^{5/2}} \\ -\dfrac{3z}{(x^2+y^2+z^2)^{5/2}} \end{pmatrix}$$

If we apply the divergence operator on this vector, we find

di = Simplify[Div(gr)]

$$\frac{6}{(x^2 + y^2 + z^2)^{5/2}}$$

We are also able to determine the curl of the gradient field:

Curl(gr) // MatrixForm

$$\begin{pmatrix} 0 \\ 0 \\ 0 \end{pmatrix}$$

---

saying that the curl of a gradient vanishes. This result, demonstrated for a specific example, has the generalization

---

`Nabla[] × Nabla[U]`

0

Another relation combining ∇ is the divergence of a curl applied to a vector field:

`Nabla[].Nabla[] × `$\vec{V}$

0

2.2.9 Integrals

The vector which results from the volume integration of a vector function $\vec{A} = \vec{A}(\vec{x})$ throughout a volume V is given by

$$\int_V \vec{A}\, dv = \left(\int_V A_1\, dv, \ \int_V A_2\, dv, \ \int_V A_3\, dv \right) \tag{2.2.55}$$

Thus, the integration of the vector $\vec{A}$ throughout V is accomplished simply by performing three separate, ordinary integrations.

The integral over the surface S of the projection of a vector function $\vec{A} = \vec{A}(\vec{x})$ onto that surface is defined to be

$$\int_S \vec{A}.d\,\vec{a} \tag{2.2.56}$$

where $d\vec{a}$ is an infinitesimal element of area of the surface. We write $d\vec{a}$ as a vector quantity since we can attribute to it not only a magnitude da but also a direction corresponding to the normal to the surface at the point in question. If the unit normal vector is $\vec{n}$, then

$$d\vec{a} = \vec{n}\,da. \qquad (2.2.57)$$

Therefore, we have

$$\int_S \vec{A}.d\vec{a} = \int_S \vec{A}.\vec{n}\,da \qquad (2.2.58)$$

or

$$\int_S \vec{A}\,d\vec{a} = \int_S \sum_i A_i\,da_i. \qquad (2.2.59)$$

Equation (2.2.58) states that the integral of $\vec{A}$ over a surface S is the integral of the normal component of $\vec{A}$ over this surface. The normal to a surface can be taken to lie in either of two possible directions (up or down); thus, the sign of $\vec{n}$ is ambiguous. If the surface is closed, we adopt the convention that the outward normal is positive.

The line integral of a vector function $\vec{A} = \vec{A}(\vec{x})$ along a given path extending from the point B to the point C is given by the integral of the component of $\vec{A}$ along the path:

$$\int_{BC} \vec{A}\,d\vec{s} = \int_{BC} \sum_i A_i\,dx_i. \qquad (2.2.60)$$

The quantity $d\vec{s}$ is an element of unit length along the given path. The direction of $d\vec{s}$ is taken to be positive along the direction in which the path is traversed.

The form (2.2.60) is exactly the same as that encountered when we calculate the work done with a force that varies along the path,

$$W = \int \vec{F}\,d\vec{r}. \qquad (2.2.61)$$

In this expression $\vec{F}$ is the force exerted on a particle.

2.210 Exercises

1. Show that the components of a vector $\vec{a}$ in the direction orthogonal to a vector $\vec{b}$ is

$$\vec{a} - (\vec{a}.\vec{b}) \frac{\vec{b}}{b^2} = \frac{1}{b^2} (\vec{b} \times (\vec{a} \times \vec{b})) \tag{2.2.62}$$

2. Show that under double reflection in two mirrors, one in the $x - z$ plane and te other in the $x - y$ plane, a axial vector transforms in the same way as a polar vector.

3. The transformation $\bar{x}_j = -x_j$ ($j = 1, 2, 3$) describes a reflection in the origin. Draw a diagram to illustrate how two radius vectors $\vec{r}_1$ and $\vec{r}_2$ and their vector product $\vec{r}_1 \times \vec{r}_2$ transform under this reflection.

4. Prove the following identities:
   a) $\nabla(\vec{B}.\vec{A}) = 2 \nabla.\vec{B}\vec{A} + \vec{B}.\nabla \vec{A} + 2 \nabla.\vec{A}\vec{B} + \vec{A}.\nabla \vec{B}$
   b) $\nabla \times (\nabla \times \vec{A})'' = \vec{A} \nabla^2 + \nabla.\vec{A} \nabla$
   c) $\nabla \times (\nabla \phi)'' = 0$

2.3 Kinematics

2.3.1 Introduction

Kinematics is concerned with the motion of a body. We assume that the motion is in itself present. At the moment, we do not ask for the origin of the motion (i.e., the forces causing the motion). The consideration of the forces will be discussed in Section 2.4, where we discuss the dynamics of a mass point.

Kinematics is concerned with the mathematical description of the path a body moves along. The body is taken as a mass m with vanishing extension. We call such an object a point mass or, in short, a particle. The location of this point is measured with respect to a second point, a reference point. This reference point is part of a fixed system. In practical applications, the fixed system is the Earth. Defining on the surface of the Earth, a fixed point allows us to introduce coordinates x, y, and z. These so-called cartesian coordinates allow us to locate a point in space by specifying the position by the triple (x, y, z) which may depend on time if the particle moves in space. In such a case, the location of the particle is given by a vector $\vec{r} = \vec{r}(t)$ given by

$$\vec{r}(t) = \begin{pmatrix} x(t) \\ y(t) \\ z(t) \end{pmatrix}. \tag{2.3.1}$$

For a certain time t, the position of the point is given by Equation (2.3.1) (see Figure 2.3.1)

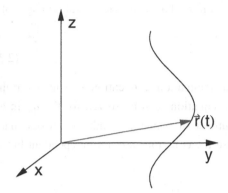

Figure 2.3.1.   A track in a coordinate system. Time is used as a parameter of the motion.

If time t changes continuously, the point moves along a track.

To characterize the particle in a more precise way, we not only need the position but also other quantities like the velocity or acceleration of the particle. The following subsections will discuss these terms in more detail.

2.3.2 Velocity

We already mentioned that the coordinates describing a particle may vary with time. Let us consider a system consisting of a single particle. The position of the particle is described by the values of its cartesian coordinates x_i at each value of the time t. The rate at which these coordinates change with time gives the velocity of a particle. Denoting the cartesian components of velocity by v_i, we have

$$v_i = \frac{dx_i}{dt} = \dot{x}_i.$$ (2.3.2)

This can be written in vector notation as

$$\vec{v} = \frac{d\vec{r}}{dt} = \dot{\vec{r}}.$$ (2.3.3)

Velocity can be described in terms of generalized coordinates of Section 2.2. From Equation (2.2.6), we see that

$$x_i = x_i(q_1, q_2, q_3, t).$$ (2.3.4)

depend on the q_i and also on t. Then, the temporal change of the coordinates is

$$\dot{x}_i = \frac{\partial x_i}{\partial q_m}\, \dot{q}_m + \frac{\partial x_i}{\partial t}, \tag{2.3.5}$$

where we used the Einstein summation convention to sum over the m components. If desired, these equations can be solved for the $\dot{q}_m$ in terms of the $\dot{x}_i$ even though the number of equations might be greater than the number of unknowns, because the equations are not independent but must satisfy the constraints.

The cartesian components of velocity are seen to be linear functions of the generalized velocity components and are, in general, nonlinear functions of the q_i's no matter how the generalized coordinates are defined. This means that it is easy to express velocities in generalized coordinates. The term $\partial x_j / \partial t$ appears only when there are moving constraints on the system or in the rare cases where it is convenient to introduce moving coordinate axes.

Example 1: Coordinate Systems

We can apply the results from above to commonly used special coordinate systems defined by Equations (2.2.1–2.2.3). In order to describe the motion of a particle in a plane, it is convenient to introduce the plane polar coordinates r and θ defined by Equations (2.2.1) and (2.2.2), r being the length of the position vector of the particle and θ the angle between the position vector and the x-axis (see Figure 2.3.2).

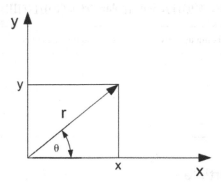

Figure 2.3.2. Polar coordinates.

We solve Equations (2.2.1) and (2.2.2) for x and y and assume that all coordinates depend on t. In *Mathematica* notation, we write

polar = {$x(t) == r(t)\cos(\theta(t))$, $y(t) == r(t)\sin(\theta(t))$}; TableForm[polar]

$x(t) == \cos(\theta(t))\, r(t)$
$y(t) == r(t)\sin(\theta(t))$

The velocity components are found by differentiating this relation with respect to time:

velocity = $\dfrac{\partial\,\text{polar}}{\partial t}$; TableForm[velocity]

$x'(t) == \cos(\theta(t))\, r'(t) - r(t)\sin(\theta(t))\, \theta'(t)$
$y'(t) == \sin(\theta(t))\, r'(t) + \cos(\theta(t))\, r(t)\, \theta'(t)$

The terms in r' give the velocity toward or away from the origin, and those in θ' give the velocity around the origin. The equations can be solved for r' and θ'

> **TableForm[Flatten[Simplify[PowerExpand[FunctionExpand[**
> **Solve[velocity, {r′(t), θ′(t)}] /. Solve[polar, {r(t), θ(t)}][[4]]]]]]]**

Solve::ifun : Inverse functions are being used by Solve, so some solutions may not be found.

$$r'(t) \rightarrow \frac{x(t)\, x'(t) + y(t)\, y'(t)}{\sqrt{x(t)^2 + y(t)^2}}$$

$$\theta'(t) \rightarrow \frac{x(t)\, y'(t) - y(t)\, x'(t)}{x(t)^2 + y(t)^2}$$

Example 2: Moving Particle

For a particle moving under the influence of a force which possesses spherical symmetry (i.e., is directed toward a fixed point and depends only on the distance of the particle from the point), it is convenient to introduce the spherical coordinates defined by Equation (2.2.3). We solved for the x's; these yield

> **spherical = {$x(t) == r(t)\sin(\theta(t))\cos(\phi(t))$, $y(t) == r(t)\sin(\theta(t))\sin(\phi(t))$,**
> **$z(t) == r(t)\cos(\theta(t))$}; TableForm[spherical]**

$x(t) == \cos(\phi(t))\, r(t)\sin(\theta(t))$
$y(t) == r(t)\sin(\theta(t))\sin(\phi(t))$
$z(t) == \cos(\theta(t))\, r(t)$

The geometrical significance of these coordinates is the following: r is the length of the radius vector of the particle; θ is the angle between the radius vector and the z-axis; ϕ is the angle between the plane containing the radius vector and the z-axis and the plane containing the x-axis and the z-axis.

The velocity components are found by differentiating with respect to t:

$\text{velocity} = \dfrac{\partial \text{spherical}}{\partial t}$
$\{x'(t) == \cos(\phi(t)) \sin(\theta(t)) r'(t) +$ $\quad \cos(\theta(t)) \cos(\phi(t)) r(t) \theta'(t) - r(t) \sin(\theta(t)) \sin(\phi(t)) \phi'(t),$ $\quad y'(t) == \sin(\theta(t)) \sin(\phi(t)) r'(t) + \cos(\theta(t)) r(t) \sin(\phi(t)) \theta'(t) +$ $\quad \cos(\phi(t)) r(t) \sin(\theta(t)) \phi'(t), \ z'(t) == \cos(\theta(t)) r'(t) - r(t) \sin(\theta(t)) \theta'(t)\}$

2.3.3 Acceleration

In general, the components of the position vector (as well as velocity and acceleration) depend on the generalized coordinates and their time derivatives. So, the velocity components of a particle can vary with time. The rate of change of the velocity components gives the *acceleration* components

$$a_i = v'_i = x''_i. \tag{2.3.6}$$

The acceleration in vector notation for a particle is

$$\vec{a} = \vec{v}' = \vec{r}''. \tag{2.3.7}$$

Acceleration components can be given in terms of the generalized coordinates. From Equation (2.3.5), we obtain

$$x''_i = \frac{\partial x_i}{\partial q_m} q''_m + \frac{\partial^2 x_i}{\partial q_m \partial q_n} q'_m q'_n + 2 \frac{\partial^2 x_i}{\partial q_m \partial t} q'_m + \frac{\partial^2 x_i}{\partial t^2}. \tag{2.3.8}$$

If the transformation from the x_i's to the q_i's does not depend explicitly on time, which is the usual situation, Equation (2.3.8) reduces to

$$x''_i = \frac{\partial x_i}{\partial q_m} q''_m + \frac{\partial^2 x_i}{\partial q_m \partial q_n} q'_m q'_n. \tag{2.3.9}$$

The cartesian acceleration components are nonlinear functions of the first derivatives of the generalized coordinates and depend linearly on the second derivatives of the generalized velocity components. The quadratic dependence on the generalized velocity coordinates disappears only if all of the second derivatives of the transformation function with respect to the q's vanish; that is, only if the x's are linear functions of the q's. The terms quadratic in the velocity components, which enter whenever the coordinate curves q_i = const are not straight lines, represent effects like the centripetal and Coriolis acceleration.

Higher derivatives of the coordinates with respect to time could be named and discussed. However, this proves unnecessary because the laws of mechanics are stated in terms of the acceleration. Even the computation of the components (2.3.8) of the acceleration in terms of generalized coordinates can become tedious for relatively simple problems. The advantage of introducing generalized coordinates would then seem to be counterbalanced by the algebraic complexity of the acceleration components which are to be inserted in the dynamic law $\vec{F} = m\vec{a}$. Fortunately, a method due to Lagrange, discussed in Section 2.7, makes it possible to avoid this difficulty and to write down equations of motion in terms of generalized coordinates without ever having to compute the second time derivatives of these coordinates.

2.3.4 Kinematic Examples

Having the fundamental quantities such as velocity and accelaration available, we are able to examine physical systems. In the following we will examine two examples demonstarting the application of the notons introduced.

Example 1: Motion on a Helix

As a first example of kinematics, let us consider the motion of a bead with constant orbital velocity confined to a helix. This motion can be divided into two parts. First, we have a circular motion of the bead in the (x, y)-plane and a linear motion in the z-direction. The motion of the bead can be described in a parametric way by using time t as a parameter. For example, the three coordinates are given by

coordinates = $\{\rho \sin(\omega t),\ \rho \cos(\omega t),\ \gamma t\}$
$\{\rho \sin(t\,\omega),\ \rho \cos(t\,\omega),\ t\,\gamma\}$

where ρ, ω, and γ are certain parameters determining the radius, the time of revolution, and the velocity along the z-direction, respectively. The velocity of this track is given by

$$\text{velocity} = \frac{\partial\,\text{coordinates}}{\partial t}$$

$$\{\rho\,\omega\cos(t\,\omega),\ -\rho\,\omega\sin(t\,\omega),\ \gamma\}$$

The amount of the velocity is determined by the three parameters α, β, and γ:

$$\text{Simplify}\left[\sqrt{\text{velocity.velocity}}\,\right]$$

$$\sqrt{\gamma^2 + \rho^2\,\omega^2}$$

The acceleration follows by

$$\text{acceleration} = \frac{\partial^2\,\text{coordinates}}{\partial t\,\partial t}$$

$$\{-\rho\,\omega^2\sin(t\,\omega),\ \ \rho\,\omega^2\cos(t\,\omega),\ 0\}$$

which has a total amount

$$\text{PowerExpand}\left[\text{Simplify}\left[\sqrt{\text{acceleration.acceleration}}\,\right]\right]$$

$$\rho\,\omega^2$$

independent of γ. If we choose these parameters in a certain way, we can plot the path of the bead in cartesian coordinates. Let us take a circle of radius $\rho = 1$, $\omega = 1/2$, and the velocity along the z-direction $\gamma = 1/10$.

```
parameterRules = {ρ → 1, ω → 1/2, γ → 1/10}
```

$$\left\{\rho \to 1, \omega \to \frac{1}{2}, \gamma \to \frac{1}{10}\right\}$$

With these values, the three coordinates of the bead simplify to

Coord = coordinates /. parameterRules

$$\left\{\sin\left(\frac{t}{2}\right), \cos\left(\frac{t}{2}\right), \frac{t}{10}\right\}$$

The related velocity and acceleration are

vel = velocity /. parameterRules;

accel = acceleration /. parameterRules;

The track of the bead can be displayed by plotting the coordinates by varying the parameter *t*:

```
track = ParametricPlot3D[Coord, {t, 0, 10 π}];
```

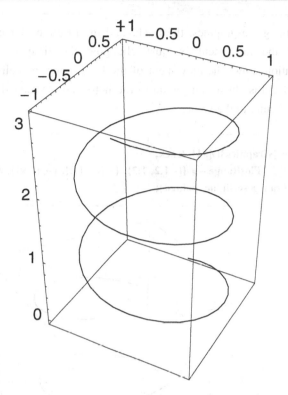

The motion of the bead itself follows by creating a table containing the coordinates along the track. Again, for each point, we change the time t in steps of 0.5:

```
points = Table[{RGBColor[0, 0, 0.996109],
        PointSize[.1], Point[Coord]}, {t, 0, 10 π, .5}];
```

We also generate a table containing the velocity of the bead

```
li = Table[{RGBColor[0.996109, 0, 0],
        Line[{Coord, Coord + vel}]}, {t, 0, 10 π, .5}];
```

and the acceleration

```
ac = Table[{RGBColor[0, 0.500008, 0],
        Line[{Coord, accel + Coord}]}, {t, 0, 10 π, .5}];
```

Combining the graphics, the track, and the location of the bead, we can follow the movement by just changing the time *t*. The following illustration shows the movement of the bead along the helix. The velocity of the bead is always tangential to the helix and the related acceleration is perpendicular to the velocity:

```
(Show[Graphics3D[#1], track,
        PlotRange -> {{-1.2, 1.2}, {-1.2, 1.2}, {-.1, 3.}}] &) /@
    Transpose[{li, ac, points}];
```

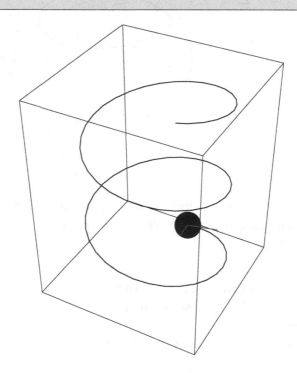

The illustration demonstrates that the bead climbs up the helix with a constant speed and revolves around the center of the (x, y)-plane. We can check that $\vec{v}$ is perpendicular to $\vec{a}$ by the scalar product:

```
velocity . acceleration
```

```
0
```

At this stage of our understanding, we described a physical system (a bead) by means of a parametric description. At the moment, we do not understand what kind of laws this motion governs. A similar situation is encountered if we describe the kinematic movement of a projectile.

Example 2: Motion of a Projectile

The motion of a projectile was and is an example of importance because in a baseball or golf play, we need to know where the ball touches down if we give it a strike. Ancient people needed also to know where the stones or bullets go if they are thrown by a bow. Applications for military purposes are evident.

In this example let us consider the motion of a projectile or a ball in the atmosphere. In our considerations, we neglect the air resistance. Furthermore, we consider only kinematics; we also demand that the projectile follows a parabolic orbit with a vertical symmetry axis and with constant horizontal velocity. The motion of the ball takes place in a three-dimensional space; thus, the velocity and the location of the ball is a certain vector with three components, respectively. These components are independent of each other and, thus, can be considered on its own. Thus we are able to separate each direction of the motion from the others. If we assume that the projectile is moving in a plane, we only need two coordinates to describe the motion. Let us further assume that the ball is thrown with a finite velocity $\vec{v}_0$ inclined by an angle α with respect to the horizontal direction. To simplify things, let us first define the origin as the starting point of the ball. Later, we will generalize this to the situation where the starting point does not coincide with the origin. The track of the ball is defined by the parametric representation in t by

$$\text{track} = \left\{t\,\text{vx} + \text{x0}, \frac{1}{2}\,(-g)\,t^2 + \text{vy}\,t + \text{y0}\right\};$$

where x_0 and y_0 are the starting point and v_x and v_y are the velocities in x- and y-direction, respectively. The assumption that the origin is the starting point of the track causes the vanishing of x_0 and y_0.

```
cond1 = Solve[Thread[{0, 0} == track /. t → 0, List],
    {x0, y0}] // Flatten
```

$\{x0 \to 0, y0 \to 0\}$

Inserting these initial conditions into the track, we find a simplified representation by

```
tracS = track /. cond1
```

$$\left\{t\,\text{vx},\ t\,\text{vy} - \frac{g\,t^2}{2}\right\}$$

The assumption that the ball is thrown in a certain direction with inclination α to the horizon and initial velocity v allows us to determine the parameters vx and vy in the track representation.

$$\text{cond2} = \text{Flatten}\Big[$$
$$\text{Solve}\Big[\text{Thread}\Big[\{v\cos(\alpha),\ v\sin(\alpha)\} == \frac{\partial\,\text{tracS}}{\partial t}\ /.\,t \to 0,\ \text{List}\Big],\ \{\text{vx, vy}\}\Big]\Big]$$

$\{\text{vx} \to v\cos(\alpha),\ \text{vy} \to v\sin(\alpha)\}$

Again inserting the results into the track coordinates, we end up with the final representation of the path by

> **tracS1 = tracS /. cond2**
>
> $$\left\{t\,v\cos(\alpha),\ t\,v\sin(\alpha) - \frac{g\,t^2}{2}\right\}$$

This representation contains two parameters v and α, the amount of the velocity and the inclination, respectively. Choosing this parameters allows us to plot the track of the ball:

> **BallTrack = ParametricPlot[Evaluate[tracS1 /. $\left\{v \to 1,\ \alpha \to \dfrac{\pi}{3},\ g \to 1\right\}$],**
>
> **{t, 0, 2}, AxesLabel $\to$ {"x", "z"}];**

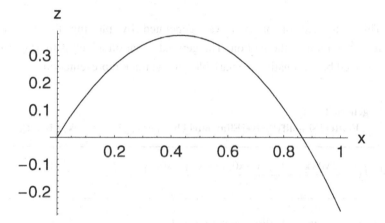

The ball itself for different times can be represented by the coordinates

> **Ball = Table[{RGBColor[0, 0, 0.62501], PointSize[.08],**
>
> **Point[tracS1 /. $\left\{v \to 1,\ \alpha \to \dfrac{\pi}{3},\ g \to 1\right\}$]}, {$t$, 0, 1.8, .1}];**

Combining both sets of data allows us to display the movement of the ball. The illustration shows that the ball moves first upward and then downward, hitting the ground at a finite distance.

> **(Show[BallTrack, Graphics[#1], PlotRange –> {−0.2, 0.5}] &) /@ Ball;**

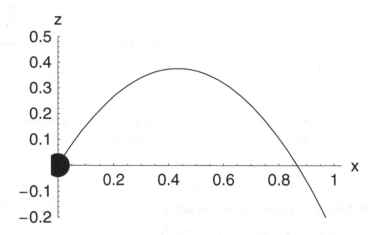

This sequence of pictures was generated by plotting the parametric representation of the motion. The general equation of the path $y(x)$ can be obtained be eliminating the variable t in the track representation:

> **generalTrack =**
> **Flatten[Simplify[Solve[Eliminate[Thread[{x, y} == track, List], t], y]]]**
>
> $$\left\{ y \rightarrow \frac{2\,y0\,vx^2 + 2\,vy\,(x - x0)\,vx - g\,(x - x0)^2}{2\,vx^2} \right\}$$

Writing out the velocity component yields

> **gTrack = Simplify[generalTrack /. cond2]**
>
> $$\left\{ y \rightarrow \frac{2\,v^2\,(y0 + (x - x0)\tan(\alpha)) - g\,(x - x0)^2\sec^2(\alpha)}{2\,v^2} \right\}$$

for the ball's path. This relation is of the form of a parabola passing through the point (x_0, y_0). The following figure shows the path of a ball:

```
Plot[y /. gTrack /. {x0 → 1, y0 → 2, v → 2, α → π/5, g → 9.81},
    {x, 1, 2.3}, AxesLabel → {"x", "y"},
    PlotStyle → RGBColor[0, 0, 0.996109], PlotRange → ( 0  2.5`
                                                        0  2.5` )];
```

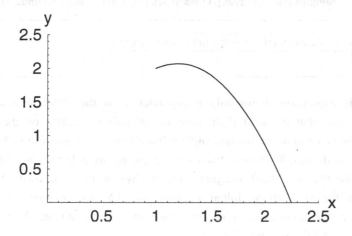

The range of the flight can be determined from the condition that the y elevation vanishes. This condition serves to determine x from the relation

```
ranges = Simplify[Solve[gTrack /. {Rule → Equal, y → 0}, x]]
```

$$\left\{\left\{x \rightarrow \frac{\cos(\alpha)\sin(\alpha)v^2 + \cos^2(\alpha)\sqrt{2gy0\sec^2(\alpha) + v^2\tan^2(\alpha)}\,v + g\,x0}{g}\right\},\right.$$
$$\left.\left\{x \rightarrow \frac{\cos(\alpha)\sin(\alpha)v^2 - \cos^2(\alpha)\sqrt{2gy0\sec^2(\alpha) + v^2\tan^2(\alpha)}\,v + g\,x0}{g}\right\}\right\}$$

The solution of the quadratic equation in x delivers two solutions. Because we are looking for positive ranges, we select the first solution:

```
range = ranges[[1]]
```

$$\left\{x \rightarrow \frac{\cos(\alpha)\sin(\alpha)v^2 + \cos^2(\alpha)\sqrt{2gy0\sec^2(\alpha) + v^2\tan^2(\alpha)}\,v + g\,x0}{g}\right\}$$

The total flight time T is gained by inserting this result into the x-component of the track and solving the resulting equation with respect to t.

T = Flatten[
** Simplify[Solve[Thread[{x} == track[[1]], List] /. range /. cond2, t]]]**

$$\left\{t \to \frac{\sqrt{2\,g\,y0\,\sec^2(\alpha) + v^2 \tan^2(\alpha)}\;\cos(\alpha) + v \sin(\alpha)}{g}\right\}$$

This expression shows only a dependence on the y initial condition, meaning that the total flight time is not only dependent on the initial velocity v and the inclination angle α but also on the height from which the ball is thrown. With these two expressions, we are able to solve problems of the following kind. Imagine a joyful physics student throwing his cap into the air with an initial velocity of 24.5 m/s at $36.9°$ from the horizontal. Find a) the total time the cap is in the air and b) the total horizontal distance traveled.

To solve this problem, we first have to convert the angle α given in degree into radians:

β = 36.92 $\dfrac{\pi}{360}$;

The other parameters are given by

---

para = {$v \to$ 24.5, $\alpha \to \beta$, $g \to$ 9.81, y0 $\to$ 0, x0 $\to$ 0};

Inserting the values into the expression for the total time of flight, we get

---

T /. para

{$t \to$ 2.99904}

---

corresponding to approximately 3 s. The range the cap transverses is given by

range /. para
$\{x \rightarrow 58.758\}$

in meters.

Another problem of the same kind is the following. A helicopter drops a supply package to soldiers in a jungle clearing. When the package is dropped, the helicopter is 100 m above the clearing and flying at 25 m/s at an angle $\alpha = 36.9°$ above the horizontal. The question is how wide must the clearing extend in one direction that the package is available for the soldiers and how long does it take to hit the ground?

We collect the numerical data in the list paraHeli and apply this rules to the expressions for the range and the total flight time derived above:

$$\text{paraHeli} = \left\{v \rightarrow 25, \alpha \rightarrow \frac{36.9\,2\,\pi}{360}, g \rightarrow 9.81, y0 \rightarrow 100, x0 \rightarrow 0\right\};$$

We find that the extension of the clearing should be

range /. paraHeli
$\{x \rightarrow 125.902\}$

in meters. The time to touch down is

T /. paraHeli
$\{t \rightarrow 6.29758\}$

in seconds. The following figure shows a graph of y versus x for supply packages dropped at various initial angles and with an initial speed of 25 m/s.

```
Plot[Evaluate[Table[y /. gTrack /. α → i /. paraHeli, {i, .2, .7, .1}]],
    {x, 1, 130}, AxesLabel → {"x", "y"},
    PlotStyle → Table[Hue[i], {i, .2, .7, .1}], PlotRange → {0, 120}];
```

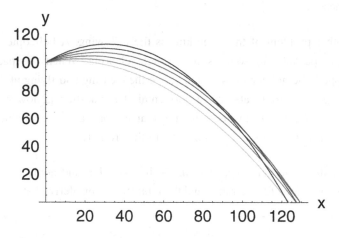

Note that the maximum range no longer occurs at 45 ° (verify this).

The examples demonstrate that once we know the path $\vec{r}(t)$ of a particle as a function of time t, we are able to answer any question related to the motion of the particle. The following sections of this chapter deal with the central problem of classical mechanics: to determine the path of one or more particles under the action of given forces.

2.3.5 Exercises

1. A particle is constrained to move with constant speed on the ellipse

$$a_{ij}\, x_i\, x_j = 1 \qquad (i, j = 1, 2).$$

Find the cartesian and polar components of its acceleration.

2. In Exercise 1, introduce as a generalized coordinate the angle θ between the radius vector of the particle and the major axis of the ellipse, and find the velocity and acceleration in terms of θ.

3. A particle is constrained to move with constant speed in the circle $r = a$. Find the cartesian and polar coordinates of its velocity and acceleration.

4. A gun is mounted on a hill of height h above a level plain. Assuming that the path of the projectile is a parabola, find the angle of elevation α for greatest horizontal range and given initial speed V,

$$\operatorname{cosec}^2(\alpha) = 2\,(1 + g\,h/V^2).$$

What physical effects are neglected in the above approximation?

2.4 Newtonian Mechanics

2.4.1 Introduction

The science of mechanics seeks to provide a precise and consistent description of the dynamics of particles and systems of particles; that is, we attempt to discover a set of physical laws that provide us with a method for mathematically describing the motion of bodies and aggregates of bodies. In order to do this, we need to introduce certain fundamental concepts. It is implicit in Newtonian theory that the concept of *distance* is intuitively understandable from a geometric viewpoint. Furthermore, *time* is considered to be an absolute quantity, capable of precise definition by an arbitrary observer. In the theory of relativity, however, we must modify these Newtonian ideas (see Chapter 6). The combination of the concepts of distance and time allows us to define the *velocity* and *acceleration* of a particle. The third fundamental concept, *mass*, requires some elaboration which we will give in connection with the discussion of Newton's laws.

The physical laws that we introduce must be based on experimental facts. A physical law can be characterized by the statement that it "might have been otherwise". Thus, there is no *a priori* reason to expect that the gravitational attraction between two bodies must vary exactly as the inverse square of the distance of them. However, experiment indicates that this is so. Once a set of data has been correlated and a postulate has been formulated regarding the phenomena to which the data refer, then various implications can be worked out. If these implications are all verified by experiment, there is reason to believe that the postulate is generally true. The postulate then assumes the status of a physical law. If some experiments are found to be in disagreement with the predictions of the law, then the theory must be modified in order to be consistent with all known facts.

Figure 2.4.1.   Sir Isaac Newton born January 4, 1643; died March 31, 1727.

Newton (Figure 2.4.1) has provided us with the fundamental laws of mechanics. We will state these laws in modern terms and discuss their meaning and then proceed to derive the implications of the laws in various situations. It must be noted, however, that the logical structure of the science of mechanics is not a straightforward issue. The line of reasoning that is followed here in interpreting Newton's laws is not the only one possible. An alternate interpretation is given by Ernst Mach (1838–1916).

Figure 2.4.2.   Ernst Mach born February 18, 1838; died February 19, 1916.

Mach expressed his views in his famous book, *The Science of Mechanics*, first published in 1883. We will not pursue in any detail the philosophy of mechanics, but will give only sufficient elaboration of Newton's laws to allow us to continue with the discussion of classical dynamics.

2.4.2 Frame of Reference

We start out by outlining the Newtonian framework. An event intuitively means something happening in a fairly limited region of space and for a short duration in time. Mathematically, we idealize this concept to become a point in space and an instant in time. Everything that happens in the universe is an event or collection of events. Consider a train traveling from one station P to another R, leaving at 10 a.m. and arriving at 11 a.m. We can illustrate this in the following way: For simplicity, let us assume that the motion takes place in a straight line, say along the x-axis; then, we can represent the motion by a space–time diagram in which we plot the position of some fixed point on the train against time. The curve in the diagram is called the history or world-line of the pointer.

P 10.00 a.m   R 11.00 a.m

We will call individuals equipped with a clock and a measuring rod or ruler observers. Had we looked out of the train window on our journey at a clock in a passing station, then we would have expected it to agree with our watch. One of the central assumptions of the Newtonian framework is that two observers will, once they have synchronized their clocks, always agree about the time of an event, irrespective of their relative motion. This implies that for all observers, time is an absolute quantity. In particular, all observers can agree on an origin of time. In order to fix an event in space, an observer may choose a convenient origin in space together with a set of three coordinate axes as a frame of reference. Then, an observer is able to locate events; that is, determine the time t an event occurs and its position (x_1, x_2, x_3) relative to the origin. We will refer to these collectively as a frame of reference.

Newton realized that in order for the laws of motion to have meaning, a reference frame must be chosen with respect to which the motion of bodies can be measured. A reference frame is called an inertial frame if Newton's laws are valid in that frame; that is, if a body subject to no external force is found to move in a straight line with constant velocity (or to remain at rest), then the coordinate system used to establish this fact is an inertial reference frame. This is a clear-cut operational definition and one that also follows from the general theory of relativity.

In Newtonian mechanics, the principle of relativity plays an outstanding role. Two bodies, for example, fall downward because they are attracted toward the Earth. Thus, position has a meaning only relative to the Earth, or to some other body. In just the same way, velocity has only a relative significance. Given two bodies moving with uniform relative velocity, it is impossible to decide which of them is at rest and which is moving.

In view of the relativity principle, the frames of reference used by different unaccelerated observers are completely equivalent. The laws of physics expressed in terms of x_1, x_2, x_3, and t must be identical with those in terms of the coordinates of another frame, x_1', x_2', x_3', and t', respectively. They are not, however, identical with the laws expressed in terms of the coordinates used by an accelerated observer. The frames used by unaccelerated observers are called inertial frames.

We have not yet said how we can tell whether a given observer is unaccelerated. We need a criterion to distinguish inertial frames from others. Formally, an inertial frame can be defined to be one with respect to which an isolated body, far removed from all other matter, would move with uniform velocity. This is, of course, an idealized definition, because in practice we never can get infinitely far away from other matter. For all practical purposes, an inertial frame is one whose orientation is fixed relative to the fixed stars and in which the Sun (the center of mass of the solar system) moves with uniform velocity. It is an essential assumption of classical mechanics that such frames exist.

It is generally convenient to use only inertial frames, but there is no necessity to do so. Sometimes, it proves convenient to use a non inertial (e.g., rotating) frame in which the laws of mechanics take on a more complicated form.

2.4.3 Time

In Newton's theory, time is an absolute quantity, capable of precise definition by an arbitrary observer. It exists and flows in a continuous way. We assume further that there is a universal timescale in the sense that two observers who have synchronized their clocks will always agree about the time of any event.

2.4.4 Mass

In order to understand the motion of a system of particles, it is necessary to consider the environment of the system – potentially all of the other particles in the universe – and learn how that environment influences the motion of the system in question. We begin by considering two particles which influence each other's motion but which move in such a manner that we may reasonably expect all other matter in the universe to have negligible effect on their relative motion. Thus, for example, we may imagine two particles connected by a small spring and free to move on a smooth horizontal table. We should expect that the matter in the Earth would not affect the motion of the masses in the plane of the table and that extraterrestial matter would be too far away to have anything but a negligible effect. It is found that under such conditions, if $\vec{a}_A$ and $\vec{a}_B$ are the accelerations of the two particles A and B, respectively, then these vectors are parallel and in the opposite sense and that the ratio of the magnitudes of $\vec{a}_A$ and $\vec{a}_B$ is a constant for a given pair of particles. This ratio is called the ratio of the masses of the two particles:

$$\frac{m_A}{m_B} = \frac{|\vec{a}_B|}{|\vec{a}_A|}. \tag{2.4.1}$$

It is also found that if mass C is allowed to interact with A in the absence of B, not only is it true that $\vec{a}_A$ and $\vec{a}_C$ are parallel and in the opposite sense but that

$$\frac{m_A}{m_C} = \frac{|\vec{a}_C|}{|\vec{a}_A|}. \tag{2.4.2}$$

is identical with the ratio $(m_A/m_B)/(m_C/m_B)$ determined by comparing A and B in the absence of C, and C and B in the absence of A.

Thus, with each particle there may be associated a mass that has a unique meaning no matter how many stages it goes through in being compared to another and which may, therefore, eventually be compared with a standard mass of platinum called the international prototype kilogram, which is preserved in Sèvres. If in an experiment the vectors $\vec{a}_A$ and $\vec{a}_B$ were found not to be parallel, the two particles would be considered as not acting on each other alone, and the discrepancy would be attributed to the influence of another particle or system.

Another way to measure a mass is by a direct comparison using a balance with arms of unequal lengths l_A and l_B. The ratio of weights w_A and w_B is given by

$$\frac{w_A}{w_B} = \frac{l_B}{l_A}. \qquad (2.4.3)$$

Because the weights are the forces exerted by gravity on the masses ($w_A = m_A\, g$, $w_B = m_B\, g$, etc.) and gravity is assumed not to vary across the balance, we have

$$\frac{m_A}{m_B} = \frac{l_B}{l_A}. \qquad (2.4.4)$$

The masses compared in this manner are sometimes referred to as *gravitational masses*, in contrast to the *inertial masses* defined by Equation (2.4.1). According to Newton's law of gravitation, the gravitational attractive force between two masses m_A and m_B separated by a distance r is

$$F = \frac{G m_A m_B}{r^2}, \qquad (2.4.5)$$

where G is the gravitational constant and m_A and m_B are the gravitational masses. If, however, the masses of two particles attracting each other by gravity are compared by the method discussed in Equation (2.4.1), taking the ratio of their accelerations, then it is the masses appearing in the equation $\vec{F} = m\,\vec{a}$ which are compared. *A priori* there is no reason for these to be identical with the masses appearing in Equation (2.4.4).

However, it has not proved possible to distinguish experimentally between these two apparently different types of mass. Galileo was the first to test the equivalence of inertial and gravitational mass in his experiment with falling weights at the Tower of Pisa. Newton also considered the problem and measured the periods of pendula of equal lengths but with bobs of different material. Neither found any difference, but the method was quite crude. Later experiments are due to R.V. Eötvös, L. Southerns, and P. Zeeman. More recent experiments by R.H. Dicke have improved the accuracy, and it has now been established that inertial and gravitational mass are identical to within a few parts in 10^{11}. In Newtonian theory, we accept this result as an empirical fact and refer to the mass of a body without specifying which method is to be used to measure it. One important feature of the general theory of relativity is that from this point of view the distinction between the two types of mass loses its meaning so that they become automatically identical. The assertion of the exact equality of inertial and gravitational mass is termed the *principle of equivalence*.

2.4.5 Newton's Laws

Newton's laws of mechanics are stated as follows:

> I. (**lex prima**)
>
> A body remains at rest or in uniform motion unless acted upon by a force.
>
> II. (**lex secunda**)
>
> A body acted upon by a force moves in such a manner that the time rate of change of momentum equals the force.
>
> III. (**lex tertia**)
>
> If two bodies exert forces on each other, these forces are equal in magnitude and opposite in direction.

These laws were enunciated by Sir Isaac Newton (1642–1727) in his *Philosophiae naturalis principia mathematica* or, in short, *Principia*, 1687. Galileo had previously generalized the results of his mechanics experiment with statements equivalent to the First and Second Laws, although he was unable to complete the description of dynamics because he did not appreciate the significance of the Third Law and therefore lacked a precise meaning of force.

These laws are so familiar that we sometimes tend to lose sight of their true significance as physical laws. The First Law, for example, is meaningless without the concept of *force*. In fact, standing alone, the First Law conveys a precise meaning only for zero force; meaning that a body which remains at rest or in uniform motion is subject to no force whatsoever. A body which moves in this manner is termed a free body or a free particle. We note that the First Law by itself provides us with only a qualitative notion regarding *force*.

An explicit statement concerning force is provided by the Second Law, in which force is related to the time rate of change of momentum. Momentum was appropriately defined by Newton. He called momentum the quantity of motion. The momentum of a particle acted upon by mechanical, gravitational, or electrical forces is defined to be the product of its mass and its velocity:

$$\vec{p} = m\,\vec{v}. \tag{2.4.6}$$

The force $\vec{F}$ acting on a particle is defined by the rate of change of momentum it produces:

$$\vec{F} = \frac{d\vec{p}}{dt} = \vec{p}\,'. \tag{2.4.7}$$

The definition of force becomes complete and precise only when mass is defined. Thus, the First and Second Laws are not really laws in the usual sense of the term as used in physics; rather, they can be considered as definitions. If the mass of the particle is constant in time, then

$$\vec{p}\,' = m\,\vec{v}\,' = m\,\vec{a}, \tag{2.4.8}$$

where $\vec{a}$ is the acceleration vector. Thus, in this case, the force on the particle can be defined by

$$\vec{F} = m\,\vec{a} = m\,\vec{r}\,''. \tag{2.4.9}$$

Thus, if $\vec{F} = 0$, the velocity of the particle is constant. This is Newton's First Law.

The Third Law, on the other hand, is indeed a law. It is a statement concerning the real physical world and contains all of the physics in Newton's laws of motion. When two particles exert forces on each other, as they are made to do in the measurement of their mass ratio, we have

$$m_A\,\vec{a}_A = -m_B\,\vec{a}_B. \tag{2.4.10}$$

Thus, when two particles exert forces on each other, these forces are equal in magnitude and opposite in direction. This is the Third Law of Newton that *action is equal and opposite to reaction.*

We must hasten to add that the Third Law is not a general law of Nature. The law applies only in the event that the force exerted by one body on another body is directed along the line connecting the two objects. Such forces are called central forces. However, the Third Law applies whether a central force is attractive or repulsive. Gravitational and electrostatic forces are central forces, so Newton's Laws can be used in problems involving these types of force. Sometimes, elastic forces originating from microscopic electrostatic forces are central in character. For example, two point masses connected by a straight spring or elastic string are subject to forces that obey the Third Law. Any force that depends on the velocities of the interacting bodies is non central in character, and the Third Law does not apply in such a situation. Velocity-dependent forces are characteristic of interactions that propagate with finite velocity. Thus, the force between moving electric charges does not obey the Third Law because the force propagates with the velocity of light. Even the gravitational force between moving bodies is velocity dependent, but the effect is small and difficult to detect; the only observable effect is the precession of the perihelia of the inner planets (see Chapter 6). This chapter is concerned exclusively with gravitational and elastic forces; the accuracy of the Third Law is quite sufficient for all such discussions.

From definition (2.4.7) of $\vec{F}$, it follows that since $\vec{p}$ is a vector, so also is $\vec{F}$. Thus, if the force $\vec{F}$ is the sum of two forces $\vec{F}_1$ and $\vec{F}_2$, this sum must be understood as a vector sum.

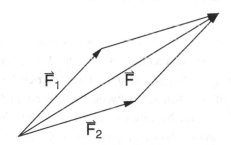

Figure 2.4.3.   Parallelogram law for forces.

This constitutes the parallelogram law for the composition of forces (see Figure 2.4.3). However, the parallelogram law is a trivial mathematical fact. It acquires physical significance in those cases where the force between two particles is independent of the presence of other particles. The parallelogram law is valid only when the various forces are independent. This independence does exist for most of the forces met within mechanics, such as gravitation and the forces between charged particles. It does not exist between polarizable molecules moving in electric fields, for the induced electric moment of a molecule depends on the field at that location of the molecules and the fields at those locations. The forces between nuclear particles can be of this many-body character rather than simple two-body forces.

The Second Law (2.4.7), on the other hand, is central in mechanics. This relation constitutes the simplest form of the *equations of motion* for a particle. Equations (2.4.7) and (2.4.9) are a set of ordinary differential equations of the second order. If the forces are given as functions of position and time, the values of the coordinates and of the velocity components at a given time t_0 determine the solution of the equations uniquely and thus determine the whole future course of the motion. We say the history of the motion is deterministic. However, in certain cases, the motion of the particle can be very complicated, if not chaotic. Chaotic means here that the final state of the motion is not predictable if the initial state is changed by a very small amount. In any case, the future states of a system are determined by the state at any given time and by the equations of motion.

2.4.6 Forces in Nature

The full power of Newton's second law emerges when it is combined with the force laws that describe the interactions of objects. For example, Newton's law for gravitation gives the gravitational force exerted by one object on another in terms of the distance between the objects and the masses of each. This combined with Newton's second law, enables us to calculate the orbits of planets around the sun, the motion of the moon, and variations with altitude of g, the acceleration due to gravity.

2.4.6.1 The Fundamental Forces

All of the different forces observed in nature can be explained in terms of four basic interactions that occur between elementary particles.

1. The Gravitational Force

The gravitational force between the Earth and an object near the Earth's surface is the weight of the object. The gravitational force exerted by the Sun keeps the planets in their orbits. Similarly, the gravitational force exerted by the galaxies in the universe generates a certain structure or distribution of galaxies. Figure 2.4.4 shows a group of galaxies interacting with each other.

Figure 2.4.4.   Group of galaxies in the cluster MS1054-03.

This galaxy cluster, called MS1054-03, is 8 billion light-years away — one of the most distant known groups of galaxies. Although hundreds of galaxies appear in this NASA/ESA Hubble Space Telescope image, a European-led team of astronomers has studied in detail 81 galaxies that certainly belong to the cluster, 13 of which are remnants of recent collisions or pairs of colliding galaxies. This is, by far, the largest number of colliding galaxies ever found in a cluster.

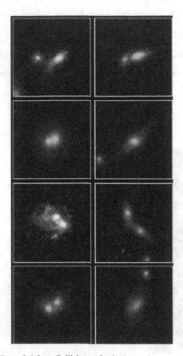

Figure 2.4.5.   Collisions of galaxies various stages.

A gallery of HST images showing distant galaxies in various stages of collision (see Figure 2.4.5). The merging galaxies have weird, distorted shapes unlike normal spiral or elliptical galaxies. Some show streams of stars apparently being pulled from one galaxy into another. All of the galaxy pairs shown here are located in a larger grouping of galaxies known as MS1054-03.

2. The Electromagnetic Force

The electromagnetic force includes both the electric and the magnetic force. A familiar example of the electric force is the attraction between bits of paper and a comb that is electrified after being run through hair. The magnetic force between a magnet and iron arises when electric charges are in motion. These two forces were recognized in the 19th century as independent forces from gravitation. The electromagnetic force between charged elementary particles is vastly greater than the gravitational force between them. For example, the electrostatic force repulsion between two protons is of order of 10^{36} times the gravitational attraction between them. The lightning shown in Figure 2.4.6 is the result of the electromagnetic force.

Figure 2.4.6.   Lightning in a thunder storm.

3. The Strong Nuclear Force (Also Called Hadronic Force)

The strong nuclear force occurs between elementary particles called hadrons, which include protons and neutrons. The strong force results from the interaction of quarks, the building blocks of hadrons, and is responsible for holding nuclei together. The magnitude of the strong force decreases rapidly with distance and is negligible beyond a few nuclear diameters. The hydrogen bomb explosion shown in Figure 2.4.7 illustrates the strong nuclear forces.

Figure 2.4.7.   Atomic explosion in 1951 at Eniwetok Atoll in the South Pacific.

In 1951, a test at Eniwetok Atoll in the South Pacific, demonstrated the release of energy from nuclear fusion. Weighing 65 tons, the apparatus was an experimental device, not a weapon, that had been constructed on the basis of the principles developed by Edward Teller and Stanislaw Ulam. On November 1, 1952, a 10.4-megaton thermonuclear explosion code-named MIKE, ushered in the thermonuclear age. The island of Elugelab in the Eniwetok Atoll was completely vaporized.

4. The Weak Nuclear Force

The weak nuclear force, which also has a short range, occurs between leptons (which include electrons and muons) and between hadrons (which include protons and neutrons). The bubble chamber photographs (Figure 2.4.8) illustrate the weak interaction.

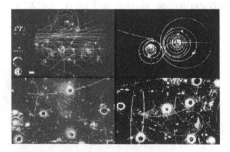

Figure 2.4.8.   Bubble chamber photographs.

2.4.7 Conservation Laws

We now turn to a detailed discussion of the Newtonian mechanics of a single particle and derive the important laws regarding conserved quantities. The background of the conservation laws are symmetries [2.9]. However, at this stage of the presentation, we do not go into a detailed examination of symmetries but use physical arguments to motivate the conserved quantities. We are merely deriving the consequences of Newton's laws of dynamics. The fact that these conservation laws have been found to be valid in many instances furnishes an important part of the proof for the correctness of Newton's laws. Today, we know that these proofs are valid for nonrelativistic systems.

2.4.7.1 Linear Momentum

The first of the conservation laws concerns the linear momentum of a particle. If the particle is free (i.e., if the particle experiences no force), then Newton's second law (2.4.7) becomes

$$\frac{\partial \vec{p}}{\partial t} == 0$$

Therefore, $\vec{p}$ is a vector constant in time, and the linear momentum of a free particle is conserved. Associated with this conservation law is the fact that Newton's equation of motion for a free particle is invariant with respect to translations in the coordinates. We also note that this result is derived from a vector equation $\partial_t \vec{p} = 0$ and therefore applies to each component of the linear momentum.

The derived result can be stated in other terms if we let $\vec{s}$ be some constant vector such that $\vec{F}.\vec{s} = 0$, independent of time. Then,

$$\frac{\partial \vec{p}}{\partial t}.\vec{s} == \vec{F}.\vec{s} == 0$$

or, integrating with respect to time,

$$\int \frac{\partial \vec{p}(t)}{\partial t} \cdot \vec{s}\, dt == \text{const}$$

$$\vec{s}\,\vec{p}(t) == \text{const}$$

which states that the component of linear momentum in a direction in which the force vanishes is constant in time.

2.4.7.2 Angular Momentum

The angular momentum $\vec{L}$ of a particle with respect to an origin from which $\vec{r}$ is measured is defined by

$$\vec{L}(t) == \vec{r}(t) \times \vec{p}(t)$$

$$\vec{L}(t) == \vec{r}(t) \times \vec{p}(t)$$

The torque or moment of force $\vec{M}$ with respect to the same origin is defined to be

$$\vec{M}(t) == \vec{r}(t) \times \vec{F}(t) == \vec{r}(t) \times \frac{\partial \vec{p}(t)}{\partial t}$$

$$\vec{M}(t) == \vec{r}(t) \times \vec{F}(t) == \vec{r}(t) \times \vec{p}'(t)$$

Now,

$$\text{momentum} = \frac{\partial \vec{L}(t)}{\partial t} == \frac{\partial (\vec{r}(t) \times \vec{p}(t))}{\partial t}$$

$$\vec{L}'(t) == \vec{r}(t) \times \vec{p}'(t) + \vec{r}'(t) \times \vec{p}(t)$$

but the product $\dot{\vec{r}} \times \vec{p} = m\dot{\vec{r}} \times \dot{\vec{r}} = 0$, so that

momentum $/. \; \vec{p}(t) \rightarrow m \; \dfrac{\partial \vec{r}(t)}{\partial t}$
$\vec{L}'(t) == \vec{r}(t) \times \vec{p}'(t)$

which is the representation of the torque. If there is no torque acting on a particle (i.e., $\vec{M} = 0$), then $\vec{L}' = 0$ and $\vec{L}$ is a vector constant in time. This is the second important conservation law: The angular momentum of a particle subject to no torque $\left(\vec{M} = 0\right)$ is conserved. We note that this conservation law is associated with the symmetry of rotation.

2.4.7.3 Work and Energy

Work and energy are important concepts in physics as well as in our everyday life. In physics, a force does work when it acts on an object that moves through a distance and there is a component of the force along the line of motion. For a constant force in one dimension, the work done equals the force times the distance. This differs somewhat from the everyday use of the word work. When you study hard for an exam, the only work you do as the term is understood in physics is in moving your pencil or turning the pages of your book.

The concept of energy is closely associated with that of work. When work is done by one system on another, energy is transferred between the two systems. For example, when you do work pushing a swing, chemical energy in your body is transferred to the swing and appears as kinetic energy of motion or gravitational potential energy of the Earth-swing system. There are many forms of energy. Kinetic energy is associated with the motion of an object. Potential energy is associated with the configuration of a system, such as the separation distance between some objects and the Earth. Thermal energy is associated with the random motion of the molecules within a system and is closely connected with the temperature of the system.

If work is done on a particle by a force $\vec{F}$ in transforming the particle from condition 1 to condition 2, then this work is defined to be

$$W_{12} = \int_{1}^{2} \vec{F}[\vec{r}] \, d\vec{r}$$

$$\int_{1}^{2} \vec{F}(\vec{r}) \, d\vec{r}$$

Now,

$$\vec{F}(\vec{r}) \, \text{DifferentialD}(\vec{r}) == m \, \frac{\partial \vec{v}(t)}{\partial t} . \frac{\partial \vec{r}(t)}{\partial t} \, \text{DifferentialD}(t) ==$$

$$m \, \frac{\partial \vec{v}(t)}{\partial t} . \vec{v}(t) \, \text{DifferentialD}(t) == \frac{1}{2} \, m \, \frac{\partial (\vec{v}(t).\vec{v}(t))}{\partial t} \, \text{DifferentialD}(t)$$

$$\text{DifferentialD}(\vec{r}) \, \vec{F}(\vec{r}) == m \, \text{DifferentialD}(t) \, \vec{v}'(t).\vec{r}'(t) ==$$
$$m \, \text{DifferentialD}(t) \, \vec{v}'(t).\vec{v}(t) == m \, \text{DifferentialD}(t) \, v(t) \, v'(t)$$

Therefore, the integrand is an exact differential and

$$W_{12} = \left(\frac{1}{2} \, m \, v^2 \right) \Big|_{1}^{2} == \frac{1}{2} \, m \, (v_2^2 - v_1^2) == T_2 - T_1$$

where $T = \frac{1}{2} m v^2$ is the kinetic energy of the particle. If $T_1 > T_2$, then $W_{12} < 0$ and the particle has done work with a resulting decrease in kinetic energy.

The total work done on a particle is equal to the change in its kinetic energy:

$$W_{\text{total}} = \Delta T = \tfrac{1}{2} m v_2^2 - \tfrac{1}{2} m v_1^2.$$

This theorem is known as the work–kinetic energy theorem. It holds whether the force is constant or variable. The theorem holds for all kinds of force. The theorem does not tell anything about where the energy ΔT goes.

2.4.7.4 Constant Forces

The work W done by a constant force $\vec{F}$ whose point of application moves through a distance $d\,\vec{r}$ is defined to be

$$W == \int \vec{F} \, d\vec{r} == \vec{F} \, \Delta\vec{r} == F \cos(\theta) \, \Delta x == F_x \, \Delta x$$

where θ is the angle between $\vec{F}$ and the x-axis, and Δx is the displacement of the particle.

Work is a scalar quantity that is positive if Δx and F_x have the same signs and negative if they have opposite signs. The dimensions of work are those of force times distance. The SI unit of work and energy is the Joule (J), which equals the product of a Newton and a meter: $1 \, J = 1 \, Nm$.

When there are several forces that do work, the total work is found by computing the work done by each force and summing:

$$W_{\text{total}} == F_{1x} \, \Delta x_1 + F_{2x} \, \Delta x_2 + \dots$$

When the forces do work on a particle, the displacement of the force Δx_i is the same for each force and is equal to the displacement of the particle Δx:

$$W_{\text{total}} == \Delta x \, F_{1x} + \Delta x \, F_{2x} + \dots == (F_{1x} + F_{2x} + \dots) \, \Delta x == F_{\text{net}} \, \Delta x$$

Thus, for a particle, the total work can be found by summing all of the forces to find the net force and then computing the work done by the net force.

Let us now examine the work integral from a different point of view. In many physical problems, the force $\vec{F}$ has the property that the work required to move a particle from one position to another without any change in kinetic energy is dependent only on the original and final positions and not upon the exact path taken by the particle. This property is exhibited, for example, by a constant gravitational force field. Thus, if a particle of mass m is raised through a height h, then an amount of work mgh has been done on the particle and the particle has the capacity to do an equal amount of work in returning to its original position. This *capacity to do work* is called then *potential energy* of the particle.

We can define the potential energy of a particle in terms of the work required to transport the particle from a position 1 to a position 2:

$$\int_1^2 \vec{F}(\vec{r})\, d\vec{r} == U_1 - U_2$$

That is, the work done in moving the particle is simply the difference in the potential energy U at the two points. This equation can be written in a different way if we represent $\vec{F}$ as the gradient of the scalar function U:

$$\vec{F}(\vec{r}) == -\nabla U(\vec{r})$$

$$\vec{F}(\vec{r}) == -\nabla U(\vec{r})$$

Then,

$$\int_1^2 \vec{F}[\vec{r}]\,.\,d\vec{r} == -\int_1^2 \nabla U[\vec{r}]\,.\,d\vec{r} == U_1 - U_2$$

In most systems of interest, the potential energy is a function of position and, probably, the time: $U = U(\vec{r})$ or $U = U(\vec{r}, t)$.

It is important to note that the potential energy is defined only to within an additive constant; that is, the force defined by $-\nabla U$ is not different from that definition by $-\nabla(U + \text{const.})$. Therefore, potential energy has no absolute meaning; only differences of potential energy are physically meaningful.

Knowing the potential and kinetic energy, we are able to define the total energy of a particle. The total energy of a particle is defined to be the sum of the kinetic energy and the potential energies:

$$H == T + U$$

Assuming that $H = H(t)$, we can ask for the time derivative of the total energy. In order to evaluate the time derivatives appearing on the right-hand side of this equation, we first note that the time derivative of the kinetic energy can be represented by

$$\mathbf{r1} = \frac{\partial T(t)}{\partial t} \rightarrow \vec{F}(\vec{r}) \frac{\partial \vec{r}(t)}{\partial t}$$

$$T'(t) \rightarrow \vec{F}(\vec{r}) \, \vec{r}'(t)$$

For the potential energy, we have

$$\mathbf{r2} = \frac{\partial U(t)}{\partial t} \rightarrow \frac{\partial U(\vec{r}(t), t)}{\partial t}$$

$$U'(t) \rightarrow U^{(0,1)}(\vec{r}(t), t) + \vec{r}'(t) \, U^{(1,0)}(\vec{r}(t), t)$$

Substituting these two expressions into the derivative of H, we find

$$\frac{\partial H(t)}{\partial t} == \frac{\partial (T(t) + U(t))}{\partial t} \, /. \, \{\mathbf{r1}, \mathbf{r2}\} \, /. \, \vec{F}(\vec{r}) \rightarrow - \frac{\partial U(\vec{r}(t), t)}{\partial \vec{r}(t)}$$

$$H'(t) == U^{(0,1)}(\vec{r}(t), t)$$

Since the force can be represented by the negative gradient of the potential, we are able to simplify the result. Now, if U is not an explicit function of the time, then the force field represented by $\vec{F}$ is said to be conservative, meaning that $\vec{F} = -\nabla U(\vec{x})$ and that $U(\vec{x})$ exists. This condition can be equivalently stated as $\nabla \times \vec{F}(\vec{x}) = 0$. Under these conditions, we have the third important conservation law: *the total energy H of a particle in a conservative force field is a constant in time.*

Note that the general law of conservation of energy was formulated in 1847 by Hermann von Helmholtz (1821–1894). His conclusion was based largely on the calorimetric experiments of James Prescott Joule (1818–1889), which were begun in 1840.

2.4.8 Application of Newton's Second Law

Newton's equation $\vec{F} = d\,\vec{p}\,/dt$, can be expressed alternatively as

$$\vec{F} = \frac{d}{dt}\,(m\,\vec{v}) = m\,\frac{d\vec{v}}{dt} = m\,\vec{r}\,'' \tag{2.4.11}$$

if we assume that the mass does not vary with time. This is a second-order differential equation for $\vec{r} = \vec{r}(t)$, which can be integrated if the function $\vec{F}$ is known. The specification of the initial values of $\vec{r}$ and $\vec{r}\,' = \vec{v}$ then allows the evaluation of the two arbitrary constants of integration. The following examples will demonstrate how Newton's equation is applied to different physical systems.

2.4.8.1 Falling Particle

The motion of a particle that has constant acceleration is common in nature. For example, near the Earth's surface, all unsupported objects fall vertically with constant acceleration (provided air resistance is negligible). The force acting on a falling particle is governed by the acceleration of gravity by

$$\vec{F}_g = m\,\vec{g}, \tag{2.4.12}$$

where $\vec{g}$ is the acceleration of gravity. Inserting relation (2.4.12) into Equation (2.4.11), we end up with the equation of motion:

$$m\,\vec{g} = m\,\vec{r}\,''. \tag{2.4.13}$$

This equation contains on the left-hand side a vector $\vec{g}$ with its direction toward the center of the Earth. This is the one and only component of $\vec{g}$. If $\vec{g}$ and $\vec{r}\,''$ are parallel, then the direction of $\vec{r}\,''$ is the same as that of $\vec{g}$. Thus, the vector equation is reducible to a single component. If we choose the coordinate r along the direction of $\vec{g}$, we get

```
equation8 = -m g == m ∂t,t r[t]
```

$$-g\,m == m\,r''(t)$$

defining Newton's equation for a falling particle. This second-order ordinary differential equation determines the motion of the particle. If we can solve this equation, we gain information on the path $r(t)$. The solution in *Mathematica* can be derived by

```
solution = DSolve[equation8, r, t]
```

$$\left\{\left\{r \to \text{Function}\left[\{t\},\ -\frac{g\,t^2}{2} + c_2\,t + c_1\right]\right\}\right\}$$

The solution represented in a pure function form tells us that the path $r(t)$ of the particle is determined by the acceleration of gravity g and in addition by two constants $C[1]$ and $C[2]$. These two constants are constants resulting from the integration process behind the function **DSolve[]**. They are determined by initial values of the motion. Incorporating the initial values of the motion right into the solution of the equation of motion, we can write

```
solution = DSolve[
    {equation8, r[0] == r0, r'[0] == v0}, r, t] // Flatten
```

$$\left\{r \to \text{Function}\left[\{t\},\ \frac{1}{2}\,(-g\,t^2 + 2\,\text{v}0\,t + 2\,\text{r}0)\right]\right\}$$

where r_0 and v_0 are the position and velocity at initial time, respectively. It is obvious by comparing the two solutions that $C[1] = r_0$ and $C[2] = v_0$.

At this stage of our examinations, we know how a particle behaves if a constant acceleration is applied to it. However, if we want to examine the motion of a particle starting at rest at a certain height, we have to specify the initial conditions that way. For example, let us assume that a ball starts from rest at a height of $r_0 = 100\,\mathrm{m}$ above the surface of the Earth. Under these conditions, the path of the particle simplifies to

```
ssol = r[t] /. solution /. {r0 → 100, v0 → 0, g → 9.81}
```

$$\frac{1}{2}\,(200 - 9.81\,t^2)$$

This special solution can be used to simulate the actual motion of the particle. To generate the animation, we have to know how long the particle needs to go before it touches down on the surface. This question can be solved by solving the equation and selecting the positive solution:

```
end = Solve[ssol == 0, t] // Flatten; Tend = t /. end[[2]]
```

4.51524

The result for the total time is used to generate different states of the falling particle:

```
track = Table[{RGBColor[0.996109, 0, 0],
    Disk[{0, ssol}, 5]}, {t, 0, Tend, .2}];
```

These states are displayed in the following sequence of pictures:

```
Map[Show[Graphics[#], PlotRange → {0, 105},
    AspectRatio → Automatic] &, track];
```

The animation of these states show how the particle increases its velocity during the time. The analytic expression for the velocity is gained by differentiating the solution with respect to time. The following plot demonstrates that the velocity v linearly increases. The negative sign of v indicates that the orientation of the velocity is parallel to the acceleration.

```
Plot[Evaluate[∂ₜ ssol],
  {t, 0, Tend}, AxesLabel → {"t", "v"},
  PlotStyle → RGBColor[0, 0, 0.996109]];
```

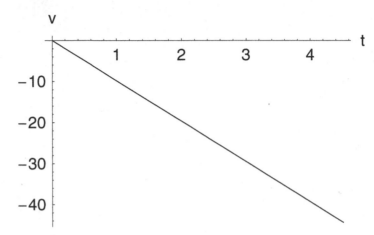

Up to the present stage of our discussion, we assumed that the particle is falling in a vacuum, meaning there is no resistance if the particle moves in the air. If we consider a real falling particle, we have to take into account that in addition to the gravitational force, several other forces act on the particle. For example, in addition to $\vec{F}_g$, there are drag forces which slow down the motion of the particle. These forces are typically functions of the velocity. Incorporating additional forces, the total force is then

$$\vec{F} = \vec{F}_g + \vec{F}_d(\vec{v}).$$ (2.4.14)

It is sufficient to consider that $\vec{F}_d$ is simply proportional to some power of the velocity. In general, real retarding forces are more complicated, but the power-law approximation is useful in many instances in which the velocity does not vary greatly. With the power-law approximation in mind, we can then write

$$\vec{F} = m\vec{g} - m\gamma v^n \frac{\vec{v}}{v},$$ (2.4.15)

where γ is a positive constant that specifies the strength of the retarding force and $\vec{v}/v$ is a unit vector in the direction of $\vec{v}$. Experimental observations indicate that for small objects moving at low velocities in air, $n \approx 1$; for larger velocities still below the velocity of sound, the retarding force is approximately proportional to the square of the velocity.

The next step of our examination is the influence of the retarding force on the path of the particle. Newton's equation for this case in one dimension is given by

$$-m g + m \gamma \left(\frac{d\, r(t)}{d\, t} \right)^n = m\, \ddot{\vec{r}}. \tag{2.4.16}$$

In *Mathematica* notation, this equation is given by

```
equation11 = -m g - m γ (∂ₜ r[t])ⁿ == m ∂ₜ,ₜ r[t]
```

$$-m \gamma\, r'(t)^n - g\, m == m\, r''(t)$$

The solution of this ordinary differential equation under the initial conditions $r(0) = r_0$ and $v(0) = v_0$ and $n = 1$ follows from

```
solutiond =
  DSolve[{equation11, r[0] == r0, r'[0] == v0} /. n -> 1,
    r, t] // Flatten
```

$$\left\{ r \rightarrow \text{Function}\left[\{t\},\ \frac{e^{-t\gamma}\,(e^{t\gamma}\,r0\,\gamma^2 - e^{t\gamma}\,g\,t\,\gamma + e^{t\gamma}\,v0\,\gamma - v0\,\gamma + e^{t\gamma}\,g - g)}{\gamma^2} \right] \right\}$$

Using the same initial conditions for the particle as in the case without drag, we find the solution

```
ssold = r[t] /. solutiond /.
  {r0 → 100, v0 → 0, g → 9.81, γ → 1/2}
```

$$4\,e^{-t/2}\,(-4.905\,e^{t/2}\,t + 34.81\,e^{t/2} - 9.81)$$

The total time of the particle needed to touch the ground is determined by solving the above relation if the particle's position equals zero.

```
endd = Solve[ssold == 0, t] // Flatten;
Tendd = t /. endd[[2]]
```

```
7.03757
```

The result is that the total falling time increases, meaning that the motion of the falling particle is slowed down by a certain factor. The following simulation shows that the motion of the particle at the end of the path reaches a constant velocity:

```
trackd = Table[{RGBColor[0, 0.500008, 0],
    Disk[{12, ssold}, 5]}, {t, 0, Tendd, .2}];
```

```
Map[Show[Graphics[#], PlotRange → {0, 105},
    AspectRatio → Automatic] &, trackd];
```

The behavior that the velocity becomes constant can be checked with

```
Limit[-∂t ssold, t → ∞]
```

19.62

The plot of the velocity versus time shows the same result.

```
Plot[Evaluate[-∂t ssold],
  {t, 0, Tendd 2}, AxesLabel → {"t", "v"},
  PlotStyle → RGBColor[0, 0, 0.996109]];
```

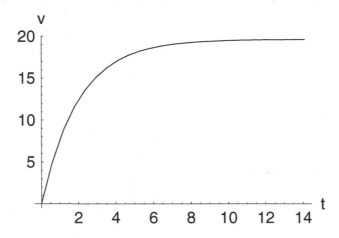

Our general observation is that a particle accelerated by gravity and influenced by additional forces change the behavior of motion.

2.4.8.2 Harmonic Oscillator

Let us consider the oscillatory motion of a particle that is constrained to move in one dimension. We assume that there exists a position of stable equilibrium for the particle and we designate its point as the origin of our coordinate system. If the particle is displaced from the origin, a certain force tends to restore the particle to its original position. This force is, in general, some complicated function of the displacement and of the particle's velocity. We consider here only cases in which the restoring force F is a function only of the displacement:

```
Force = F[x];
```

We will assume that the function $F(x)$ describing the restoring force possesses continuous derivatives of all orders so that the function can be expanded in a Taylor series:

> f = **Series[Force, {x, 0, 2}]**
>
> ---
>
> $F(0) + F'(0)\,x + \dfrac{1}{2}\,F''(0)\,x^2 + O(x^3)$

where $F(0)$ is the value of F at the origin ($x = 0$), and $F^{(n)}(0)$ is the value of the nth derivative at the origin. Because the origin is defined to be the equilibrium point, $F(0)$ must vanish. Then, if we confine our attention to displacements of the particle that are sufficiently small, we can neglect all terms involving x^2 and higher powers of x. We have, therefore, the approximate relation:

> f = $-k\,x$
>
> ---
>
> $-k\,x$

where we have replaced $F'(0) = -k$. Since the restoring force is always directed toward the equilibrium position, the derivative $F'(0)$ is negative and, therefore, k is a positive constant. Only the first power of the displacement occurs in $F(x)$, so that the restoring force in this approximation is a linear force. Physical systems that can be described in terms of linear forces are said to obey Hooke's law.

The equation of motion for the simple harmonic oscillator can be obtained by substituting Hooke's law force into the Newtonian equation $F = m\,a$. Thus,

> **equation1** = $m\,\dfrac{\partial^2 x(t)}{\partial t\,\partial t}$ == $-k\,x(t)$
>
> ---
>
> $m\,x''(t)$ == $-k\,x(t)$

If we define $\omega_0^2 = k/m$, the equation of motion becomes

> **equation1 = (equation1 /. $k \to \omega_0^2\, m$) / m**
>
> ---
>
> $x''(t) == -\omega_0^2\, x(t)$

The solution of this equation can be found by

> **solution1 = DSolve[equation1, x, t]**
>
> ---
>
> {{$x \to$ Function[{t}, $c_1 \cos(t\,\omega_0) + c_2 \sin(t\,\omega_0)$]}}

where $C[1]$ and $C[2]$ are constants of integration determining the amplitude of the oscillation. Thus, the solution for the harmonic oscillator are trigonometric functions with period $T = 2\pi/\omega_0$.

> **Plot[$x(t)$ /. solution1 /. {$\omega_0 \to 2, c_1 \to 1, c_2 \to 1$}, {$t$, 0, 4π},**
>   **AxesLabel $\to$ {"t", "x"}, PlotStyle $\to$ RGBColor[0, 0, 0.996109]];**

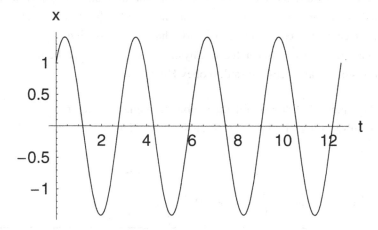

The relationship between the total energy of the oscillator and the amplitude of its motion can be obtained as follows. Using the derived solution, we find for the kinetic energy

$$T = \text{Simplify}\left[\frac{1}{2} m \left(\frac{\partial x(t)}{\partial t}\right)^2 /. \text{solution1}\right]$$

$$\left\{\frac{1}{2} m \left(\omega_0 c_2 \cos(t \, \omega_0) - \omega_0 c_1 \sin(t \, \omega_0)\right)^2\right\}$$

The potential energy for the harmonic oscillator can be calculated in the same way following the definition of work; that is, the work required to displace the particle a distance x is equivalent with the potential difference. The incremental amount of work dW that is necessary to move the particle by an amount dx against the restoring force F is

$$dW = -k \, x \, dx;$$

Integrating from 0 to x and setting the work done on the particle equal to the potential energy, we have

$$U = \int_0^{x(t)} k \, x \, dx$$

$$\frac{1}{2} k \, x(t)^2$$

Then,

$$U = \text{Simplify}[U \, /. \text{solution1}]$$

$$\left\{\frac{1}{2} k \left(c_1 \cos(t \, \omega_0) + c_2 \sin(t \, \omega_0)\right)^2\right\}$$

Combining the expressions for T and U to find the total energy E, we have

> **Energy = Simplify[$T + U$ /. $k \to \omega_0^2\, m$]**
>
> $$\left\{ \frac{1}{2}\, m\, \omega_0^2\, (c_1^2 + c_2^2) \right\}$$

so that the total energy is proportional to the square of the amplitude; this is a general result for linear systems. Notice also that the energy is independent of time; that is, energy is conserved. The conservation of energy must be expected because the potential U does not depend explicitly on time.

2.4.8.3 The Phase Diagram

So far, few attempts have been made to visualize the nature of a solution. We only plotted the *position variable* $x = x(t)$ oscillating periodically in time. A most valuable description of a solution is gained by examining its behavior in the *phase plane* or, more generally, in *phase space*.

Returning to the harmonic oscillator, the state of motion of a one-dimensional oscillator will be completely specified as a function of time if two quantities are given: the displacement $x(t)$ and the velocity $v(t) = x'(t)$. The quantities $x(t)$ and $x'(t)$ can be considered to be the coordinates of a point in a two-dimensional space, called the *phase space*. As the time varies, the point (x, x'), which describes the state of the oscillating particle, will move along a certain phase path in the phase space. For different initial conditions of the oscillator, the motion will be described by different phase paths. Any given path represents the complete time history of the oscillator for a certain set of initial conditions. The totality of all possible phase paths constitutes the phase portrait or the phase diagram of the oscillator.

According to the results of subsection 2.4.8.2, we can represent the point (x, x') in the phase plane for the single harmonic oscillator by

$$pt = \text{Flatten}\Big[\Big\{x(t), \frac{\partial x(t)}{\partial t}\Big\} \, /. \, \text{solution1}\Big]$$

$$\{c_1 \cos(t\,\omega_0) + c_2 \sin(t\,\omega_0), \, \omega_0\, c_2 \cos(t\,\omega_0) - \omega_0\, c_1 \sin(t\,\omega_0)\}$$

This point is a two-dimensional parametric representation of the path for all initial conditions. The initial conditions are chosen by specifying the values for $C[1]$ and $C[2]$. Knowing this, we can plot for different initial conditions a phase portrait by continuously changing the time.

$$\Big(\text{ParametricPlot}\Big[\text{Evaluate}\Big[\text{Table}\Big[pt\, /. \, \Big\{c_1 \to i, c_2 \to 1, \omega_0 \to \frac{1}{2}\Big\}, \{i, 1, 5\}\Big]\Big],$$

$$\{t, 0, \#1\}, \text{AxesLabel} \to \{\text{"x"}, \text{"x'"}\},$$

$$\text{PlotStyle} \to \text{RGBColor}[0, 0, 0.996109], \text{AspectRatio} \to \text{Automatic},$$

$$\text{PlotRange} \to \begin{pmatrix} -7 & 7 \\ -4 & 4 \end{pmatrix}\Big]\,\&\Big)\,/@\,\text{Table}[te, \{te, .5, 4.2\,\pi, .5\}];$$

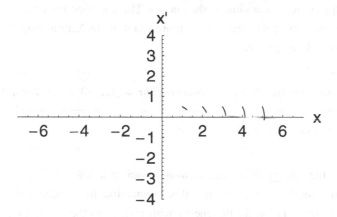

The complete phase portrait is generated by

ParametricPlot$\left[\text{Evaluate}\left[\text{Table}\left[\text{pt} \, /. \, \{c_1 \to i, c_2 \to 1, \omega_0 \to \frac{1}{2}\}, \{i, 1, 5\}\right]\right]\right.$,

$\{t, 0, 4\pi\}$, AxesLabel $\to$ {"x", "x'"},
PlotStyle $\to$ RGBColor[0, 0, 0.996109], AspectRatio $\to$ Automatic$\Big]$;

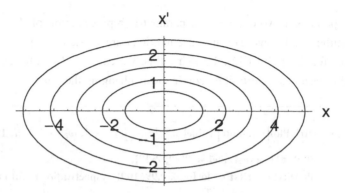

We are going to show that the paths in the phase plane are ellipses with the center as origin. The graphical representation given is based on the parametric representation of the curves. The equation governing the paths is derived from this parametric representation by eliminating the time t from the defining equations:

curve = First[FullSimplify[Eliminate[Thread[{x, xd} == pt, List], t]]]

$\text{xd}^2 == \omega_0^2 \, (-x^2 + c_1^2 + c_2^2)$

Since the energy H of the harmonic oscillator was connected with the initial conditions, we can use this connection to eliminate them. First, solving the relation for the energy with respect to the integration constant $C[1]$,

sh = Solve$\left[H == \frac{1}{2} \, m \, (c_1^2 + c_2^2) \, \omega_0^2, c_1\right]$

$$\left\{\left\{c_1 \to -\frac{i \sqrt{2} \, \sqrt{\frac{1}{2} \, m \, \omega_0^2 \, c_2^2 - H}}{\sqrt{m} \, \omega_0}\right\}, \left\{c_1 \to \frac{i \sqrt{2} \, \sqrt{\frac{1}{2} \, m \, \omega_0^2 \, c_2^2 - H}}{\sqrt{m} \, \omega_0}\right\}\right\}$$

we are able to replace the two constants $C[1]$ and $C[2]$ by the energy H:

Union[Simplify[curve /. sh]]
$\left\{ \text{xd}^2 + x^2\, \omega_0^2 == \dfrac{2\,H}{m} \right\}$

The result is the representation of an ellipses for the coordinates x and x'. Because the derived expression contains the total energy as a parameter, we know that each phase path corresponds to a definite total energy of the oscillator. This result is expected because the system is conservative (i.e., H = const.).

We observe that the phase paths do not cross. This is a general feature of the trajectories; for if they could cross, this would imply that for a given set of initial conditions $x(t_0)$ and $x'(t_0)$, the motion could proceed along different phase paths. However, this is impossible since the solution of the equation of motion is unique.

2.4.8.4 Damped Harmonic Oscillator

The motion represented by the simple harmonic oscillator is termed a free oscillation; once set into oscillation, the motion would never cease. This is, of course, an oversimplification of the actual physical case in which dissipative or frictional forces would eventually damp the motion to the point that the oscillations would no longer occur. It is possible to analyze the motion in such a case by incorporating into the differential equation a term that represents the damping force. It is frequently assumed that the damping force is a linear function of the velocity $\vec{F}_d = -\gamma\,\vec{v}$. In this subsection, we will consider only one-dimensional damped oscillations so that we can represent the damping term by $-\gamma\, x'$. The damping constant γ must be positive in order that the force indeed is resisting.

Thus, if a particle of mass m moves under the combined influence of a linear restoring force $-k\,x$ and a resisting force $-\gamma\,x'$, the differential equation which describes the motion is

$$m\,x'' + \gamma\,x' + k\,x = 0,\tag{2.4.17}$$

which we can write as

> $$\text{equation12} = x(t)\,\omega_0^2 + 2\,\beta\,\frac{\partial x(t)}{\partial t} + \frac{\partial^2 x(t)}{\partial t\,\partial t} == 0$$
>
> $$x(t)\,\omega_0^2 + 2\,\beta\,x'(t) + x''(t) == 0$$

Here, $\beta = \gamma / 2\,m$ is the damping parameter and $\omega_0 = (k/m)^{1/2}$ is the characteristic frequency in the absence of damping. The solution of this equation follows by

> **solution = Flatten[DSolve[equation12, x, t]]**
>
> $$\left\{ x \to \text{Function}\!\left[\{t\},\; e^{t\left(-\beta - \sqrt{\beta^2 - \omega_0^2}\right)} c_1 + e^{t\left(\sqrt{\beta^2 - \omega_0^2} - \beta\right)} c_2 \right] \right\}$$

There are three general cases of interest distinguished by the radicand $\beta^2 - \omega_0^2$:

---

a) Underdamping: $\omega_0^2 > \beta^2$

b) Critical damping: $\omega_0^2 = \beta^2$

c) Overdamping: $\omega_0^2 < \beta^2$

As we shall see, only the case of underdamping results in oscillatory motion.

Underdamped Motion

For this case, it is convenient to define

$$\omega_1^2 = \omega_0^2 - \beta^2,\tag{2.4.18}$$

where $\omega_1^2 > 0$; then, the exponents of the solution becomes imaginary and the solution reduces to

> **underdampedSolution = PowerExpand[$x(t)$ /. solution /. $\omega_0^2 \to \beta^2 + \omega_1^2$]**
>
> ---
>
> $e^{t(-\beta - i\,\omega_1)}\, c_1 + e^{t(i\,\omega_1 - \beta)}\, c_2$

We call the quantity ω_1 the frequency of the damped oscillator. Strictly speaking, it is not possible to define a frequency when damping is present because the motion is not periodic (i.e., the oscillator never passes twice through a given point with the same velocity). If the damping β is small, then

> **Series$\left[\sqrt{\omega_0^2 - \beta^2}\, , \{\beta, 0, 1\} \right]$**
>
> ---
>
> $\sqrt{\omega_0^2} + O(\beta^2)$

the term frequency may be used, but the meaning is not precise unless $\beta = 0$. Nevertheless, for simplicity, we will refer to ω_1 as the frequency of the damped oscillator, and we note that this quantity is less than the frequency of the oscillator in the absence of damping.

The maximal elongation of the motion of the damped oscillator decreases with time because of the factor $e^{-\beta t}$, where $\beta > 0$, and the envelope of the displacement versus time curve is given by

> **envelope = underdampedSolution /. $\omega_1 \to 0$**
>
> ---
>
> $e^{-t\beta}\, c_1 + e^{-t\beta}\, c_2$

This envelope as well as the displacement curve is shown in the following plot.

```
Plot[Evaluate[
   {envelope, -envelope, underdampedSolution} /.
   {ω₁ → 1, β → 1 / 7, C[1] → 1, C[2] → 1}],
  {t, 0, 15}, AxesLabel → {"t", "x(t)"},
  PlotStyle → {RGBColor[0, 0, 0.996109],
    RGBColor[0, 0, 0.996109],
    RGBColor[0.996109, 0, 0]}, PlotRange → All];
```

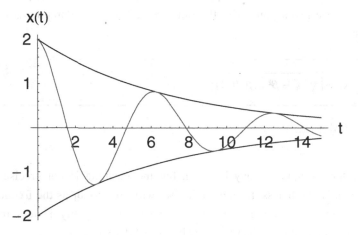

In contrast to the simple harmonic oscillator, the energy of the damped oscillator is not constant in time. Rather, energy is continually given up to the damping medium and dissipated as heat.

The "energy" of the damped oscillator is defined by

$$H = \tfrac{1}{2} m (x')^2 + \tfrac{1}{2} k x^2,$$ (2.4.19)

whereas the loss rate of the energy is dH / dt; both quantities are given in the following plot for a specific choice of parameters:

```
Plot[Evaluate[{ 1/2 (∂_t underdampedSolution)² +

    1/2 (ω₁² + β²) underdampedSolution²,

    ∂_t ( 1/2 (∂_t underdampedSolution)² +

    1/2 (ω₁² + β²) underdampedSolution²)} /.

    {ω₁ → 1, β → 1/7, C[1] → 1, C[2] → 1}],

    {t, 0, 15}, AxesLabel → {"t", "H, dH/dt"},

    PlotStyle → {RGBColor[0, 0, 0.996109],

    RGBColor[0, 0.500008, 0]}, PlotRange → All];
```

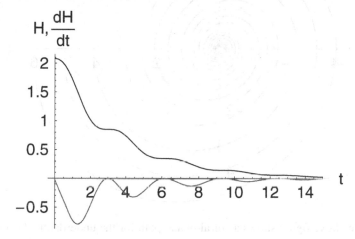

The rate of energy loss is proportional to the square of the velocity so the decrease of energy does not take place uniformly. The loss rate will be a maximum when the particle attains its maximum velocity near the equilibrium position and it will instantaneously vanish when the particle is at maximum amplitude and has zero velocity.

The phase diagram for the damped oscillator can be generated by plotting the coordinates x and x' for different choices of the integration constants $C[1]$ and $C[2]$.

ParametricPlot[

Evaluate[Table[Re({underdampedSolution, $\dfrac{\partial\,\text{underdampedSolution}}{\partial t}$} /.

$\{\omega_1 \to 1, \beta \to \dfrac{1}{7}, c_1 \to i, c_2 \to 1\}$), {i, 1, 5}]], {t, 0, 25},

AxesLabel → {"x", "x'''"}, PlotStyle → RGBColor[0, 0, 0.996109],
AspectRatio → Automatic, PlotRange → All];

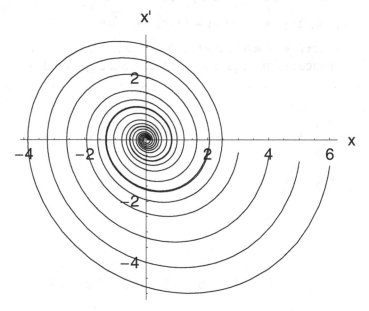

The above figure shows a spiral phase path for the underdamped oscillator. The continually decreasing magnitude of the radius vector and the decrease of the velocity affect the path in such a way that the terminal point of the motion ends in the origin.

Critically Damped Motion

In the case that the damping force is sufficiently large (i.e., if $\beta^2 > \omega_0^2$), the system is prevented from undergoing oscillatory motion. If there is zero initial velocity, the displacement decreases monotonically from its initial value to the equilibrium position $x = 0$. The case of critical damping occurs when β^2 is just equal to ω_0^2. For this choice of parameters, we have to solve the original equation of motion a second time because this special choice of parameters generates a bifurcation of the solution. A bifurcation of the solution means that the nature of the solution changes if we change the parameters in a special way in the equation of motion. We note that the reason behind this bifurcation is a change of the symmetry group of the equations of motion. The solution for the critical damping case is derived by

```
criticallydampedSolution = x(t) /.
    Flatten[DSolve[{equation12 /. ω₀² → β², x(0) == x0, x'(0) == v0}, x, t]]
```

$e^{-t\,\beta}\,(t\,\mathrm{v0} + \mathrm{x0} + t\,\mathrm{x0}\,\beta)$

where x_0 and v_0 are the initial values for the position and the velocity, respectively. Let us assume that the oscillator starts with a finite elongation of $x_0 = 1$ at a vanishing velocity $v_0 = 0$; we get the displacement by

plcritical =

$\text{Plot}\left[\text{Evaluate}\left[\text{criticallydampedSolution} /. \left\{\beta \to \frac{1}{5}, \text{x0} \to 1, \text{v0} \to 0\right\}\right],\right.$

$\left.\{t, 0, 25\}, \text{AxesLabel} \to \{\text{"t"}, \text{"x"}\},\right.$

$\left.\text{PlotStyle} \to \text{RGBColor}[0, 0, 0.996109]\right];$

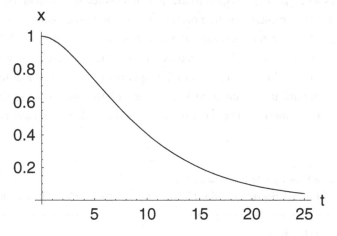

For a given set of initial conditions a critically damped oscillator will approach equilibrium at a rate more rapid than that for either an overdamped or an underdamped oscillator. This fact is of importance in the design of certain practical oscillatory systems when it is desired that the system return to equilibrium as rapidly as possible.

Overdamped Motion

If the damping parameter β is even larger than ω_0, the overdamping results. Since $\beta^2 > \omega_0^2$, it is convenient to define

$$\omega_2^2 = \omega_0^2 + \beta^2 \tag{2.4.20}$$

where $\omega_2^2 > 0$; then the exponents of the solution become real and the solution reduces to

overdampedSolution = PowerExpand[$x(t)$ /. solution /. $\omega_0^2 \to \beta^2 - \omega_2^2$]

$e^{t(-\beta-\omega_2)} c_1 + e^{t(\omega_2-\beta)} c_2$

The solution derived from the calculation contains two constants of integration c_1 and c_2, which are connected with the initial values for the elongation and the velocity by

---

gl1 = x0 == overdampedSolution /. $t \to 0$

$x0 == c_1 + c_2$

The defining equation for the velocity is

---

gl2 = v0 == $\dfrac{\partial \, \text{overdampedSolution}}{\partial t}$ /. $t \to 0$

$v0 == c_1 \, (-\beta - \omega_2) + c_2 \, (\omega_2 - \beta)$

The solution of these two equations with respect to the constants c_1 and c_2 is given by

---

sh = Simplify[Solve[{gl1, gl2}, $\{c_1, c_2\}$]]

$\left\{\left\{c_1 \to -\dfrac{v0 + x0\,\beta - x0\,\omega_2}{2\,\omega_2}, \; c_2 \to \dfrac{v0 + x0\,\beta + x0\,\omega_2}{2\,\omega_2}\right\}\right\}$

Inserting this relations into the solution for the overdamped oscillator, we get

---

os = overdampedSolution /. sh

$\left\{\dfrac{e^{t\,(\omega_2 - \beta)}\,(v0 + x0\,\beta + x0\,\omega_2)}{2\,\omega_2} - \dfrac{e^{t\,(-\beta - \omega_2)}\,(v0 + x0\,\beta - x0\,\omega_2)}{2\,\omega_2}\right\}$

Note that ω_2 does not represent a frequency because the motion is not periodic; the displacement asymptotically approaches the equilibrium position as shown in the following plot:

ploverd = Plot$\left[\text{Evaluate}\left[\text{os} /. \left\{\beta \to \dfrac{1}{5}, \omega_2 \to \dfrac{1}{10}, x0 \to 1, v0 \to 0\right\}\right],\right.$

$\{t, 0, 25\}$, AxesLabel $\to$ {"t", "x"},

PlotStyle $\to$ RGBColor[0.996109, 0, 0], PlotRange $\to \left.\begin{pmatrix} 0 & 25 \\ 0 & 1.1^{\text{`}} \end{pmatrix}\right]$;

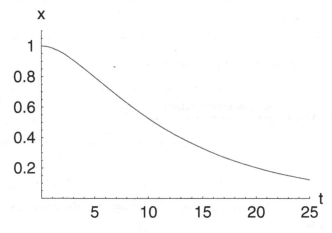

Comparing this plot with the plot of the critical damped oscillator, we observe that the displacement of the overdamped oscillator is always greater than the displacement of the critically damped oscillator. This behavior is important for the construction of certain practical oscillatory systems (e.g., galvanometers).

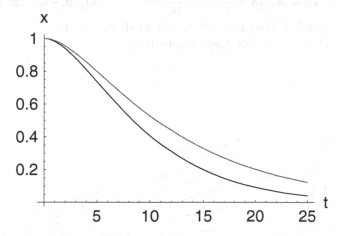

```
Show[plcritical, ploverd];
```

The case of overdamping results in a non oscillatory asymptotic decrease of the amplitude to zero. However, depending on the initial value of the velocity, there might be a change of sign of x before the displacement approaches zero. If we limit our considerations to initial positive displacements $x(0) = x_0 > 0$, there are three cases of interest for the initial velocity $x'(0) = v_0$.

a)	$v_0 > 0$,	so that x(t) reaches a maximum at some t>0 before approaching zero.
b)	$v_0 < 0$,	with x(t) monotonically approaching zero.
c)	$v_0 < 0$,	but sufficiently large so that x(t) changes sign, reaches a minimum value, and then approaches zero.

These three cases are illustrated in the following plot where we used a positive initial displacement and nine different initial values for the velocity. We observe that all three cases occur.

Plot[Evaluate[

    Table[os /. $\left\{\beta \to \dfrac{1}{5}, \omega_2 \to \dfrac{1}{10}, \text{x0} \to 1, \text{v0} \to i\right\}, \{i, -1, 1, .25\}$]],

    {t, 0, 25}, AxesLabel → {"t", "x"}, PlotRange → All,

    PlotStyle → RGBColor[0, 0, 0.996109]];

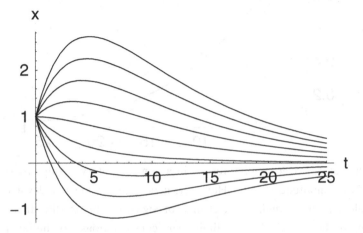

The phase plane of the overdamped oscillator is constructed by plotting x' versus x. This is possible if we first determine the velocity of the overdamped oscillator:

dos $= \dfrac{\partial \, \text{os}}{\partial t}$

$$\left\{ \frac{e^{t(\omega_2 - \beta)}\,(\omega_2 - \beta)\,(\text{v0} + \text{x0}\,\beta + \text{x0}\,\omega_2)}{2\,\omega_2} - \right.$$

$$\left. \frac{e^{t(-\beta - \omega_2)}\,(-\beta - \omega_2)\,(\text{v0} + \text{x0}\,\beta - \text{x0}\,\omega_2)}{2\,\omega_2} \right\}$$

The parametric plot of the displacement and the velocity for different values of the initial velocity generates a characteristic picture for the damped oscillator.

ParametricPlot[Evaluate[

Table[Flatten[{os, dos}] /. $\{\beta \to \frac{1}{5}, \omega_2 \to \frac{1}{10}, x0 \to 1, v0 \to i\}$,

$\{i, -2, 2, .25\}$]], $\{t, 0, 25\}$, AxesLabel $\to$ {"x", "x'"},

PlotRange $\to$ All, PlotStyle $\to$ RGBColor[0, 0, 0.996109]];

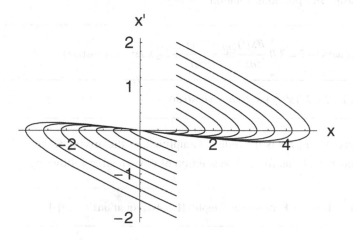

2.4.8.5 Driven Oscillations

In the preceding sections, we found that a particle undergoing free oscillations would remain in motion forever. In every real system, however, there is always a certain amount of friction that eventually damps the motion to rest. This damping of the oscillations may be prevented if there exists some mechanism for supplying the system with energy from an external source at a rate equal to that at which it is absorbed by the damping medium. Motions of this type are called driven oscillations.

The simplest case of driven oscillations is that in which an external driving force varying harmonically with time is applied to the oscillator. The total force on the particle is then

$$F = -k\,x - \gamma\,x' + F_0 \cos(\omega\,t) \qquad (2.4.21)$$

where we consider a linear restoring force and a viscous damping force in addition to the driving force. The equation of motion becomes

$$\text{equation17} = \gamma \, \frac{\partial x(t)}{\partial t} + m \, \frac{\partial^2 x(t)}{\partial t \, \partial t} + k \, x(t) == F_0 \cos(\omega t)$$

$$k \, x(t) + \gamma \, x'(t) + m \, x''(t) == \cos(t \, \omega) \, F_0$$

or using our previous notation

$$\text{equation17} = 2 \, \beta \, \frac{\partial x(t)}{\partial t} + \frac{\partial^2 x(t)}{\partial t \, \partial t} + \omega_0^2 \, x(t) == A \cos(\omega t)$$

$$x(t) \, \omega_0^2 + 2 \, \beta \, x'(t) + x''(t) == A \cos(t \, \omega)$$

where $A = F_0 / m$ is the reduced amplitude of the driving force and ω is the frequency of that force. The solution of this equation follows by

solution17 = Flatten[FullSimplify[DSolve[equation17, x, t]]]

$$\left\{ x \rightarrow \text{Function}\left[\{t\}, \; e^{t\left(-\beta - \sqrt{\beta^2 - \omega_0^2}\right)} c_1 + e^{t\left(\sqrt{\beta^2 - \omega_0^2} - \beta\right)} c_2 + \right.$$

$$\left(-A \sqrt{(\beta - \omega_0)(\beta + \omega_0)} \; \sqrt{\beta^2 - \omega_0^2} \; \cos(t \, \omega) \, \omega^2 + \right.$$

$$2 \, A \, \beta \, \sqrt{(\beta - \omega_0)(\beta + \omega_0)} \; \sqrt{\beta^2 - \omega_0^2} \; \sin(t \, \omega) \, \omega +$$

$$\left. A \, \omega_0^2 \, \sqrt{(\beta - \omega_0)(\beta + \omega_0)} \; \sqrt{\beta^2 - \omega_0^2} \; \cos(t \, \omega)\right) \Big/$$

$$\left((\beta - \omega_0)(\beta + \omega_0) \left(\beta - i \, \omega - \sqrt{\beta^2 - \omega_0^2} \right) \left(\beta + i \, \omega - \sqrt{\beta^2 - \omega_0^2} \right) \right.$$

$$\left.\left. \left(\beta - i \, \omega + \sqrt{\beta^2 - \omega_0^2} \right) \left(\beta + i \, \omega + \sqrt{\beta^2 - \omega_0^2} \right) \right) \right] \right\}$$

We observe that the solution consists of two parts. The first part represents the complementary solution containing initial conditions denoted by the constants of integration c_1 and c_2. The second part is the particular solution free of any constant of integration. This part is present in any case independent of the initial conditions. To separate the two parts from each other, we first extract the particular solution from the total solution by

---

particularSolution = $x(t)$ /. solution17 /. $\{c_1 \to 0, c_2 \to 0\}$

$$\left(-A \sqrt{(\beta - \omega_0)(\beta + \omega_0)}\ \sqrt{\beta^2 - \omega_0^2}\ \cos(t\,\omega)\,\omega^2 + \right.$$
$$2\,A\,\beta\,\sqrt{(\beta - \omega_0)(\beta + \omega_0)}\ \sqrt{\beta^2 - \omega_0^2}\ \sin(t\,\omega)\,\omega +$$
$$\left. A\,\omega_0^2\,\sqrt{(\beta - \omega_0)(\beta + \omega_0)}\ \sqrt{\beta^2 - \omega_0^2}\ \cos(t\,\omega)\right) \Big/$$
$$\left((\beta - \omega_0)(\beta + \omega_0)\left(\beta - i\,\omega - \sqrt{\beta^2 - \omega_0^2}\right)\left(\beta + i\,\omega - \sqrt{\beta^2 - \omega_0^2}\right)\right.$$
$$\left.\left(\beta - i\,\omega + \sqrt{\beta^2 - \omega_0^2}\right)\left(\beta + i\,\omega + \sqrt{\beta^2 - \omega_0^2}\right)\right)$$

The complementary solution incorporating the initial conditions follows by

---

complementarySolution =
   Simplify[$(x(t)$ /. solution17) − particularSolution]

$$e^{-t\left(\beta + \sqrt{\beta^2 - \omega_0^2}\right)}\left(c_1 + e^{2\,t\,\sqrt{\beta^2 - \omega_0^2}}\,c_2\right)$$

This solution is just the result derived for a damped oscillator. The general solution for the driven oscillator is

$$x(t) = x_p(t) + x_c(t). \tag{2.4.22}$$

The complementary solution containing the initial conditions c_1 and c_2 is responsible for the transient effects in the solution (i.e., effects that depend upon the initial conditions). The complementary solution damps out with time because of the factor $e^{-\beta t}$. The partial solution x_p represents the steady state effects and contains all of the information for time t large compared to $1/\beta$. For this reason let us first examine the particular solution.

The important part of the particular solution x_p is its amplitude. To extract the amplitude from the variable *particularSolution*, we use the property that the amplitude is independent of time. Thus we can set $t \to 0$ and make sure that the radicand $\beta^2 - \omega_0^2$ is a positive quantity (i.e., we replace $\omega_0^2 \to \beta^2 + \omega_1^2$ with $\omega_1^2 > 0$). After the simplification we rewrite the result in the original parameters β^2 and ω_0^2.

```
amplitude = (particularSolution /. {t → 0, ω₀² → β² + ω₁²} //
    PowerExpand // Simplify) /.
  {ω₁² → ω₀² - β², ω₁⁴ → (ω₀² - β²)²} // FullSimplify
```

$$\frac{A \sqrt{\beta - \omega_0} \sqrt{\beta + \omega_0} \ (\omega_0^2 - \omega^2)}{\sqrt{(\beta - \omega_0)(\beta + \omega_0)} \ (4 \beta^2 \omega^2 + (\omega - \omega_0)^2 (\omega + \omega_0)^2)}$$

The result shows that the total amplitude Δ of the particular solution depends on the driving frequency ω, the damping constant β, the frequency of the undamped oscillator ω_0, and on the reduced amplitude of the applied driving force A. We can reduce this four parameter relations to a three parameter expression if we introduce the scaled amplitude $\delta = \Delta / A$:

scaledAmplitude = $\dfrac{\text{amplitude}}{A}$ **// PowerExpand // Simplify**

$$\frac{\omega_0^2 - \omega^2}{4 \beta^2 \omega^2 + (\omega^2 - \omega_0^2)^2}$$

A plot of this relation reveals, that the scaled amplitude of the driven oscillator encounters a zero at the frequency of the undamped oscillator.

```
Plot[Evaluate[
    Table[scaledAmplitude /. {A → 1, ω₀ → 1, β → i}, {i, .1, 1.2, .1}]],
    {ω, 0, 3}, AxesLabel → {"ω", "Δ/A"},
    PlotStyle → RGBColor[0, 0, 0.996109]];
```

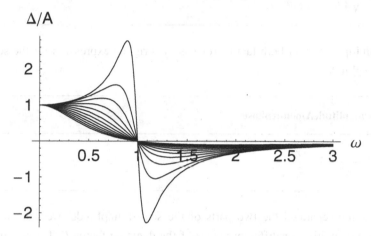

This behavior indicates that the scaled amplitude is a combination of two parts. These parts are the amount of the amplitude and the phase factor Φ of the amplitude. The phase occurs because the driving frequency ω is different from the frequency of the undriven oscillator ω_0. The scaled amplitude contains this difference in the numerator. Dividing the numerator by the square root of the denominator, we get the phase shift factor of the amplitude

$$\text{phase} = \frac{\text{Numerator[scaledAmplitude]}}{\sqrt{\text{Denominator[scaledAmplitude]}}}$$

$$\frac{\omega_0^2 - \omega^2}{\sqrt{4\,\beta^2\,\omega^2 + (\omega^2 - \omega_0^2)^2}}$$

The amplitude itself is given by the inverse square root of the denominator

$$\text{amplitudeAmount} = \frac{1}{\sqrt{\text{Denominator}[\text{scaledAmplitude}]}}$$

$$\frac{1}{\sqrt{4\,\beta^2\,\omega^2 + (\omega^2 - \omega_0^2)^2}}$$

Multiplication of both factors reveals the original expression for the scaled amplitude

amplitudeAmount phase

$$\frac{\omega_0^2 - \omega^2}{4\,\beta^2\,\omega^2 + (\omega^2 - \omega_0^2)^2}$$

Having separated the two parts of the scaled amplitude, we can plot the two quantities for different values of the damping factor β. The amount of the amplitude looks like

```
Plot[Evaluate[
    Table[amplitudeAmount /. {A → 1, ω₀ → 1, β → i}, {i, .1, 1.2, .1}]],
    {ω, 0, 4}, AxesLabel → {"ω", "Δ/A"},
    PlotStyle → RGBColor[0, 0, 0.996109]];
```

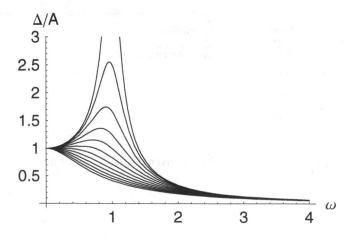

The graphical representation of the phase factor for different values of the damping constant is given by

```
Plot[Evaluate[Table[phase /. {A → 1, ω₀ → 1, β → i}, {i, .1, 1.2, .1}]],
    {ω, 0, 4}, AxesLabel → {"ω", "Φ"},
    PlotStyle → RGBColor[0, 0, 0.996109]];
```

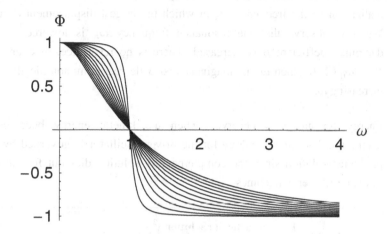

That is, there is a real delay between the action of the driving force and the response of the system.

The amplitude, and therefore the energy, of the system in the steady state depends not only on the amplitude of the driver, but also on its frequency. In the two plots above, we observe that if the driving frequency is approximately equal to the natural frequency of the system, the system will oscillate with a very large amplitude. This phenomenon is called resonance. When the driving frequency equals the natural frequency of the oscillator, the energy absorbed by the oscillator is maximum. Thus we have to distinguish two frequencies when resonances occur: First the amplitude resonance with its largest elongation and second an energy resonance with the largest energy transfer.

In order to find the resonance frequency ω_R at which the amplitude Δ / A is a maximum, we solve the defining equation for the maximal deviation

$$\omega_R = \text{Solve}\left[\frac{\partial\,\text{amplitudeAmount}}{\partial\omega} == 0, \omega\right]$$

$$\left\{\{\omega \to 0\},\ \left\{\omega \to -\sqrt{\omega_0^2 - 2\beta^2}\right\},\ \left\{\omega \to \sqrt{\omega_0^2 - 2\beta^2}\right\}\right\}$$

The result contains three roots. Only the positive root is a physical realization of the frequency ω_R at which the largest displacement occurs. We also observe that the resonance frequency ω_R is lowered as the damping coefficient β is increased. There is no resonance, of course, if $\beta^2 > \omega_0^2/2$, for then ω_R is imaginary and Δ decreases monotonically with increasing ω.

Energy resonance is observed when the kinetic energy becomes a maximum. The kinetic energy for the driven oscillator is governed by the particular solution since the complementary solution dies out for large t. The kinetic energy becomes

$$T = \text{Simplify}\left[\frac{1}{2}\,m\left(\frac{\partial\,\text{particularSolution}}{\partial t}\right)^2\right]$$

$$\frac{A^2\,m\,\omega^2\,(2\,\beta\,\omega\,\cos(t\,\omega) + (\omega^2 - \omega_0^2)\sin(t\,\omega))^2}{2\left(4\,\beta^2\,\omega^2 + (\omega^2 - \omega_0^2)^2\right)^2}$$

In order to obtain a value of T which is independent of the time, we compute the average of T over one complete period of oscillation. Thus,

$$\text{TMean} = \frac{\omega}{2\pi}\int_0^{\frac{2\pi}{\omega}} T\,dt\ //\ \text{Simplify}$$

$$\frac{A^2\,m\,\omega^2}{4\left(4\,\beta^2\,\omega^2 + (\omega^2 - \omega_0^2)^2\right)}$$

The value of ω for $\langle T \rangle$ a maximum is labeled ω_E and is obtained from

$$\omega_E = \text{Solve}\left[\frac{\partial \text{TMean}}{\partial \omega} == 0, \omega\right]$$

$$\{\{\omega \to 0\}, \{\omega \to -\omega_0\}, \{\omega \to -i\,\omega_0\}, \{\omega \to i\,\omega_0\}, \{\omega \to \omega_0\}\}$$

Since the trivial, negative, and complex solutions of this condition are of minor physical importance, we get as a result

$$\omega_E = \{\omega \to \omega_0\}$$

$$\{\omega \to \omega_0\}$$

so that the kinetic energy resonance occurs at the natural frequency of the system for undamped oscillations.

We see therefore that the amplitude resonance occurs at a frequency $\sqrt{\omega_0^2 - 2\,\beta^2}$ whereas the kinetic energy resonance occurs at ω_0. Since the potential energy is proportional to the square of the amplitude, the potential energy resonance must also occur at $\sqrt{\omega_0^2 - 2\,\beta^2}$. That the kinetic and potential energies resonate at different frequencies is a result of the fact that the damped oscillator is not a conservative system; energy is continuously exchanged with the driving mechanism and energy is being transferred to the damping medium.

Although we have emphasized the steady-state motion of the driven oscillator, the transient effects are often of considerable importance. The details of the motion during the period of time before the transient effects have disappeared (i.e., $t \lesssim 1/\beta$) are strongly dependent on the conditions of the oscillator at the time that the driving force is first applied and also on the relative magnitude of the driving frequency ω and the damping frequency $\sqrt{\omega_0^2 - \beta^2}$.

2.4.8.6 Solution Procedures of Liner Differential Equations

This subsection discuses two methods useful for solving linear ordinary a well as partial differential equations. The discussed methods are especially useful for solving initial value problems. The presented methods are the Laplace transform method and the Green's function method.

The Laplace Transform Method

In the preceding sections, we have mainly used straightforward methods in solving the differential equations that describe oscillatory motion. The procedure has been to obtain a general solution and then to impose the initial conditions in order to obtain the desired particular solution. The procedure discussed in this subsection is the Laplace transform method. This technique, which is generally useful for obtaining solutions to linear differential equations, allows the reduction of a differential equation to an algebraic equation. This is accomplished by defining the Laplace transform $f(p)$ of a function $F(t)$ according to

$$f(p) = \int_0^\infty e^{-pt} F(t)\, dt. \tag{2.4.23}$$

The Laplace transform of a function $F(t)$ exists if $F(t)$ is sectionally continuous in every finite interval $0 < t < \infty$ and if $F(t)$ increases at a rate less than exponential as t becomes infinitely large. In general, the parameter p may be complex, but we will not have occasion to consider such a case here. The Laplace transform of a function $F(t)$ will be denoted by $\mathcal{L}_t^p[F[t]]$, where the lower index denotes the original variable t and the upper index refers to the Laplace variable p.

For example, if $F(t) = 1$, the Laplace transform is given by

$\mathcal{L}_t^p[1]$
$\dfrac{1}{p}$

Similarly, for $F(t) = e^{-\alpha t}$, we find

$\mathcal{L}_t^p[e^{-\alpha t}]$
$\dfrac{1}{p+\alpha}$

Some important properties of Laplace transforms are the following:

The Laplace transform is linear. If α and β are constants, then

$\mathcal{L}_t^p[\alpha H(t) + \beta G(t)]$
$\beta \mathcal{L}_t^p[G(t)] + \alpha \mathcal{L}_t^p[H(t)]$

The Laplace transform of the derivative of $H(t)$ is given by

$\mathcal{L}_t^p\left[\dfrac{\partial H(t)}{\partial t}\right]$
$p\,\mathcal{L}_t^p[H(t)] - H(0)$

The transforms of higher derivatives can be calculated similarly; for example,

$\mathcal{L}_t^p\left[\dfrac{\partial^2 H(t)}{\partial t\,\partial t}\right]$
$\mathcal{L}_t^p[H(t)]\,p^2 - H(0)\,p - H'(0)$

The substitution of $p + a$ for the parameter p in the transform corresponds to multiplying $F(t)$ by $e^{-\alpha t}$. For example,

$$\mathcal{L}_t^p[\cos(\omega t)]$$

$$\frac{p}{p^2 + \omega^2}$$

so that

$$\mathcal{L}_t^p[e^{-\alpha t}\cos(\omega t)] \text{ // Simplify}$$

$$\frac{p + \alpha}{p^2 + 2\alpha p + \alpha^2 + \omega^2}$$

Knowing some of the main properties of the Laplace transform, let us apply this method to solve the problem of a driven damped oscillator. The equation of motion is just *equation17*. Let us assume that the initial conditions are $x(0) = 0$ and $x'(0) = 0$. The Laplace transform of this equation is

lpTr $= \mathcal{L}_t^p$**[equation17]**

$$\mathcal{L}_t^p[x(t)]\, p^2 - x(0)\, p + \omega_0^2\, \mathcal{L}_t^p[x(t)] + 2\,\beta\,(p\,\mathcal{L}_t^p[x(t)] - x(0)) - x'(0) == \frac{A\,p}{p^2 + \omega^2}$$

Applying the initial conditions to the Laplace representation

lpTr $=$ **lpTr** /. $\{x(0) \to 0,\ x'(0) \to 0\}$

$$\mathcal{L}_t^p[x(t)]\, p^2 + 2\,\beta\,\mathcal{L}_t^p[x(t)]\, p + \omega_0^2\, \mathcal{L}_t^p[x(t)] == \frac{A\,p}{p^2 + \omega^2}$$

we get a simplified version of the Laplace representation. Solving this expression with respect to the Laplace representation of $x(p) = \mathcal{L}_t^p[x(t)]$, we find

---

slpTr = Simplify[Solve[lpTr, $\mathcal{L}_t^p[x(t)]$]]

$$\left\{\left\{\mathcal{L}_t^p[x(t)] \to \frac{A\,p}{(p^2+\omega^2)\,(p^2+2\,\beta\,p+\omega_0^2)}\right\}\right\}$$

This is the solution of our initial value problem represented in Laplace space. The inversion of this expression will provide the solution

---

solution = InverseLaplaceTransform[slpTr, p, t] // Simplify

$$\left\{\left\{x(t) \to \right.\right.$$

$$\left(A\,e^{-t\left(\beta+\sqrt{\beta^2-\omega_0^2}\right)}\left(-e^{2t\sqrt{\beta^2-\omega_0^2}}\,\beta\,\omega^2 + \beta\,\omega^2 + e^{2t\sqrt{\beta^2-\omega_0^2}}\,\sqrt{\beta^2-\omega_0^2}\,\omega^2 + \right.\right.$$

$$\sqrt{\beta^2-\omega_0^2}\,\omega^2 + 4\,e^{t\left(\beta+\sqrt{\beta^2-\omega_0^2}\right)}\,\beta\,\sqrt{\beta^2-\omega_0^2}\,\sin(t\,\omega)\,\omega -$$

$$e^{2t\sqrt{\beta^2-\omega_0^2}}\,\beta\,\omega_0^2 + \beta\,\omega_0^2 + 2\,e^{t\left(\beta+\sqrt{\beta^2-\omega_0^2}\right)}\,\sqrt{\beta^2-\omega_0^2}\,(\omega_0^2-\omega^2)$$

$$\cos(t\,\omega) - e^{2t\sqrt{\beta^2-\omega_0^2}}\,\omega_0^2\,\sqrt{\beta^2-\omega_0^2} - \omega_0^2\,\sqrt{\beta^2-\omega_0^2}\right)\right)\Big/$$

$$\left.\left(2\,\sqrt{\beta^2-\omega_0^2}\,\left(4\,\beta^2\,\omega^2 + (\omega^2-\omega_0^2)^2\right)\right)\right\}\right\}$$

The graphical representation of this solution for different damping factors β is given below

```
Plot[Evaluate[
    Table[x(t) /. solution /. {A → 1, ω₀ → 2, β → i, ω → 1}, {i, .1, 4, .5}]],
    {t, 0.1, 25}, AxesLabel → {"t", "x(t)"},
    PlotStyle → RGBColor[0, 0, 0.996109]];
```

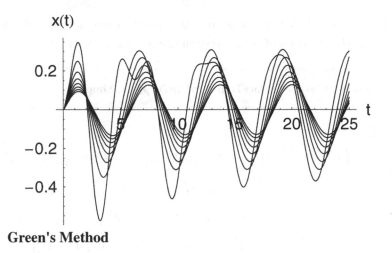

Green's Method

Green's method is generally useful for the solution of linear, inhomogeneous differential equations. The main advantage of the method lies in the fact that Green's function $G(t, \tau)$, which is the solution of the equation for an infinitesimal element of the inhomogeneous part, already contains the initial conditions. To demonstrate these facts, let us consider the linear ordinary differential equation

$$L_t[u] = f(t) \tag{2.4.24}$$

where L_t is a linear differential operator. If this linear differential operator has an inverse $L_t^{-1} = G$, the solution can be written as

$$u(t) = \int_{-\infty}^{\infty} G(t, \tau) f(\tau) \, d\tau \tag{2.4.25}$$

where the integration is over the range of definition of the functions involved. Once we know $G(t, \tau)$, Equation (2.4.25) gives the solution $u(t)$ in an integral form. However, how do we find $G(t, \tau)$? If L_t is a local differential operator, we obtain

$$L_t \, G(t, \tau) = \delta(t - \tau). \tag{2.4.26}$$

$G(t, \tau)$ is called Green's function for the differential operator L_t. Thus, Green's function is nothing more than the solution of an linear ordinary differential equation under the condition that at $t = \tau$, a unique force acts on the system. Let us examine this behavior for the damped harmonic oscillator. The linear operator L_t for this physical system is defined by

$$L(f\_) := f\,\omega_0^2 + 2\,\beta\,\frac{\partial f}{\partial t} + \frac{\partial^2 f}{\partial t\,\partial t}$$

Taking this definition into account, Green's function follows from relation (2.4.26) by

Green = $L(G(t, \tau))$ == DiracDelta[$t - \tau$]

$$G(t, \tau)\,\omega_0^2 + 2\,\beta\,G^{(1,0)}(t, \tau) + G^{(2,0)}(t, \tau) == \delta(t - \tau)$$

We assume that the system starts from rest, meaning $G(t, \tau) = 0$ for $t < \tau$, so that $G(t, \tau)$ is the response of the system on a unit force action at $t = \tau$. For times $t > \tau$, there is no force acting on the equation. Thus, Green's function is determined by the homogeneous equation

Green = $L(G(t, \tau))$ == 0

$$G(t, \tau)\,\omega_0^2 + 2\,\beta\,G^{(1,0)}(t, \tau) + G^{(2,0)}(t, \tau) == 0$$

The solution of this equation follows by applying the Laplace transform method to this equation. The Laplace transform is

lpGreen = $\mathcal{L}_t^p$[Green]

$$\mathcal{L}_t^p[G(t, \tau)]\,p^2 - G(0, \tau)\,p + \omega_0^2\,\mathcal{L}_t^p[G(t, \tau)] +$$
$$2\,\beta\,(p\,\mathcal{L}_t^p[G(t, \tau)] - G(0, \tau)) - G^{(1,0)}(0, \tau) == 0$$

Solving this relation with respect to the Laplace variable $\mathcal{L}_t^p[G(t, \tau)]$, we get

> **lpSolution = Solve[lpGreen, $\mathcal{L}_t^p[G(t, \tau)]$]**
>
> $$\left\{\left\{\mathcal{L}_t^p[G(t, \tau)] \rightarrow \frac{p\,G(0, \tau) + 2\,\beta\,G(0, \tau) + G^{(1,0)}(0, \tau)}{p^2 + 2\,\beta\,p + \omega_0^2}\right\}\right\}$$

The inversion of the Laplace transform provides us with the solution

> **GreenF = InverseLaplaceTransform[lpSolution, p, t] // FullSimplify**
>
> $$\left\{\left\{G(t, \tau) \rightarrow \frac{1}{\sqrt{(\beta - \omega_0)\,(\beta + \omega_0)}}\right.\right.$$
> $$\left(e^{-t\beta}\left(\sqrt{(\beta - \omega_0)\,(\beta + \omega_0)}\,\cosh\!\left(t\,\sqrt{(\beta - \omega_0)\,(\beta + \omega_0)}\right)G(0, \tau) + \right.\right.$$
> $$\left.\left.\left.\left.\sinh\!\left(t\,\sqrt{(\beta - \omega_0)\,(\beta + \omega_0)}\right)(\beta\,G(0, \tau) + G^{(1,0)}(0, \tau))\right)\right)\right\}\right\}$$

A transformation to a pure function representation allows us to use Green's function in symbolic expressions:

> **r1 = $G \rightarrow$ Function[{t, τ}, \$w] /. (\$w $\rightarrow$ $G(t, \tau)$ /. GreenF)**
>
> $$G \rightarrow \text{Function}\!\left[\{t, \tau\}, \frac{1}{\sqrt{(\beta - \omega_0)\,(\beta + \omega_0)}}\right.$$
> $$\left(e^{-t\beta}\left(\sqrt{(\beta - \omega_0)\,(\beta + \omega_0)}\,\cosh\!\left(t\,\sqrt{(\beta - \omega_0)\,(\beta + \omega_0)}\right)G(0, \tau) + \right.\right.$$
> $$\left.\left.\left.\sinh\!\left(t\,\sqrt{(\beta - \omega_0)\,(\beta + \omega_0)}\right)(\beta\,G(0, \tau) + G^{(1,0)}(0, \tau))\right)\right)\right]$$

> **res = $\displaystyle\int_{\tau-\epsilon}^{\epsilon+\tau}$ Green[1] dt == $\displaystyle\int_{\tau-\epsilon}^{\epsilon+\tau}$ DiracDelta[$t - \tau$] dt /.**
>
> **$\{G(\tau - \epsilon, \tau) \rightarrow 0,\ G^{(1,0)}(\tau - \epsilon, \tau) \rightarrow 0\}$**
>
> $$\int_{\tau-\epsilon}^{\epsilon+\tau} (G(t, \tau)\,\omega_0^2 + 2\,\beta\,G^{(1,0)}(t, \tau) + G^{(2,0)}(t, \tau))\,dt == \theta(\epsilon) - \theta(-\epsilon)$$

To estimate the terms in the above relation, we assume that the maximum of G is finite $\text{Max}(|G|) < \infty$, so that we can estimate the integral term by $I \leq \text{Max}(|G|)\,2\,\epsilon$, meaning that for $\epsilon \rightarrow 0$, the integral term vanishes. If we, in addition, assume that the time derivative of G is finite, $\text{Max}\big(\big|\dot{G}\big|\big) < \infty$, then we can estimate the behavior of Green's function as $|G| \leq \text{Max}\big(\big|\dot{G}\big|\big) 2\epsilon$. This again means that G vanishes if $\epsilon \to 0$. These two properties allow us to define the following conditions for the Green's function:

$$
\begin{aligned}
G(\tau + 0, \tau) &= 1, \\
G(\tau - 0, \tau) &= 0.
\end{aligned}
\qquad (2.4.27)
$$

These two conditions represent the behavior that the particle right after the application of a unit force stays at the same position but gets a unique momentum. Conditions (2.4.27) allow us to determine the initial conditions for Green's function. The first equation reads

---

eq1 = (G(τ, τ) /. r1) == 0

$$
\frac{1}{\sqrt{(\beta - \omega_0)(\beta + \omega_0)}}
$$
$$
\left(e^{-\beta \tau} \left(\sqrt{(\beta - \omega_0)(\beta + \omega_0)} \cosh\!\left(\tau \sqrt{(\beta - \omega_0)(\beta + \omega_0)} \right) G(0, \tau) + \right.\right.
$$
$$
\left.\left. \sinh\!\left(\tau \sqrt{(\beta - \omega_0)(\beta + \omega_0)} \right) (\beta\, G(0, \tau) + G^{(1,0)}(0, \tau)) \right) \right) == 0
$$

The second Equation of (2.4.27) reads

---

eq2 = $\left(\dfrac{\partial G(t, \tau)}{\partial t} \; /.\, t \to \tau \; /.\, \text{r1} \right)$ == 1

$$
\frac{1}{\sqrt{(\beta - \omega_0)(\beta + \omega_0)}}
$$
$$
\left(e^{-\beta \tau} \left((\beta - \omega_0)(\beta + \omega_0)\, G(0, \tau) \sinh\!\left(\tau \sqrt{(\beta - \omega_0)(\beta + \omega_0)} \right) + \right.\right.
$$
$$
\sqrt{(\beta - \omega_0)(\beta + \omega_0)} \cosh\!\left(\tau \sqrt{(\beta - \omega_0)(\beta + \omega_0)} \right)
$$
$$
\left.\left. (\beta\, G(0, \tau) + G^{(1,0)}(0, \tau)) \right) \right) - \frac{1}{\sqrt{(\beta - \omega_0)(\beta + \omega_0)}}
$$
$$
\left(e^{-\beta \tau} \beta \left(\sqrt{(\beta - \omega_0)(\beta + \omega_0)} \cosh\!\left(\tau \sqrt{(\beta - \omega_0)(\beta + \omega_0)} \right) G(0, \tau) + \right.\right.
$$
$$
\left.\left. \sinh\!\left(\tau \sqrt{(\beta - \omega_0)(\beta + \omega_0)} \right) (\beta\, G(0, \tau) + G^{(1,0)}(0, \tau)) \right) \right) == 1
$$

Solving these two equations for the initial conditions of Green's function, we get

---

sol = Simplify[Solve[{eq1, eq2}, {$G^{(1,0)}(0, \tau)$, $G(0, \tau)$}]]

$$\left\{\left\{G^{(1,0)}(0, \tau) \to e^{\beta\tau}\left(\cosh\left(\tau\sqrt{\beta^2 - \omega_0^2}\right) + \frac{\beta\sinh\left(\tau\sqrt{\beta^2 - \omega_0^2}\right)}{\sqrt{\beta^2 - \omega_0^2}}\right),\right.\right.$$

$$\left.\left. G(0, \tau) \to -\frac{e^{\beta\tau}\sinh\left(\tau\sqrt{\beta^2 - \omega_0^2}\right)}{\sqrt{\beta^2 - \omega_0^2}}\right\}\right\}$$

Inserting these results into the original representation of the solution, we gain

---

GreenF = Simplify[$G(t, \tau)$ /. r1 /. sol]

$$\left\{\frac{e^{\beta(\tau-t)}\sinh\left((t-\tau)\sqrt{\beta^2 - \omega_0^2}\right)}{\sqrt{\beta^2 - \omega_0^2}}\right\}$$

---

representing the Green's function for $t > \tau$. For $t \leq 0$ the Green's function vanishes. Knowing the Green's function, we are able to solve the inhomogeneous differential equation by integrating the product of the inhomogenity and the Green's function

---

lh = $\left\{\dfrac{e^{\beta(\tau-t)}\sinh\left(I(t-\tau)\sqrt{-\beta^2 + \omega_0^2}\right)}{I\sqrt{-\beta^2 + \omega_0^2}}\right\}$

$$\left\{\frac{e^{\beta(\tau-t)}\sin\left((t-\tau)\sqrt{\omega_0^2 - \beta^2}\right)}{\sqrt{\omega_0^2 - \beta^2}}\right\}$$

Knowing the Green's function also allows us, for example, to calculate the solution for a constant force of unit strength by

$$\text{Simplify}\left[\int_0^t \text{PowerExpand}[\text{TrigReduce}[\text{lh}[\![1]\!]]]\, d\tau\right]$$

$$-\frac{e^{-t\beta}\cos\!\left(t\sqrt{\omega_0^2 - \beta^2}\right) + \dfrac{e^{-t\beta}\,\beta\sin\!\left(t\sqrt{\omega_0^2-\beta^2}\right)}{\sqrt{\omega_0^2-\beta^2}} - 1}{\omega_0^2}$$

The solution for a harmonic force $g\cos(\omega_0 t)$ in the case of vanishing damping is given by

$$\text{solution} =$$
$$\text{Simplify}\left[\int_0^t \text{PowerExpand}[\text{TrigReduce}[g\cos(\omega_0 \tau)\,\text{lh}[\![1]\!]\, /.\, \beta \to 0]]\, d\tau\right]$$

$$\frac{g\, t\sin(t\,\omega_0)}{2\,\omega_0}$$

Another example is an exponential decaying force for the damped harmonic oscillator resulting in

$$\text{h1} =$$
$$\text{Simplify}\left[\int_0^t \text{PowerExpand}[\text{TrigReduce}[h\, e^{-\gamma\tau}\, \text{lh}[\![1]\!]]]\, d\tau\right]$$

$$\left(e^{-t(\beta+\gamma)}\, h\!\left(-e^{t\gamma}\sqrt{\omega_0^2-\beta^2}\,\cos\!\left(t\sqrt{\omega_0^2-\beta^2}\right) + e^{t\gamma}(\gamma-\beta)\sin\!\left(t\sqrt{\omega_0^2-\beta^2}\right) + \right.\right.$$
$$\left.\left. e^{t\beta}\sqrt{\omega_0^2-\beta^2}\right)\right)\Big/\left(\sqrt{\omega_0^2-\beta^2}\,(\gamma^2 - 2\beta\gamma + \omega_0^2)\right)$$

All of these solutions are solutions free of any transient effects.

2.4.8.7 Nonlinear Oscillation

Solutions of certain nonlinear oscillation problems can be expressed in closed form in terms of elliptic integrals. The pendulum is one example of a nonlinear model exhibiting elliptic functions as solutions. A pendulum is a system with mass m which is kept in orbit by a massless supporting rod of length l (see Figure 2.4.9). The pendulum moves within the gravitational field of the Earth and is thus exposed to the vertical gravitational force mg. The dynamic force F is perpendicular to the supporting rod and takes the form $F(\phi) = -mg \sin(\phi)$.

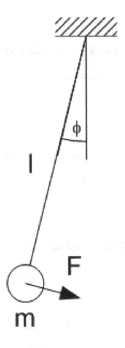

Figure 2.4.9.   Pendulum as a nonlinear system.

For small amplitudes, we can model the pendulum in terms of a linear system which is equivalent to a harmonic oscillator. The accuracy of this approximation will be determined in the course of our calculations. Taking the angle of libration to be ϕ (see Figure 2.4.9), the equation of motion for an oscillating particle of unit mass is

$$\phi'' + \omega_0^2 \sin(\phi) = 0, \tag{2.4.28}$$

with $\omega_0^2 = g/l$ being the ratio between the gravitational acceleration g and the length of the pendulum l. If the amplitudes around the equilibrium position are small, then $\sin(\phi)$ in Equation (2.4.28) can be approximated by $\sin(\phi) \approx \phi$.

Series[sin(ϕ), {ϕ, 0, 1}]

$\phi + O(\phi^2)$

As a result, the equation of motion is reduced to an equation of a harmonic oscillator

$$\phi'' + \omega_0^2 \phi = 0. \tag{2.4.29}$$

Within this approximation, the oscillation period T is given by $T = 2\pi/\omega = 2\pi\sqrt{l/g}$ and is independent of the amplitude.

If we wish to determine the oscillation period for larger amplitudes, we need to start with Equation (2.4.28). Since we have neglected damping in our equations, the total energy of the system can be written as the sum of the potential and kinetic energy (conservation of energy):

$$T_{\text{kin}} + V = E = \text{const.} \tag{2.4.30}$$

This formulation allows us to easily construct the solution to Eqation (2.4.28). Equation (2.4.30) gives a first integral of motion. Due to the explicit time independence of the equation of motion (2.4.28), the second step of the integration process can be done by a quadrature. The duration of oscillation can be expressed in the form of an integral.

If we choose the origin of the potential energy to be at the lowest point in the orbit, then we get for the potential energy

```
V = m g l (1 - Cos[φ[t]])
```

$$g\,l\,m\,(1 - \cos(\phi(t)))$$

$$V = m\,g\,l\,(1 - \cos(\phi)). \tag{2.4.31}$$

A graphical representation of the potential energy is given in the following plot. In addition to the potential energy, we also plotted three different energy values of the pendulum. As we will see, these values correspond to three different kinds of motion of the pendulum.

```
Plot[{1 - Cos[φ], 4/3, 2, 7/3},
    {φ, -2 π, 2 π}, AxesLabel → {"φ", " V[φ]/mgl "},
    PlotStyle → {Hue[0], Hue[0.2], Hue[0.4], Hue[0.6]}];
```

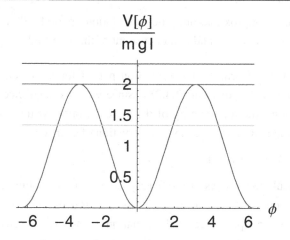

The kinetic energy is derived from the equation

```
Tkin = 1/2 m l² (∂t φ[t])²
```

$$\frac{1}{2}\,l^2\,m\,\phi'(t)^2$$

$$T_{\text{kin}} = \tfrac{1}{2}\,m\,l^2(\phi')^2. \tag{2.4.32}$$

The total energy of the pendulum then follows by adding up the kinetic and potential energy as

```
H = Tkin + V
```

$$\frac{1}{2} \, l^2 \, m \, \phi'(t)^2 + g \, l \, m \, (1 - \cos(\phi(t)))$$

A phase space portrait of the pendulum is generated next by specifying the parameters *l*, *m*, and *g*.

```
<< Graphics`ImplicitPlot`;
ImplicitPlot[
   Evaluate[Table[H == e /. {l -> 10, g -> 10, m -> .01,
       φ[t] -> φ, ∂t φ[t] -> p}, {e, 1/3, 5, 1/3}]],
   {φ, -2 π, 2 π}, {p, -7, 7}, PlotPoints -> 41,
   AxesLabel -> {"φ", "φ'"},
   PlotStyle -> Table[Hue[i/20], {i, 0, 15}]];
```

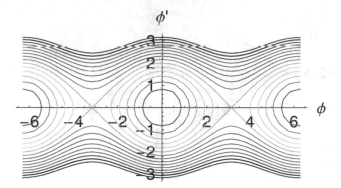

The phase space diagram shows that three different kinds of motion are possible. Near the center, there exist oscillations. For larger energies, we find revolutions, and for a certain energy, there is an asymptotic motion starting at one point and terminating at the upper turning point of the pendulum. This third kind of motion separates the two other motions. The phase space curve is thus called a separatrix.

Combining the energy plot with the phase space plot, we get an impression
of how the motion in the different region of the potential takes place.

```
<< Graphics`Graphics3D`;
ShadowPlot3D[Evaluate[
    H /. {l → 10, g → 10, m -> 0.01, φ[t] → φ, ∂_t φ[t] → p}],
    {φ, -2 π, 2 π}, {p, -3, 3}, PlotPoints → 45,
    AxesLabel → {"φ", "φ'", "H"}, Axes → True,
    SurfaceMesh → False, ShadowMesh → False,
    ViewPoint -> {1.756, -2.089, 2.000}];
```

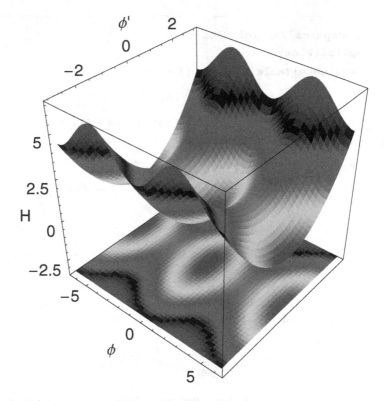

If we designate the angle at the highest orbital point as ϕ_1, the potential
and the kinetic energies at this point are given by

$$V(\phi = \phi_1) = E = m\,g\,l\,(1 - \cos(\phi_1)), \tag{2.4.33}$$
$$T_{\text{kin}}(\phi = \phi_1) = 0. \tag{2.4.34}$$

By means of the trigonometric identity $\cos(\phi) = 1 - 2\sin^2(\phi/2)$, the total energy at the upper reversal point can be expressed in the form

$$E = 2\,m\,g\,l\sin^2\left(\tfrac{\phi_1}{2}\right). \tag{2.4.35}$$

Because E is constant in time, this expression is also valid for amplitudes smaller than ϕ_1. The potential energy takes the form

$$V = 2\,m\,g\,l\sin^2\left(\tfrac{\phi}{2}\right); \tag{2.4.36}$$

we used the trigonometric identity $\cos(\phi) = 1 - 2\sin^2(\phi/2)$ to simplify the relation. In accordance with Equation (2.4.30), the kinetic energy is given as the difference between the total energy and the potential energy by

$$\tfrac{1}{2}\,m\,l^2\,\phi'^2 = 2\,m\,g\,l\left(\sin^2\left(\tfrac{\phi_1}{2}\right) - \sin^2\left(\tfrac{\phi}{2}\right)\right). \tag{2.4.37}$$

In other words, we get

$$\phi' = 2\,\omega_0\left(\sin^2\left(\tfrac{\phi_1}{2}\right) - \sin^2\left(\tfrac{\phi}{2}\right)\right)^{1/2}. \tag{2.4.38}$$

Separating the variables, we find

$$dt = \frac{d\phi}{2\,\omega_0\sqrt{\sin^2(\tfrac{\phi_1}{2}) - \sin^2(\tfrac{\phi}{2})}}. \tag{2.4.39}$$

We can obtain the oscillation period T of the pendulum by integrating both sides over a complete period

$$\int_0^T dt = 4\int_0^{\phi_1} \frac{d\phi}{2\,\omega_0\sqrt{\sin^2(\tfrac{\phi_1}{2}) - \sin^2(\tfrac{\phi}{2})}}. \tag{2.4.40}$$

The left hand side of (2.4.35) can be directly integrated and we find

$$T = \frac{2}{\omega_0}\int_0^{\phi_1} \frac{d\phi}{\sqrt{\sin^2(\tfrac{\phi_1}{2}) - \sin^2(\tfrac{\phi}{2})}}. \tag{2.4.41}$$

Thus, the oscillation period is reduced to a complete elliptic integral. By substituting $z = \sin(\phi/2)/\sin(\phi_1/2)$ and $k = \sin(\phi_1/2)$, the integral on the right-hand side of Equation (2.4.41) is transformed to the standard form

$$T = \frac{4}{\omega_0}\int_0^1 \frac{dz}{\sqrt{(1-z^2)(1-k^2 z^2)}} \tag{2.4.42}$$

$$= \frac{4}{\omega_0}\,K(k^2).$$

$K(k^2)$ denotes the complete elliptic integral of the first kind and $k^2 = E / (2\,m\,g\,l)$ denotes the modulus of the elliptic function.

By calling **EllipticK[]**, *Mathematica* executes $K(k^2)$. **Integrate[]** executes the integration of Equation (2.4.42):

$$\textbf{PowerExpand}\Big[\int_0^1 \frac{1}{\sqrt{(1-z^2)\,(1-k^2\,z^2)}}\; dz\Big]$$

$$\text{If}\Big[\text{Im}\Big(\frac{1}{k}\Big)\neq 0 \bigvee \text{Im}(k)\neq 0 \bigvee$$

$$1+\frac{1}{k}=0 \bigwedge \text{Re}\Big(\frac{1}{k}\Big)>1 \bigvee 1+\frac{1}{k}=0 \bigwedge \text{Re}(k)<0 \bigvee$$

$$\frac{1}{k}=1 \bigwedge \text{Re}(k)\geq 0 \bigvee \frac{1}{k}=1 \bigwedge \text{Re}\Big(\frac{1}{k}\Big)+1<0 \bigvee$$

$$\text{Re}\Big(\frac{1}{k}\Big)>1 \bigwedge \text{Re}(k)\geq 0 \bigvee \text{Re}\Big(\frac{1}{k}\Big)+1<0 \bigwedge \text{Re}(k)<0,$$

$$K(k^2),\ \text{Integrate}\Big[\frac{\sqrt{z^2-1}\,\sqrt{k^2\,z^2-1}}{(z^2-1)\,(k^2\,z^2-1)},\ \{z,0,1\},$$

$$\text{Assumptions}\to \neg\,\Big(\text{Im}\Big(\frac{1}{k}\Big)\neq 0 \bigvee \text{Im}(k)\neq 0 \bigvee$$

$$1+\frac{1}{k}=0 \bigwedge \text{Re}\Big(\frac{1}{k}\Big)>1 \bigvee 1+\frac{1}{k}=0 \bigwedge \text{Re}(k)<0 \bigvee$$

$$\frac{1}{k}=1 \bigwedge \text{Re}(k)\geq 0 \bigvee \frac{1}{k}=1 \bigwedge \text{Re}\Big(\frac{1}{k}\Big)+1<0 \bigvee$$

$$\text{Re}\Big(\frac{1}{k}\Big)>1 \bigwedge \text{Re}(k)\geq 0 \bigvee \text{Re}\Big(\frac{1}{k}\Big)+1<0 \bigwedge \text{Re}(k)<0\Big)\Big]\Big]$$

Once we know the length of the pendulum and its initial angular displacement, the oscillation period is completely determined. Since *Mathematica* recognizes all elliptic integrals as well as all Jacobian elliptic functions, we can straightforwardly determine the dependence of the period on the initial amplitude. A graphical representation of $K(k)$ via ϕ_1 can be found in Figure 2.4.10. We are now able to evaluate the period T with the following function:

```
T[omega_, phi1_] := Block[{k, duration},
    k = Sin[ phi1 / 2 ];
    duration = 4 EllipticK[k²] / omega
]
```

Our input values are the angle of displacement ϕ_1 and the frequency $\omega_0 = \sqrt{g/l}$. We first calculate the modulus k^2 in accordance with the above definition and then determine the period in accordance with Equation (2.4.42). As we see from Figure 2.4.10, $K(k)$ with $k = 1$ tends toward ∞ (i.e., at the upper point of reversal $\phi_1 = \pi$, the period is infinitely large).

Approximated equations are often cited in the literature for the period. To obtain a valid comparison between exact and approximated oscillation periods, we use the approximation procedure described below. If the pendulum oscillates, we know that $k < 1$. Using this condition, we can expand the second part of the integrand in Equation (2.4.42) into a Taylor series:

$$\frac{1}{\sqrt{1-k^2 z^2}} = 1 + \frac{k^2 z^2}{2} + \frac{3 k^4 z^4}{8} + \dots \qquad (2.4.43)$$

We execute this procedure using

$$\text{res} = \text{Series}\left[\frac{1}{\sqrt{1 - k^2 z^2}}, \{k, 0, 8\}\right]$$

$$1 + \frac{z^2 k^2}{2} + \frac{3 z^4 k^4}{8} + \frac{5 z^6 k^6}{16} + \frac{35 z^8 k^8}{128} + O(k^9)$$

We have expanded the expression $1/\sqrt{1 - k^2 z^2}$ around $k = 0$ up to the eighth order by calling the function **Series[]**, which yields a Taylor expansion. The period is expressed by using the Taylor representation

$$T_N = \frac{4}{\omega_0} \int_0^1 \frac{1 + \frac{k^2 z^2}{2} + \frac{3 k^4 z^4}{8} + \dots}{\sqrt{(1 - z^2)}} \, dz, \qquad (2.4.44)$$

which in *Mathematica* looks as follows:

$$\text{TN = Expand}\Big[\dfrac{4\displaystyle\int_0^1 \frac{\text{Normal[res]}}{\sqrt{1-z^2}}\,dz}{\omega}\Big]$$

$$\dfrac{1225\,\pi\,k^8}{8192\,\omega} + \dfrac{25\,\pi\,k^6}{128\,\omega} + \dfrac{9\,\pi\,k^4}{32\,\omega} + \dfrac{\pi\,k^2}{2\,\omega} + \dfrac{2\,\pi}{\omega}$$

By calling **Normal[]**, we eliminate the symbol $O(k^9)$ from the variable *res*. After executing the integration of the truncated expression *res* with **Integrate[]** and applying **Expand[]** to simplify the result, we get the same result as given by Landau with respect to the first set of orders:

$$T_N \approx \frac{2\pi}{\omega_0}\Big(1 + \frac{k^2}{4} + \frac{9\,k^4}{64} + \dots\Big). \tag{2.4.45}$$

To use the same independent variables in a graphical representation, we replace k by $\sin(\phi_1/2)$. *Mathematica* executes such a replacement with the operator **ReplaceAll[]** (*/.*) .

$$\text{tn = TN } /. \, k \to \sin\Big(\frac{\phi 1}{2}\Big)$$

$$\dfrac{1225\,\pi\,\sin^8(\frac{\phi 1}{2})}{8192\,\omega} + \dfrac{25\,\pi\,\sin^6(\frac{\phi 1}{2})}{128\,\omega} + \dfrac{9\,\pi\,\sin^4(\frac{\phi 1}{2})}{32\,\omega} + \dfrac{\pi\,\sin^2(\frac{\phi 1}{2})}{2\,\omega} + \dfrac{2\,\pi}{\omega}$$

In order to get a graphical representation of this approximation, we now need to specify a value for ω in T_N to obtain an expression void of any parameter. To keep it simple, we choose $\omega = 4$. The replacement is executed by

```
tn = tn /. ω→4;
```

T and T_N can now be graphically presented as follows:

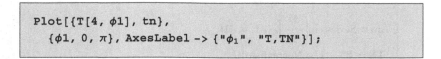

```
Plot[{T[4, φ1], tn},
    {φ1, 0, π}, AxesLabel -> {"φ1", "T,TN"}];
```

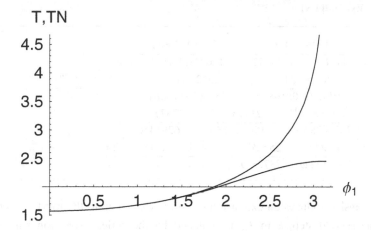

Figure 2.4.10.   Comparison between the exact period T (upper curve) and the approximation T_N with an expansion up to the eighth order with $\omega_0 = 4$.

Plot[] here is used together with a list of functions pertaining to the first argument. The second argument contains the range of representations. The third argument contains the axis labels.

Figure 2.4.10 shows that for small ϕ_1, the amplitudes between the exact period and its approximations are negligible. However, the difference between the exact theory and the approximation becomes larger and larger for angular displacement larger than $\phi_1 \approx 2$. In other words, for large ϕ_1 (i.e., for large amplitudes), a larger number of higher-order Taylor components is needed to obtain an accurate representation of the period.

If, however, we make the period dependent on the initial displacement ϕ_1 and note that k is connected to the initial condition via $k = \sin(\phi_1/2) \approx \phi_1/2 - \phi_1^3/48$. ..., the range of agreement is further reduced by

$$T_N \approx \frac{2\pi}{\omega_0} \left(1 + \frac{1}{16}\phi_1^2 + \frac{11}{3072}\phi_1^4 + ...\right). \tag{2.4.46}$$

The steps in *Mathematica* for this formulation are

```
sin = Series[sin(φ1/2), {φ1, 0, 4}];

TN = TN /. k → Normal[sin];
Expand[TN]
```

$$\frac{1225\,\pi\,\phi1^{24}}{230844665274826752\,\omega} - \frac{1225\,\pi\,\phi1^{22}}{1202315964973056\,\omega} +$$

$$\frac{8575\,\pi\,\phi1^{20}}{100192997081088\,\omega} - \frac{25625\,\pi\,\phi1^{18}}{6262062317568\,\omega} + \frac{4675\,\pi\,\phi1^{16}}{38654705664\,\omega} -$$

$$\frac{8075\,\pi\,\phi1^{14}}{3623878656\,\omega} + \frac{21773\,\pi\,\phi1^{12}}{905969664\,\omega} - \frac{757\,\pi\,\phi1^{10}}{6291456\,\omega} +$$

$$\frac{9\,\pi\,\phi1^{8}}{2097152\,\omega} + \frac{25\,\pi\,\phi1^{6}}{73728\,\omega} + \frac{11\,\pi\,\phi1^{4}}{1536\,\omega} + \frac{\pi\,\phi1^{2}}{8\,\omega} + \frac{2\,\pi}{\omega}$$

Series[] produces an expansion of sin at $\phi_1 = 0$ up to the fourth order. In the second step, k in T_N is replaced by the series expansion sin and is simplified by **Expand[]** in the last step.

Despite the limited accuracy, we can see from this approximation procedure that the period of a nonlinear problem depends on the initial conditions. In case of linear approximation, however, the period is independent of initial conditions.

Solutions for Different Values of Energy

When we look at the potential $V(x) = 1 - \cos(x)$ for the mathematical pendulum, we observe that three forms of motion are possible. For a total energy smaller than the maximum value of the potential energy, oscillations occur (bound motion). For energy values of $E > V_{max}$, we get rotations. Finally, for $E = V_{max}$, we get the asymptotic behavior of the pendulum (see Figure 2.4.11). The solutions for the different values of energy result from (2.4.38) in the form of

$$\phi' = \pm\sqrt{\tfrac{2}{m\,l^2}\,\left(E - m\,l^2\,\omega_0^2\,(1 - \cos\phi)\right)}. \tag{2.4.47}$$

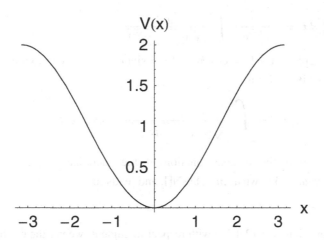

Figure 2.4.11.   Scaled potential $V(x)$ for the mathematical pendulum.

Scaling the energy with $E^* = E / (m \, l^2 \, \omega_0^2)$, we get

$$\phi' = \pm \omega_0 \sqrt{2 \, (\cos \phi - 1 + E^*)} \, . \tag{2.4.48}$$

Different forms of motion occur for different values of the scaled energy

$$
\begin{array}{lll}
E^* > 2 & \text{rotation} & \\
E^* = 2 & \text{asymptotic motion} & \text{(2.4.49)} \\
0 \leq E^* < 2 & \text{oscillations} &
\end{array}
$$

In the following, we will investigate a case that is characterized by its fixed energy $E^* = 2$. For this case, Equation (2.4.48) takes the form

$$\dot{\phi} = \pm \omega_0 \sqrt{2 \, (\cos \phi + 1)} \, . \tag{2.4.50}$$

Substituting $\cos \phi = y$, we get

$$\sqrt{2} \; \omega_0 \int_0^t dt' = \int_1^y \frac{dy'}{\sqrt{(1 - y'^2) \, (1 + y')}} \, . \tag{2.4.51}$$

The integration of this equation yields

$$\pm \omega_0 \sqrt{2} \, t = -\sqrt{2} \; \text{Arctanh}\!\left(\frac{\sqrt{(1+y) \, (1-y^2)}}{\sqrt{2} \, (1+y)} \right) . \tag{2.4.52}$$

By inverting these functions, the solution for the angle ϕ is obtained:

$$\phi = \arccos(1 - 2 \tanh^2 \omega_0 \, t) \, . \tag{2.4.53}$$

From Equation (2.4.48), we get for $0 < E^* < 2$,

$$\int_0^t d\,t' = \pm \frac{1}{\sqrt{2}\,\omega_0} \int_0^\phi \frac{d\phi'}{\sqrt{\cos\phi' - (1 - E^*)}}. \tag{2.4.54}$$

If we replace $1 - E^* = \cos\phi_1$ and $k = \sin(\phi_1/2)$, we can express Equation (2.4.54) in the form

$$\pm \omega_0 \int_0^t d\,t' = \int_0^y \frac{d\,y'}{\sqrt{(1 - y'^2)(1 - k^2\,y'^2)}} = sn^{-1}(y, k), \tag{2.4.55}$$

where sn is the inverse function of the Jacobian elliptic function, in *Mathematica* known as **JacobiSN[]**, and leads to

$$y = sn(\omega_0\,t, k). \tag{2.4.56}$$

Solving Equation (2.4.55) with respect to angle ϕ, we get the expression

$$\phi = 2\arcsin(k\,sn(\omega_0\,t, k)). \tag{2.4.57}$$

If we choose $E^* > 2$, we obtain the solution for the angle by applying a similar strategy to the one above. The solution is

$$\phi = 2\,am\left(\frac{\omega_0\,t}{k}, k\right), \tag{2.4.58}$$

where am denotes the **JacobiAmplitude[]**. The course of the solutions for the various k values is $k = \{0.1, 0.5, 0.9\}$; different initial amplitudes and $\omega_0 = 4$ are shown in Figures 2.4.12, 2.4.13, and 2.4.14. The figures are produced with **Plot[]** as well as with **ArcSin[]**, **JacobiSN[]**, and **JacobiAmplitude[]**. The Jacobi elliptic functions have two arguments: the independent variable $\omega_0\,t$ and the modulus k.

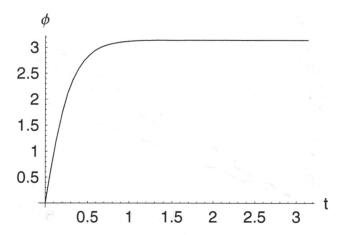

Figure 2.4.12. Solution for $E^* = 2$.

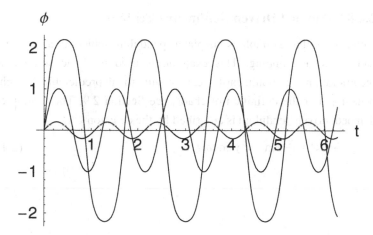

Figure 2.4.13. Solutions for $0 < E^* < 2$. The amplitudes of the solution increase by increasing the values of the modulus $k = \{0.1, 0.5, 0.9\}$.

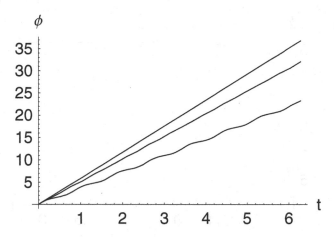

Figure 2.4.14.   Solutions of the mathematical pendulum for $E^* > 2$. The slope of the solution decreases by increasing the modulus k. The three values for k are {0.1, 0.5, 0.9}.

2.4.8.8 Damped Driven Nonlinear Oscillator

Another familiar example is the planar pendulum subject to a driving force and frictional damping. This example is used to demonstrate that the incorporation of nonlinearity can result to unpredictable or chaotic behavior. For the definition of chaos, see Section 2.9. The motion of the damped, driven pendulum is described by the equation

$$x'' + \alpha x' + \frac{g}{l} \sin x = \gamma \cos \omega t. \qquad (2.4.59)$$

Apart from its application to the pendulum, this equation is used to describe a Josephson tunneling junction. In a Josephson junction, two superconducting materials are separated by a thin nonconducting oxide layer. Among the practical applications of such junctions are high-precision magnetometers and standards of voltage elements. The ability of these Josephson junctions to switch rapidly and with very low dissipation from one current-carrying state to another might provide microcircuit technologies for, say, supercomputers, which are more efficient than those based on conventional semiconductors. Hence, the nature of the dynamic response of a Josephson junction to the external driving force — the $\cos \omega t$ term — is a matter of technological as well as of fundamental interest.

One of the characteristics of this equation is the occurrence of chaotic states. These states depend on the choice of parameters for damping and driving force. Since standard analytical techniques are of limited use in the chaotic regime, we demonstrate the existence of chaos by relying on graphical results from numerical simulations.

We first note that because there is an external time dependence in the equation of motion, the system really involves three first-order differential equations. In a normal dynamic system, each degree of freedom results in two first-order equations and such a system is said to correspond to one-and-a-half degrees of freedom. To see this explicitly, we introduce the variable $z = \omega t$ and rewrite the equation of motion (2.4.59) resulting in

$$x' = v(t), \tag{2.4.60}$$
$$v' = -\alpha\, v(t) - \tfrac{g}{l}\, \sin(x(t)) + \gamma \cos(z(t)), \tag{2.4.61}$$
$$z' = \omega. \tag{2.4.62}$$

The equations show how the system depends on the three generalized coordinates x, v, and z. Note further that the presence of damping implies that the system is no longer conservative but is dissipative and, thus, can have attractors.

Analysis of the damped driven pendulum illustrates two separate but related aspects of chaos: first, the existence of a strange attractor and, second, the presence of several different attracting sets and the resulting extreme sensitivity of the asymptotic motion to initial conditions.

To identify the signature of chaos, we use the Poincaré technique to represent a section of phase space. A Poincaré section is a plot showing only the phase plane variables x and x'. A stroboscopic snapshot of the motion is taken during each cycle of the driving force. The obtained complicated attracting set of points shown in Figure 2.4.20 is, in fact, a strange attractor and describes a never-repeating, non periodic motion in which the pendulum oscillates and flips over its pivot point in an irregular, chaotic manner. Before we examine this chaotic behavior, let us first discuss the regular motion of the system.

Regular Motion

We use for the numerical integration Equations (2.4.60) and (2.4.61). The relevant system of equations reads

$$\text{eq1} = \frac{\partial x(t)}{\partial t} == v(t)$$

$$x'(t) == v(t)$$

$$\text{eq2} = \frac{\partial v(t)}{\partial t} == \gamma \cos(\omega t) - \alpha v(t) + \sin(x(t))\,(-\omega_0^2)$$

$$v'(t) == -\sin(x(t))\,\omega_0^2 + \gamma \cos(t\,\omega) - \alpha v(t)$$

where we abbreviated $g/l = \omega_0^2$, and α and γ are the damping constant and the amplitude of the driving force, respectively. Since we cannot access the solution of the driven nonlinear pendulum by analytic procedures, we are forced to carry out numerical integrations. For that reason, we have to select specific numerical values for the parameters:

$$\text{parameterRules} = \{\omega_0 \to 1,\ \alpha \to 0.2,\ \gamma \to 0.52,\ \omega \to 0.694\}$$

$$\{\omega_0 \to 1,\ \alpha \to 0.2,\ \gamma \to 0.52,\ \omega \to 0.694\}$$

To generate the numerical solution, we select 30 cycles of the driving frequency for the endpoint in time.

```
cycles = 30;
```

The numerical solution then follows from

```
pts = NDSolve[
    {eq1, eq2, x[0] == 0.8, v[0] == 0.8} /. parameterRules,
    {x, v}, {t, 0, cycles (2 π) / 0.694}, MaxSteps → 20000]
```

$\{\{x \rightarrow \text{InterpolatingFunction}[(\ 0.\quad 271.607\ ), <>],$
$\quad v \rightarrow \text{InterpolatingFunction}[(\ 0.\quad 271.607\ ), <>]\}\}$

The result of the integration procedure is now displayed in phase space by a parametric plot (see Figure 2.4.15):

```
ParametricPlot[Evaluate[{x[t], v[t]} /. pts],
    {t, 0, 271}, AxesLabel → {"x", "x'"},
    PlotStyle → RGBColor[0, 0, 0.996109]];
```

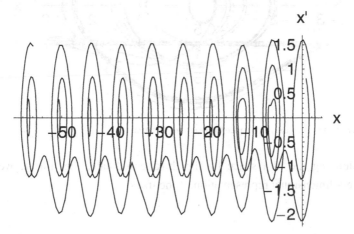

Figure 2.4.15.   Phae space representation of a trajectory for the driven pendulum.

The solution we gain by **NDSolve[]** is in principle defined for any value of x (i.e., $x \in (-\infty, \infty)$). However, the real motion of a pendulum is restricted to the range $x \in (-\pi, \pi)$. Thus, we can reduce the total integration time to the interval $(-\pi, \pi)$. To find the motion modulo 2π, we define the function

```
red[x_] := Mod[x,2 π]/; Mod[x,2 π] ≤ π;

red[x_] := (Mod[x,2 π]-2 π) /; Mod[x,2 π] > π;
```

Mapping this function onto the first argument of each of the solutions *pts,* we generate a reduced representation of the phase space modulo 2π (see Figure 2.4.16).

```
ParametricPlot[Evaluate[{red[x[t]], v[t]} /. pts],
   {t, 0, 271}, AxesLabel → {"x", "x'"},
   PlotStyle → RGBColor[0, 0, 0.996109]];
```

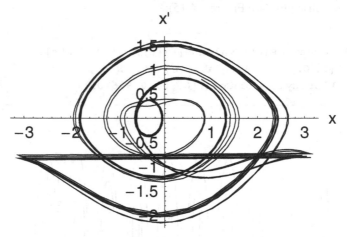

Figure 2.4.16.   Reduced phase space for the driven pendulum.

Extending the plot space by the third coordinate, the time t, we get a three-dimensional representation of the track.

```
ParametricPlot3D[
  Evaluate[Flatten[{red[x[t]], v[t], red[0.694 t],
    RGBColor[0, 0, 0.996109]} /. pts]],
  {t, 0, 271}, AxesLabel → {"x", "x'", "t"},
  PlotPoints → 1700];
```

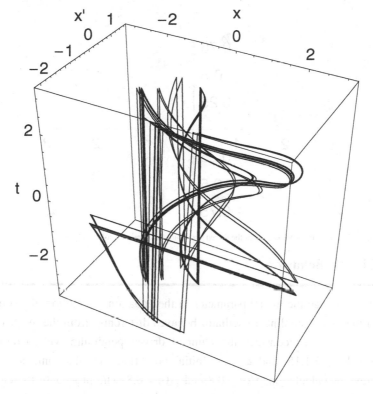

To show the oscillating behavior of this solution, a Poincaré section is created by a stroboscopic map (see Figure 2.4.17). We extract only those points of the solution which are commensurate with the driving frequency:

```
ListPlot[Table[Flatten[{red[x[t]], v[t]} /. pts],
   {t, 16, 271, 2 ( 2 π / 0.694 )}]],
 PlotStyle → {RGBColor[1, 0, 0], PointSize[0.025]},
 AxesLabel → {"x", "v"}];
```

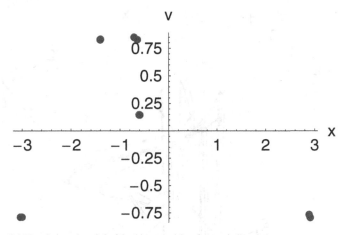

Figure 2.4.17.   Poincaré section of the driven pendulum for a periodic solution.

Chaotic Behavior

If we change the model parameter in the equations of motion, the solution of the driven nonlinear oscillator behaves differently from the result found earlier. Let us consider the damped driven pendulum with parameters $\alpha = \frac{1}{2}$, $\gamma = 1.15$, and $\omega = \frac{2}{3}$. Initial conditions are the same as in the previous calculation: $x(0) = 0.8$ and $v(0) = 0.8$. The procedure to generate the solution is the same as earlier. First, we define the parameters by

```
cycles = 300;
parameterRules = {ω₀ → 1, α → 0.5, γ → 1.15, ω → 0.6666}
```

$\{\omega_0 \to 1, \alpha \to 0.5, \gamma \to 1.15, \omega \to 0.6666\}$

The next step generates the numerical solution

```
ptsChaos = NDSolve[
    {eq1, eq2, x[0] == 0.8, v[0] == 0.8} /. parameterRules,
    {x, v}, {t, 0, cycles (────── )}, MaxSteps → 200000]
                          ( 2 π  )
                          ( 0.6666)
```

{{*x* → InterpolatingFunction[(0.   2827.72), <>],
    v → InterpolatingFunction[(0.   2827.72), <>]}}

The representation of the solution in phase space is given by

```
ParametricPlot[
    Evaluate[{x[t], v[t]} /. ptsChaos], {t, 0, 2827},
    AxesLabel → {"x", "x'"}, PlotPoints → 120,
    PlotStyle → RGBColor[0, 0, 0.996109]];
```

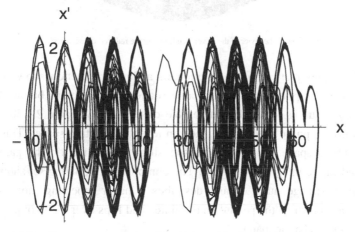

Figure 2.4.18.   Phase space representation of the driven pendulum in a chaotic state.

Comparing Figure 2.4.18 with Figure 2.4.15, we observe that the phase plane picture is more complicated. A reduction of the phase space to the interval $(-\pi, \pi)$ reveals the impression of a chaotic entanglement (see Figure 2.4.19):

```
ParametricPlot[
  Evaluate[{red[x[t]], v[t]} /. ptsChaos], {t, 0, 2827},
  AxesLabel → {"x", "x'"}, PlotPoints → 100,
  PlotStyle → RGBColor[0, 0, 0.996109]];
```

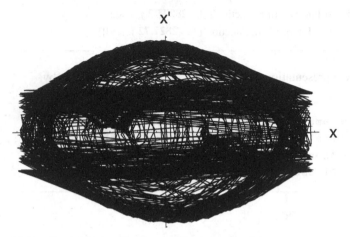

Figure 2.4.19.   Chaotic behavior of the driven pendulum in the reduced phase space.

The representation of the solution in a Poincaré section shows that the intersecting points are not randomly scattered in the plane but are located along a strange entangled curve. We observe from Figure 2.4.20 that the motion in phase space takes place on a finite attracting subset. This subset of phase space has a characteristic shape depending on the parameters used in the integration process. The complicated attracting set shown is in fact a strange attractor and describes a never repeating, non periodic motion in which the pendulum oscillates and flips over its pivot point in an irregular, chaotic manner.

```
ListPlot[Table[Flatten[{red[x[t]], v[t]} /. ptsChaos],
    {t, 16, 2827, (2 π / 0.6666)}],
    PlotStyle → {RGBColor[1, 0, 0], PointSize[0.012]},
    AxesLabel → {"x", "v"}];
```

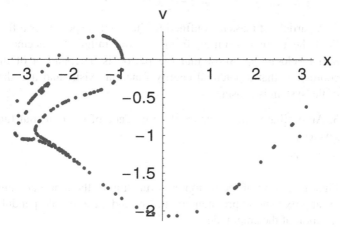

Figure 2.4.20.   Strange attractor of the driven pendulum.

A convenient way to delineate the dynamics of a system is given by the Poincaré section. The Poincaré section represents a slice of the phase space of the system. For the three-dimensional case under examination, a slice can be obtained from the intersection of a continuous trajectory with a two-dimensional plane in the phase space. One method of creating a Poincaré section is to check the system over a full cycle of the driving frequency. If we are dealing with a periodic evolution of period n, then this sequence consists of n dots being indefinitely repeated in the same order (compare Figure 2.4.17). If the evolution is chaotic, then the Poincaré section is a collection of points that show interesting patterns with no obvious repetition (compare Figure 2.4.20). The process of obtaining a Poincaré section can be compared to sampling the state of the system randomly instead of continuously.

2.4.9 Exercises

1. A system of particles moves in a uniform gravitational field g in the z-direction. Show that g can be eliminated from the equations of motion by a transformation of coordinates given by

$$\bar{x} = x, \quad \bar{y} = y, \quad \bar{z} = z - \tfrac{1}{2} g \, t^2.$$

2. A particle of mass m confined to the x-axis experiences a force $-k\,x$. Find the motion resulting from a given initial displacement x_0 and initial velocity v_0. Show that the period is independent of the initial coordiates, that a potential energy function exists, and that the energy of the system is constant.

3. An oscillator moves under the influence of the potential function V given by

$$V = \tfrac{1}{2} k \, x^2 + \kappa \, x^4.$$

Find the period of the moton as a function of the amplitude and derive an approximate expression for the period of a simple pendulum as a function of the amplitude.

4. A particle is attracted toward a center of force according to the relation $F = -m\,k^2/x^3$. Show that the time required for the particle to reach the force center from a distance d is d^2/k.

5. A particle is projected with an initial velocity v_0 up a slope which makes an angle α with the horizontal. Assume frictionless motion and find the time required for the particle to return to its starting position.

2.4.10 Packages and Programs

This subsection contains some declarations for notations used in the text. We also made some extensions of functions **Cross[]** in connection with the cross-product, the function **Dot[]** for the scalar product, the function **Derivative[]** in connection with vector multiplications, and the functions **Times[]** and **Equal[]** related to the multiplication of equations. The definitions introduced below allow a more convenient to use of mathematical expressions in the text. The idea was to generate an environment for the reader which is very similar to traditional textbooks.

Notations

> `Symbolize[`ω_0`]`

> `Symbolize[`ω_0`, WorkingForm → TraditionalForm]`

Symbolize::bsymbexs :
> Warning: The box structure attempting to be symbolized has a similar or identical
> symbol already defined, possibly overriding previously symbolized box structure.

> `Symbolize[`ω_R` ]`

> `Symbolize[`ω_R`, WorkingForm → TraditionalForm]`

> `Symbolize[`ω_{ε}` ]`

> `Symbolize[`ω_{ε}`, WorkingForm → TraditionalForm]`

LaplaceTransform

> `Notation[`$\mathcal{L}_x^p[f\_]$ $\Longleftrightarrow$ `LaplaceTransform[f_, x_, p_],`
>   `WorkingForm → TraditionalForm]`

> `Notation[`$(\mathcal{L}^{-1})_f^p[f\_]$ $\Longleftrightarrow$ `InverseLaplaceTransform[f_, x_, p_],`
>   `WorkingForm → TraditionalForm]`

> `Notation[`$\mathcal{L}_{x\_}^p[f\_]$ $\Longleftrightarrow$ `LaplaceTransform[f_, x_, p_]]`

> `Notation[`
> $(\mathcal{L}^{-1})_{x\_}^p[f\_]$ $\Longleftrightarrow$ `InverseLaplaceTransform[f_, x_, p_]]`

$$\begin{pmatrix} \mathcal{L}_\Box^\Box[\Box] \\[2mm] (\mathcal{L}^{-1})_\Box^\Box[\Box] \end{pmatrix}$$

Integrate

```
Unprotect[Integrate];
```

```
Integrate[f_ , {x_ , x0_ , xe_}] :=
 Map[Integrate[#, {x, x0, xe}] &, f] /; ! FreeQ[f, Plus]
```

```
Protect[Integrate];
```

Cross Product

```
a1 = Attributes[Cross]
```

{Protected, ReadProtected}

```
Unprotect[Cross]
```

{Cross}

```
ClearAttributes[Cross, a1]
```

```
Attributes[Cross]
```

{}

```
Cross[a_, b_] := 0 /;
  a == b ∧ ! FreeQ[a, OverVector] ∧ ! FreeQ[b, OverVector]
```

```
Cross[a_, b_OverVector] := 0 /; FreeQ[a, OverVector]
```

```
Cross[b_OverVector, a_] := 0 /; FreeQ[a, OverVector]
```

```
Cross[c_ a_, b_] :=
 c Cross[a, b] /; FreeQ[c, OverVector] ∧
   ! FreeQ[a, OverVector] ∧ ! FreeQ[b, OverVector]
```

```
Cross[a_, c_ b_] :=
 c Cross[a, b] /; FreeQ[c, OverVector] ∧
   ! FreeQ[a, OverVector] ∧ ! FreeQ[b, OverVector]
```

```
Cross[b_, a_OverVector] := Map[Cross[#, a] &, b] /;
  Head[b] == Plus ∧ ! FreeQ[b, OverVector]
```

```
Cross[a_OverVector, b_] := Map[Cross[a, #] &, b] /;
  Head[b] == Plus ∧ ! FreeQ[b, OverVector]
```

```
Cross[a_, b_] :=
 (c[x_] := Map[Cross[x, #] &, b];
   Fold[Plus, 0, Map[c[#] &, Level[a, 1]]]) /;
  Head[b] == Plus ∧ Head[a] == Plus ∧
   ! FreeQ[b, OverVector] ∧ ! FreeQ[a, OverVector]
```

```
Cross[a_OverVector,
  Cross[b_OverVector, c_OverVector]] :=
 Dot[a, c] b - Dot[a, b] c
```

```
Cross[Cross[a_OverVector, b_OverVector],
  Cross[c_OverVector, d_OverVector]] :=
 (Dot[Cross[a, b], d] c - Dot[Cross[a, b], c] d)
```

```
SetAttributes[Cross, a1]
```

```
Attributes[Cross]
```

{Protected, ReadProtected}

```
Protect[Cross]
```

{}

Dot Product

```
a2 = Attributes[Dot]
```

{Flat, OneIdentity, Protected}

```
Unprotect[Dot]
```

{Dot}

```
ClearAttributes[Dot, a2]
```

```
Attributes[Dot]
```

{}

```
Dot[a_, b_OverVector] := 0 /; FreeQ[a, OverVector]
```

```
Dot[b_OverVector, a_] := 0 /; FreeQ[a, OverVector]
```

```
Dot[a_OverVector, b_OverVector] := HoldForm[Dot[a, b]]
```

```
Dot[a_, b_] := (a /. OverVector[x_][y___] → x[y]^2) /;
  a == b ∧ ! FreeQ[a, OverVector] ∧
   ! FreeQ[b, OverVector] ∧ FreeQ[a, Plus] ∧
  FreeQ[b, Plus] ∧ FreeQ[a, Cross] ∧ FreeQ[b, Cross]
```

```
Dot[a_OverVector, b_OverVector] :=
  (a /. OverVector[x_] → x^2) /; a == b ∧ FreeQ[a, Plus] ∧
   FreeQ[b, Plus] ∧ FreeQ[a, Cross] ∧ FreeQ[b, Cross]
```

```
Dot[c_ a_, b_] := c Dot[a, b] /; FreeQ[c, OverVector] ∧
   ! FreeQ[a, OverVector] ∧ ! FreeQ[b, OverVector]
```

```
Dot[a_, c_ b_] := c Dot[a, b] /; FreeQ[c, OverVector] ∧
   ! FreeQ[a, OverVector] ∧ ! FreeQ[b, OverVector]
```

```
Dot[b_, a_OverVector] := Map[Dot[#, a] &, b] /;
  Head[b] == Plus ∧ ! FreeQ[b, OverVector]
```

```
Dot[a_OverVector, b_] := Map[Dot[a, #] &, b] /;
  Head[b] == Plus ∧ ! FreeQ[b, OverVector]
```

```
Dot[a_, b_] :=
  (c[x_] := Map[Dot[x, #] &, b];
   Fold[Plus, 0, Map[c[#] &, Level[a, 1]]]) /;
  Head[b] == Plus ∧ Head[a] == Plus ∧
   ! FreeQ[b, OverVector] ∧ ! FreeQ[a, OverVector]
```

```
Dot[Cross[a_OverVector, b_OverVector],
  Cross[c_OverVector, d_OverVector]] :=
 Dot[a, Cross[b, Cross[c, d]]]
```

```
Dot[a_OverVector, Cross[c_OverVector,
   d_OverVector]] := 0 /; c == a ∨ a == d
```

```
SetAttributes[Dot, a2]
```

```
Attributes[Dot]
```

{Flat, OneIdentity, Protected}

```
Protect[Dot]
```

{}

Derivative

```
a3 = Attributes[D]
```

{Protected, ReadProtected}

```
Unprotect[D]
```

{D}

```
ClearAttributes[D, a3]
```

```
Attributes[D]
```

{}

```
D[Equal[a_, b_], t_] := Equal[D[a, t], D[b, t]]
```

```
D[Cross[a_, b_], t_] :=
 Cross[D[a, t], b] + Cross[a, D[b, t]] /;
  ! FreeQ[a, OverVector] ∧ ! FreeQ[b, OverVector] ∧
   ! FreeQ[a, t] ∧ ! FreeQ[b, t]
```

```
D[Times[c_, Cross[a_, b_]], t_] :=
 c (Cross[D[a, t], b] + Cross[a, D[b, t]]) /;
  ! FreeQ[a, OverVector] ∧ ! FreeQ[b, OverVector] ∧
   ! FreeQ[a, t] ∧ ! FreeQ[b, t]
```

```
D[Dot[a_, b_], t_] := Dot[D[a, t], b] + Dot[a, D[b, t]] /;
  ! FreeQ[a, OverVector] ∧ ! FreeQ[b, OverVector] ∧
   ! FreeQ[a, t] ∧ ! FreeQ[b, t]
```

```
D[f_, t_] := Map[D[#, t] &, f] /;
  (! FreeQ[f, Cross] ∨ ! FreeQ[f, Dot]) ∧
   ! FreeQ[f, OverVector] ∧ Head[f] == Plus
```

```
SetAttributes[D, a3]
```

```
Attributes[D]
```

```
{Protected, ReadProtected}
```

```
Protect[D]
```

```
{}
```

Times

```
a4 = Attributes[Times]
```

{Flat, Listable, NumericFunction, OneIdentity, Orderless, Protected}

```
Unprotect[Times]
```

{Times}

```
ClearAttributes[Times, a4]
```

```
Attributes[Times]
```

{}

```
Times[Dot[a_ , b_], c_] := Times[Dot[a, c], b] /;
  ! FreeQ[b, Dot] ∧ ! FreeQ[c, OverVector]
```

```
SetAttributes[Times, a4]
```

```
Attributes[Times]
```

{Flat, Listable, NumericFunction, OneIdentity, Orderless, Protected}

```
Protect[Times]
```

{}

Equal

```
a5 = Attributes[Equal]
```

{Protected}

```
Unprotect[Equal]
```

{Equal}

```
ClearAttributes[Equal, a5]
```

```
Attributes[Equal]
```

{}

```
(**********************************************)
Equal /: Integrate[left_ == right_, limits__] :=
(**********************************************)
 Block[{lhs = Expand[left],        (*left hand side*)
        rhs = Expand[right]        (*right hand side*)},
  (*hint:There is no other need of an
     integration constant or*)(*of another
    lower integration level instead of Zero:*)
   (*--------------------------------------------------
                            ------------*)
  Off[Integrate::gener];
  If[!AtomQ[lhs], If[Head[lhs] === Plus,
     lhs = Map[Integrate[#, limits] &, lhs];,
     lhs = Integrate[lhs, limits];];,
   lhs = Integrate[lhs, limits];];
  If[!AtomQ[rhs], If[Head[rhs] === Plus,
     rhs = Map[Integrate[#, limits] &, rhs];,
     rhs = Integrate[rhs, limits];];,
   rhs = Integrate[rhs, limits];];
  On[Integrate::gener];
  (*return result*)
  (*---------------*)
  lhs == rhs]
```

```
Equal /: Plus[left_ == right_, term__] :=
 Plus[left, term] == Plus[right, term]
```

```
Equal /: Times[left_ == right_, term__] :=
 Times[left, term] == Times[right, term]
```

```
(*Equal/:f_[left_==right_]:=
  f[left]==f[right]/;Fold[And,True,Map[
     FreeQ[f,#]&,{List,Rule,RuleDelayed,ToRules}]]*)
```

```
Fold[And, True,
  Map[FreeQ[f, #] &, {List, Rule, RuleDelayed, ToRules}]]
```

True

```
SetAttributes[Equal, a5]
```

```
Attributes[Equal]
```

{Protected}

```
Protect[Equal]
```

{}

```
RHSToLHS = Equal[a_, b_] :> Equal[a - b, 0]
```

$a\_ == b\_ :\to a - b == 0$

```
LHSToRHS = Equal[a_, b_] :> Equal[0, b - a]
```

$a\_ == b\_ :\to 0 == b - a$

```
Plus[a == b, c]
```

$a + c == b + c$

```
a + (b == c)
```

$a + b == a + c$

```
Times[a == b, c]
```

$$a\,c == b\,c$$

```
d (c == b)
```

$$c\,d == b\,d$$

```
∫ (f[x] == g[x²]) dx
```

$$\int (-k\,x)[x]\,dx == \int g(x^2)\,dx$$

```
√a == b
```

$$\sqrt{a == b}$$

```
Log[a == b]
```

$$\log(a) == \log(b)$$

```
f[a == b]
```

$$(-k\,x)[a] == (-k\,x)[b]$$

2.5 Central Forces

2.5.1 Introduction

This section discusses the two-body problem in a central field. We restrict our discussion mainly on planet movements. The nonintegrable problems in central fields are briefly discussed and examined.

The motion of a two-body problem with central forces is important with respect to its applications. This kind of model is applicable to macroscopic as well as microscopic systems. An important macroscopic example governed by these laws is the motion of planets around the Sun. A microscopic example from atomic physics is the movement of electrons around a nucleus. An example in between the macroscopic and the microscopic range is the scattering of α-particles on gold atoms, so called Rutherford scattering.

We mentioned in Section 2.4.6 that gravitation is the weakest force of the four fundamental forces. This kind of force is negligible in considerations concerning nuclear components such as neutrons and protons. It also is of no importance if we examine interactions of molecules and atoms.

In our daily life, gravitation is omnipresent but does not influence our actions. For example, a sky scraber with its mass has some gravitational influence on a car standing in front of such a building. However, the strength with which the building interacts with the car is much smaller than the interaction of the car with the Earth. Gravitation is an important factor if we consider the interaction of planets. It is only gravitation which holds us to Earth, which determines the movement of Earth around the Sun, and which determines the motion of planets in the solar system. Gravitation also is responsible for the development, creation, and history of stars, galaxies, and the whole universe. Gravitation determines the evolution of our life and the development of our universe.

2.5.2 Kepler's Laws

The dark sky with its myriads of stars always impressed mankind. At the end of the 16th century, Tycho Brahe (1546–1601) examined the sky with great accuracy. These experimental data were the basis for his co-worker and successor, the imperial mathematician Johannes Kepler (1571–1630) (see Figure 2.5.1).

Figure 2.5.1.   Johannes Kepler born December 27, 1571 in Leonberg/Württemberg and died Nvember 15, 1630 in Regensburg.

In a laborious work, Kepler extracted from these observations his three general planetary laws. In his famous Rudolphine tables, he summarized his work, which took him 20 years to the completion. He demonstrated in his *Astronomia nova* that the planetary tracks are ellipses slightly deviating from a circle. Also in this work, he discussed the velocity of planets, which is highest in the perihelion and lowest in the aphelion. In his extensive calculations, Kepler derived a mathematical expression connecting the mean diameter of a track with the period of revolution around the Sun. The last law was given by him in his 1619 published book *Hamonices mundi* 10 years after the formulation of his first and second law. These three laws were the basics for Newton's theory on gravitation. The three laws by Kepler read as follows:

I.   All planets move on ellipses around the Sun.

II.   In equal times, equal areas are scanned by a planet.

III.   The square of the period is proportional to the third power of the mean radius.

Kepler, for example, determined that the Earth's track is nearly circular with its shortest distance in the perihelion of about 1.48×10^{11}m and the largest distance in the aphelion of about 1.52×10^{11}m. The mean radius of the track around the sun is approximately 1.5×10^{11}m. This quantity is today defined as an astronomical unit (AU).

Later, Newton demonstrated mathematically that the planets of the solar system move on ellipses, parabolas, or hyperbolas in a r^{-2}-force field. This kind of curves also occur in conic sections. This is the reason why Kepler's paths are also called conic sections. Figure 2.5.2 demonstrates the four types of conic section.

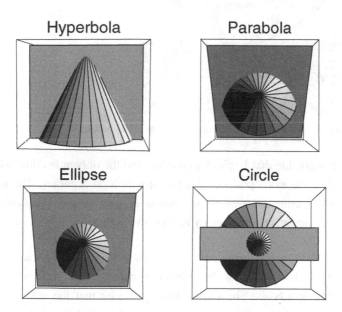

Figure 2.5.2.   Conic sections. The sections are created by intersecting a cone with a plane. Different section angles between the center line of the cone and the plane result to different intersecting curves.

This figure demonstrates that circles also occur as a deviation from ellipses. Circles and ellipses are those paths on which planets move periodically around the Sun. On parabolas and hyperbolas, objects move only once in the direction of the force center and then depart from it to infinity. Kepler was a harmony-loving man who connected the different planet paths of the solar system with the platonic bodies known at that time. His idea was that each platonic body is connected with the period of a planet (see Figure 2.5.3).

Figure 2.5.3.   Planet model by Kepler represented by the platonic bodies.

It is remarkable that Kepler was the one and the only at his time who could calculate the exact position of a planet with high accuracy. The main tool for his calculations was his collection of data in the Rudolphin tables. These tables were published by Kepler after a long journey in 1628 to Ulm.

Later, Newton demonstrated that an ellipse is a possible track in a $1/r^2$ potential. The first law by Kepler becomes with Newton's theory a mathematical basis. The second law by Kepler that the areas of scanned arcs are equal is supported by the central action of forces between the Sun and a planet. These forces are called central forces.

The following illustration shows a consequence of Kepler's second law. The planet moves in the vicinity of the Sun faster than far away from it. As we will see, this behavior is closely related to the conservation of the angular momentum.

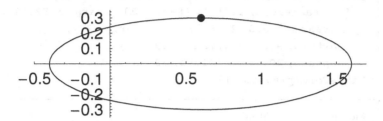

The third law by Kepler relates the time of revolution with the mean distance between a planet and the Sun. If we denote the mean distance of the planet from the Sun by r and the time of revolution by T, we are able to mathematically formulate the third Kepler law by

$$T^2 = C r^3, \tag{2.5.1}$$

where C is a universal constant for the planet system. This relation is a direct consequence of the $1/r^2$ force law. If we are interested in the period of revolution of Jupiter around the Sun, we can use Kepler's third law. The unknown constant C is determined from the Earth's period of revolution by

$c =$ **Solve**$[T_{\text{Er}}^2 == C \, r_{\text{Er}}^3, C]$ **// Flatten**

$$\left\{ C \to \frac{T_{\text{Er}}^2}{r_{\text{Er}}^3} \right\}$$

For Jupiter's period, we find

Solve$[T_J^2 == C \, r_J^3 \, /. \, c, T_J]$

$$\left\{ \left\{ T_J \to -\frac{r_J^{3/2} \, T_{\text{Er}}}{r_{\text{Er}}^{3/2}} \right\}, \left\{ T_J \to \frac{r_J^{3/2} \, T_{\text{Er}}}{r_{\text{Er}}^{3/2}} \right\} \right\}$$

where we used C from the calculation for the Earth. This demonstrates that the knowledge of the mean distances allows us to determine the times of revolution. The mean distances for our solar system in astronomical units (AU) are known to be

```
planetList = {{Mercury, 0.387`}, {Venus, 0.723`},
   {Eros asteroid, 1.45`}, {Earth, 1}, {Mars, 1.523`},
   {Ceres asteroid, 2.767`}, {Jupiter, 5.2`},
   {Sarturn, 9.57`}, {Uranus, 19.28`},
   {Neptune, 30.14`}, {Pluto, 39.88`}};
TableForm[planetList]
```

Mercury	0.387
Venus	0.723
asteroid Eros	1.45
Earth	1
Mars	1.523
asteroid Ceres	2.767
Jupiter	5.2
Sarturn	9.57
Uranus	19.28
Neptune	30.14
Pluto	39.88

A graphical representation of these data in connection with Kepler's third law shows a linear dependence with slope $\alpha = 3/2$ in a log-log plot:

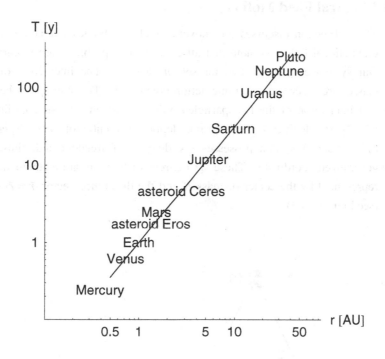

This double logarithmic representation of data shows that a scaling law between time and distance exists. This characteristic behavior relates time and distance via a finite transformation as

$$\tilde{t} = a\,t,$$
$$\tilde{r} = a^{2/3}\,r,$$

(2.5.2)

where a = const. Eliminating the constant a, it follows that

$$\frac{\tilde{t}^2}{\tilde{r}^3} = \frac{t^2}{r^3}.$$

(2.5.3)

Scaling time by a and the orbit by $a^{2/3}$, we get another orbit and another time of revolution. Both orbits are related by the relation $\tilde{t}^2/\tilde{r}^3 = t^2/r^3$. In fact, this relation is, in essence, the third law by Kepler.

2.5.3 Central Field Motion

This subsection discusses the movement of two bodies interacting via a gravitational field. We note that all central force problems are integrable. Our system consists of two masses m_1 and m_2. The interaction of the masses are described by an interaction potential U. The assumption here is that interaction of the two particles only depend on relative coordinates $\vec{r}_1 - \vec{r}_2$ or velocities $\vec{r}'_1 - \vec{r}'_2$ (prims denote differentiation with respect to time). Such a system possesses six degrees of freedom and, thus, six generalized coordinates. These six degrees of freedom are mathematically represented by the center of mass $\vec{R}$ and the difference vector $\vec{r} = \vec{r}_1 - \vec{r}_2$ (see Figure 2.5.4).

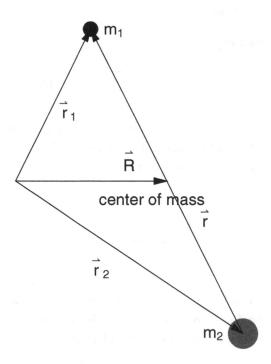

Figure 2.5.4.   Characteristic configuration of a two-body problem. The two masses m_1 and m_2 are a distance r away from each other. The center of mass is given by $\vec{R}$.

The center of mass in a two-body system moves like a single particle. The forces acting on the single particles are transformed to the center of mass. The equations of motion for the single particles with masses m_1 and m_2 are given by

$$\text{particle1} = m_1 \frac{\partial^2 \vec{r}_1(t)}{\partial t \, \partial t} == \vec{F}_1(t)$$

$$m_1 \vec{r}_1''(t) == \vec{F}_1(t)$$

and

$$\text{particle2} = m_2 \frac{\partial^2 \vec{r}_2(t)}{\partial t \, \partial t} == \vec{F}_2(t)$$

$$m_2 \vec{r}_2''(t) == \vec{F}_2(t)$$

Adding up both equations, we find

$$\text{cmMotion} = \text{Thread[particle1 + particle2, Equal]}$$

$$m_1 \vec{r}_1''(t) + m_2 \vec{r}_2''(t) == \vec{F}_1(t) + \vec{F}_2(t)$$

Defining the center of mass by

$$\text{cm} = \vec{R}(t) == \frac{m_1 \vec{r}_1(t) + m_2 \vec{r}_2(t)}{M}$$

$$\vec{R}(t) == \frac{m_1 \vec{r}_1(t) + m_2 \vec{r}_2(t)}{M}$$

with $M = m_1 + m_2$ and replacing $\vec{r}_1$ by $\left(M\vec{R} - m_2 \vec{r}_2\right)/m_1$ in the center of mass equation, we get

$$\text{cmMotion} = \text{cmMotion} \, /. \, \left\{ \vec{r}_1 \rightarrow \text{Function}\left[t, \, \frac{M \, \vec{R}(t) - m_2 \, \vec{r}_2(t)}{m_1} \right] \right\}$$

$$M \, \vec{R}''(t) == \vec{F}_1(t) + \vec{F}_2(t)$$

Since the forces in the system are central forces and since the masses m_1 and m_2 are interchangeable, we must consider

$$\vec{F}_1 = -\vec{F}_2 \tag{2.5.4}$$

by Newton's second law. Thus, the center of mass moves in a force-free state:

$$\text{cmMotion} = \text{cmMotion} \, /. \, \vec{F}_1(t) \rightarrow -\vec{F}_2(t)$$

$$M \, \vec{R}''(t) == 0$$

Taking into account Newton's first law, the center of mass is at rest or travels with constant velocity.

On the other hand, subtracting both equations of motion, we get

$$\text{rel} = \text{Thread}\left[(\text{Thread}[\#1, \text{Equal}] \, \&) \, /@ \left(\frac{\text{particle1}}{m_1} - \frac{\text{particle2}}{m_2} \right), \text{Equal} \right]$$

$$\vec{r}_1''(t) - \vec{r}_2''(t) == \frac{\vec{F}_1(t)}{m_1} - \frac{\vec{F}_2(t)}{m_2}$$

We introduce the reduced mass μ by

$$\text{reducedMass} = \mu == \frac{m_1 \, m_2}{m_1 + m_2}$$

$$\mu == \frac{m_1 \, m_2}{m_1 + m_2}$$

Here, μ is always smaller than the smallest mass. Inserting this relation into the difference of the equations of motion and transforming to relative coordinates $\vec{r}$, we find

---

rel = Simplify[rel /. {$\vec{r}_1$ → Function[t, $\vec{r}(t) + \vec{r}_2(t)$],
  $\vec{F}_1$ → Function[t, $\vec{F}(t)$], $\vec{F}_2$ → Function[t, $-\vec{F}(t)$]}]

$$\vec{r}''(t) == \left(\frac{1}{m_2} + \frac{1}{m_1}\right)\vec{F}(t)$$

With the reduced mass replaced, we get

---

relEquation = Simplify[rel /. Flatten[Solve[reducedMass, m_1]]]

$$\vec{r}''(t) == \frac{\vec{F}(t)}{\mu}$$

The introduction of center of mass and relative coordinates allowed us to separate the two-body problem into two independent problems. First, the center of mass moves force-free and, second, the fictitious particle with mass μ is governed due to the central force $\vec{F}$ in direct connection to the masses.

The equation of motion for the center of mass delivers

```
DSolve[cmMotion, R̃[t], t]
```

$$\{\{\vec{R}(t) \to c_1 + t\, c_2\}\}$$

---

meaning a center of mass at rest ($c_2 = 0$) or a movement with a constant velocity ($c_2 \neq 0$). The constants c_1 and c_2 are determined by the initial conditions of the motion.

The relative movement is described by a fictitious particle. The force $\vec{F}$ governing this movement can be directed toward the center of mass or in the opposite direction. The direction of the force determines some properties of the movement. The instrumental behavior is that the force is a central force. The following observations summarize these properties.

First, we observe the following:

> The movement under the action of a central force always is bound to a plane.

This property is obviously governed by the direction of the force, the direction of the location vector, and the acceleration. The central force and the acceleration are parallel to the position vector $\vec{r}$. Thus, $\vec{r}''$, $\vec{r}'$, and $\vec{r}$ all belong to the same plane. The particle will never leave this plane because there is no force component directing outward this plane.

Second, we observe the following:

> The angular momentum is a conserved quantity.

The angular momentum $\vec{L}$ along the track is

angularMomentum $= \vec{L}(t) == \vec{r}(t) \times \vec{p}(t)$
$\vec{L}(t) == \vec{r}(t) \times \vec{p}(t)$

with $\vec{p}$ the linear momentum given by $\vec{p} = \mu \vec{r}'$. Replacing $\vec{p}$ by this expression in the representation of the angular momentum, we obtain

angularMomentum = angularMomentum /. $\vec{p}(t) \rightarrow \mu \dfrac{\partial \vec{r}(t)}{\partial t}$
$\vec{L}(t) == \mu \vec{r}(t) \times \vec{r}'(t)$

Differentiating this expression with respect to time, it follows that

$$\textbf{timeDerivativeOfL} = \frac{\partial\,\text{angularMomentum}}{\partial t}$$

$$\vec{L}'(t) == \mu\,\vec{r}(t) \times \vec{r}''(t)$$

Since $\vec{r}$ is parallel to $\vec{r}''$ (i.e., $\vec{r}'' = \alpha\,\vec{r}$) the temporal changes in $\vec{L}$ are thus

$$\textbf{timeDerivativeOfL} = \textbf{timeDerivativeOfL} \;/.\; \frac{\partial^2 \vec{r}(t)}{\partial t\, \partial t} \to \alpha\,\vec{r}(t)$$

$$\vec{L}'(t) == 0$$

This relation shows that $\vec{L}$ is a conserved quantity:

$$\textbf{DSolve}\big[\textbf{timeDerivativeOfL},\, \vec{L},\, t\big]$$

$$\cdot\ \big\{\big\{\vec{L} \to \text{Function}[\{t\},\, c_1]\big\}\big\}$$

$\vec{L}$ is fixed for all times in direction as well as in total.

These two properties are major consequences of the central character of the acting force. In each two-particle system with central forces, these properties hold.

Because the force in direct connection between the particles is only dependent on the radial distance, we restrict our considerations to the case where the interaction potential $U = U(r)$ is a pure function of the distance r. Note that the force is derivable from U by the gradient. The behavior of radial dependence only establishes a spherical symmetry of the problem, meaning that an arbitrary rotation around any axis will not change the solution of the problem. The spherical symmetry simplifies the problem because there are conserved quantities related to this symmetry. Especially, the angular momentum is such a quantity.

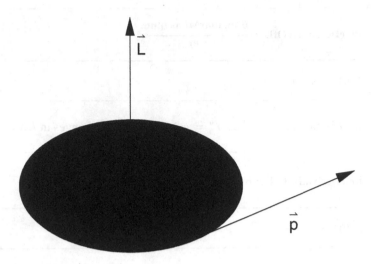

Figure 2.5.5. Geometrical relations between momentum $\vec{p}$ and radius $\vec{r}$ definig the angular momentum $\vec{L}$.

It is natural to use spherical coordinates (r, θ, ψ) for a spherical symmetric problem. r is the radial coordinate, ψ is the zenith angle, and θ describes the azimutal angle. If we chose the polar axis as the direction of $\vec{L}$, then the movement always is perpendicular to $\vec{L}$ (see Figure 2.5.5).

The mathematical description of the movement can be based on cartesian coordinates. The position vector $\vec{r}$ is represented by

$\vec{x} = \{x(t), y(t), z(t)\}; \vec{x}$ // **MatrixForm**

$$\begin{pmatrix} x(t) \\ y(t) \\ z(t) \end{pmatrix}$$

The kinetic energy in cartesian coordinates is given by

$T = \dfrac{1}{2} \mu \, (\partial_t \vec{x}) \cdot (\partial_t \vec{x})$

$$\frac{1}{2} \mu (x'(t)^2 + y'(t)^2 + z'(t)^2)$$

Now, the transformation to spherical coordinates can be carried out by the following transformations:

```
coordinates =
 {x → Function[t, r[t] Sin[θ[t]] Cos[ψ[t]]],
  y → Function[t, r[t] Sin[θ[t]] Sin[ψ[t]]],
  z → Function[t, r[t] Cos[θ[t]]]};
coordinates // TableForm
```

$x \to$ Function$[t, r(t)\sin(\theta(t))\cos(\psi(t))]$
$y \to$ Function$[t, r(t)\sin(\theta(t))\sin(\psi(t))]$
$z \to$ Function$[t, r(t)\cos(\theta(t))]$

Since ψ is a fixed quantity ($\psi = \frac{\pi}{2}$), the kinetic energy is simplified to

kineticEnergy = Simplify$\left[T\ /.\ \text{coordinates}\ /.\ \psi \to \text{Function}\left[t, \frac{\pi}{2}\right]\right]$

$$\frac{1}{2}\,\mu\,(r'(t)^2 + r(t)^2\,\theta'(t)^2)$$

This expression represents the kinetic energy in polar coordinates. The constant angular momentum $p_\theta = l$ is determined from this expression by

angularMomentum $= \dfrac{\partial\,\text{kineticEnergy}}{\partial\,\frac{\partial\theta(t)}{\partial t}} == l$

$\mu\,r(t)^2\,\theta'(t) == l$

The fact that l is a constant has a geometrical interpretation. The position vector $\vec{r}$ overrides in a time interval dt a certain area dA (see Figure 2.5.6):

$$dA = \frac{r^2\,d\theta}{2}$$

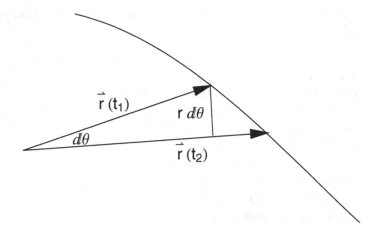

Figure 2.5.6.   The position vector $\vec{r}$ overrides in a time interval $dt = t_2 - t_1$ an area dA.

This expression divided by dt generates the area velocity:

$$\frac{dA(t)}{dt} == \frac{r(t)^2 \, d\theta(t)}{2 \, dt} == \frac{l}{2\mu}$$

$$A'(t) == \frac{1}{2} \, r(t)^2 \, \theta'(t) == \frac{l}{2\mu}$$

We observe that the area velocity is a constant of motion. This relation was first established by Kepler in 1609. He derived this relation on an empirical basis by studying Brahe's (died 1601) observations. It is of fundamental importance that the second law by Kepler is not related to the $1/r^2$ dependence of the Newtonian force field. However, it only resides on the existence of central force. Thus, this law exists for any central force problem independent of the structure of the force field.

I addition to the conservation of the linear momentum of the center of mass and the conservation of the angular momentum, the kinetic energy of the relative movement is conserved:

totalEnergy = *H* == kineticEnergy + *U*(*r*(*t*))

$$H == U(r(t)) + \frac{1}{2}\,\mu\,(r'(t)^2 + r(t)^2\,\theta'(t)^2)$$

or with

sangular = Flatten[Solve[angularMomentum, $\dfrac{\partial\,\theta(t)}{\partial t}$]]

$$\left\{\theta'(t) \to \frac{l}{\mu\,r(t)^2}\right\}$$

we find

totalEnergy = Expand[totalEnergy /. sangular]

$$H == U(r(t)) + \frac{1}{2}\,\mu\left(\frac{l^2}{\mu^2\,r(t)^2} + r'(t)^2\right)$$

2.5.3.1 Equations of Motion

Knowing the total energy and the interaction potential $U(r)$ of the two-body problem allows us to derive the equations of motion. The equation depends on the two conserved quantities H and l, the total energy and the angular momentum, respectively. Solving the total energy with respect to r', we find the equation of motion for the radial coordinate:

> **eq1 =**
> **Flatten[Simplify[Solve[totalEnergy /. $(\partial_t r[t])^2 \to \kappa, \kappa$]]] /. $\kappa \to (\partial_t r[t])^2$ /.**
> **Rule $\to$ Equal // Flatten[Solve[#, $\partial_t r[t]$]]] &**

$$\left\{ r'(t) \to -\frac{i\sqrt{2}\sqrt{\frac{l^2}{2\mu r(t)^2} - H + U(r(t))}}{\sqrt{\mu}}, \right.$$

$$\left. r'(t) \to \frac{i\sqrt{2}\sqrt{\frac{l^2}{2\mu r(t)^2} - H + U(r(t))}}{\sqrt{\mu}} \right\}$$

We select the second solution because of the plus sign:

> **equationOfMotion = eq1[[2]]**

$$r'(t) \to \frac{i\sqrt{2}\sqrt{\frac{l^2}{2\mu r(t)^2} - H + U(r(t))}}{\sqrt{\mu}}$$

Since the result is separable, we solve this expression with respect to dt and carry out an integration on both sides:

$$\int 1\, dt == \sqrt{\mu} \int \frac{1}{\sqrt{2\left(-\frac{l^2}{2\mu r^2} + H - U(r)\right)}}\, dr$$

The above integral delivers an expression for $t = t(r)$ as a function of time. If we can invert this expression, we get the radial distance as a function of time. An alternative representation is gained by eliminating time as an parameter by the relation

$$\text{pathEquation} = d\theta == \frac{d\theta\, dt\, dr}{dt\, dr} == \frac{\frac{\partial\theta(t)}{\partial t}\, dr}{\frac{\partial r(t)}{\partial t}}$$

$$\frac{dr\,\theta'(t)}{r'(t)} == d\theta$$

Using the conservation of the angular momentum by the definition $\theta' = l/(\mu\, r^2)$, we can write

$$d\theta = \frac{l}{\mu\, r^2\, \dot{r}}\, dr. \tag{2.5.5}$$

In addition, the total energy delivers r' and thus we get

pathEquation /. sangular /. equationOfMotion /. $r(t) \to r$

$$-\frac{i\,l\,dr}{\sqrt{2}\; r^2\, \sqrt{\mu}\, \sqrt{\frac{l^2}{2r^2\mu} - H + U(r)}} == d\theta$$

Integrating both sides, we find

$$\int 1\, d\theta == \int \frac{l}{\sqrt{2}\; r^2\, \sqrt{\mu}\, \sqrt{-\frac{l^2}{2r^2\mu} + H - U(r)}}\, dr$$

$$\theta == \frac{l \int \frac{1}{r^2\, \sqrt{-\frac{l^2}{2r^2\mu} + H - U(r)}}\, dr}{\sqrt{2}\; \sqrt{\mu}}$$

Since l is a constant of motion, the sign of θ' cannot change. Thus, the angle $\theta(t)$ is an monotonous increasing function in time.

So far, we gained a formal solution of the equation of motion. The explicit solution of the problem depends mainly on the interaction potential U(r). Such solutions are symbolically accessible for a certain kind of forces

$F(r) = -\partial_r U(r)$. In cases where the potential $U(r) \sim r^{n+1}$ is represented by a power law relation with n an integer or rational expression the solution is given by elliptic integrals. For the specific cases n = 1, -2 and -3 the solutions are known symbolically.

2.5.3.2 Orbits in a Central Force Field

The radial velocity of a fictitious particle with mass μ is determined by the relation

```
equationOfMotion /. Rule → Equal
```

$$r'(t) == \frac{i\sqrt{2}\sqrt{\frac{l^2}{2\mu r(t)^2} - H + U(r(t))}}{\sqrt{\mu}}$$

It is obvious that the radial velocity vanishes if the particle comes to rest. This situations occurs at a turning point when the particle changes its direction. If the radial velocity vanishes, then the following relation must hold:

$$\textbf{turningPoints} = -\frac{l^2}{2\mu r^2} + H - U(r) == 0$$

$$-\frac{l^2}{2r^2\mu} + H - U(r) == 0$$

Because this relation is at least quadratic in r, we can expect that under certain conditions, there exist two turning points. These two points can be finite r_{min} and r_{max} or one of these points is located at infinity. Under certain conditions determined by $U(r)$, H, and l, there exists only one turning point. A detailed discussion is given below. In such a case, we have

$$r' = 0 \tag{2.5.6}$$

for any time t. This property means r = const. or the orbit is a circle.

If the motion of the particle is periodic in the potential $U(r)$, then we find two turning points. If, in addition, the radial oscillations are rational commensurable with the angular oscillations, then we find closed orbits. The following illustration shows two examples of such orbits.

If, however, the radial and angular frequencies are rational incommensurable, then the orbits are not closed. The particle now sweeps out the complete space without any recurrence of the orbit. Two examples of this behavior are given in the following illustrations.

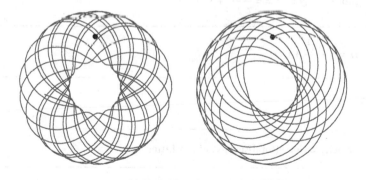

Mathematically, this behavior is determined by the formula

$$\Delta\theta == -2\,i\,l\,\frac{\displaystyle\int_{r_{min}}^{r_{max}}\frac{1}{r^2\sqrt{\frac{l^2}{r^2\mu}-2\,H+2\,U(r)}}\,dr}{\sqrt{\mu}}$$

$$\Delta\theta = \begin{cases} 2\,\pi\,\dfrac{m}{n}, & \text{closed orbits} \\ \text{any}, & \text{open orbits.} \end{cases} \tag{2.5.7}$$

2.5.3.3 Effective Potential

Up to now we discussed different principal forms of orbits. However, we did not solve the problem by integration. This subsection discusses under which conditions a solution is derivable and what kinds of solution are allowed.

For example, we know that the radial velocity v can be determined by the total energy H and the angular momentum l. The radial velocity $r' = v$ is gained from energy conservation:

totalEnergy = H $==$ $\dfrac{\mu\,v^2}{2} + U(r)$

$$H == \frac{\mu\,v^2}{2} + U(r)$$

or

velocity = Solve[totalEnergy, v] // Flatten

$$\left\{ v \rightarrow -\frac{\sqrt{2}\,\sqrt{H-U(r)}}{\sqrt{\mu}}, v \rightarrow \frac{\sqrt{2}\,\sqrt{H-U(r)}}{\sqrt{\mu}} \right\}$$

In the case of planetary motion, we already know the radial velocity:

---

equationOfMotion /. Rule → Equal

$$r'(t) == \frac{i \sqrt{2} \sqrt{\frac{l^2}{2 \mu r(t)^2} - H + U(r(t))}}{\sqrt{\mu}}$$

The right-hand side of this expression follows from the total energy. In addition to the total energy H and the potential $U(r)$, this expression contains a term expressed by

---

$$-\frac{l^2}{2 r^2 \mu} == \frac{1}{2} \mu r^2 \left(\frac{\partial \theta(t)}{\partial t}\right)^2$$

$$-\frac{l^2}{2 r^2 \mu} == \frac{1}{2} r^2 \mu \theta'(t)^2$$

This relation expresses the rotational energy on the orbit. Because the left-hand side shows a radial dependence, we can interpret this term as a sort of effective potential. The part of the total potential is given by

---

$$U_c = \frac{l^2}{2 \mu r^2}$$

$$\frac{l^2}{2 r^2 \mu}$$

The related force corresponding to the orbit potential is

---

cForce $= -\partial_r U_c$

$$\frac{l^2}{r^3 \mu}$$

This kind of force is known as centrifugal force. The conventional representation of this force is written as

$$F_c = m r \omega^2, \tag{2.5.8}$$

where m is mass and ω is the frequency of revolution. This kind of force was first introduced by Christian Huygens (1629–1695). If we identify $\omega = \theta'$ and $\mu = m$, we are able to write

$$F_c = \frac{d}{dr} \left(\frac{1}{2} \mu r^2 \theta'^2 \right) = \mu r \theta'^2. \tag{2.5.9}$$

This allows us to identify $\frac{l^2}{2\mu r^2}$ as a centrifugal potential. Because U_c is a pure function in r, we can combine the interaction potential $U(r)$ with U_c to an effective potential. This potential is

$$\text{effectivePotential} = \frac{l^2}{2\,\mu r^2} + U(r)$$

$$\frac{l^2}{2\,r^2\,\mu} + U(r)$$

The effective potential is an fictitious potential consisting of the real interaction potential and a part containing the energy of rotation.

In Newton's theory of the two body problem the central force is assumed to decrease quadratic in the radial coordinate

$$\text{force} = -\frac{k}{r^2}$$

$$-\frac{k}{r^2}$$

The related potential is thus given by

$$U(r) = -\int \text{force}\, dr$$

$$-\frac{k}{r}$$

The effective potential U_{eff} then takes the explicit form:

effectivePotential

$$\frac{l^2}{2\,r^2\,\mu} - \frac{k}{r}$$

A graphical representation of the effective potential is given in figure 2.5.7.

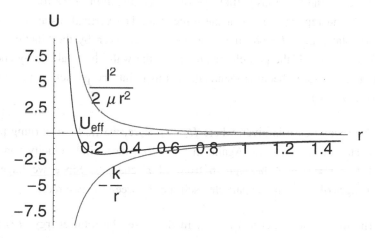

Figure 2.5.7.   Effective potential for central forces.

In this representation of the effective potential U_{eff}, we assume the vanishing asymptotic behavior for $r \to \infty$.

Figure 2.5.8 shows the effective potential with three different values for total energy. The three values for the total energy H_1, H_2, and H_3 characterize three different regimes of orbits.

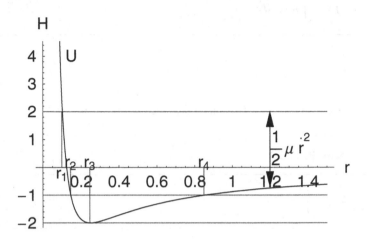

Figure 2.5.8.   Three regimes of motion (circular, elliptic, hyperbolic).

First, if the total energy $H_1 \geq 0$, then the motion on the orbit is infinite. In this case, the fictitious particle moves in the direction of the force center at $r = 0$ and repells at $r = r_1$ at the force wall. The vertical distance between the total energy H_1 and the potential $U(r)$ is given by the kinetic energy $T = \frac{1}{2}\mu r'^2$. If the particle hits the potential wall, the total energy and the potential energy become identical. At this point, the particle comes to rest (i.e., $r' = 0$).

The second case is given where $H_2 < 0$. Here, we find two turning points located at $r_2 < r < r_4$. Again, in r_2 and r_4, the radial velocity vanishes; that is $r' = 0$ and the sign in front of r' changes. Since we have two changes of the sign, the particle oscillates between the two radii.

The third case is defined by H_3. In this case, the total energy $H = H_3$ is always equal to the potential energy at the potential minimum $H_3 = U_{\text{eff}}(r_{\text{min}})$. The radial velocity is always zero; that is the radius is a finite constant. Thus, the particle moves on a circle around the force center at $r = 0$. Energies smaller than $U_{\text{eff}}(r_{\text{min}}) = -\mu k^2/(2 l^2)$ are of no physical relevance because here $r'^2 < 0$ (i.e., imaginary velocities).

2.5.3.4 Planet Motions

Taking into account the forces derived in the previous subsections, we can use Newton's equation of motion to write down the second-order equation for the radial component:

$$\text{KeplersEquation} = \mu \, \frac{\partial^2 r(t)}{\partial t \, \partial t} == -\frac{\partial \left(\frac{l^2}{2 r(t)^2 \mu} - \frac{k}{r(t)} \right)}{\partial r(t)}$$

$$\mu \, r''(t) == \frac{l^2}{\mu \, r(t)^3} - \frac{k}{r(t)^2}$$

The acting forces are the gravitation force and the centrifugal force. The aim of this subsection is to solve this equation of motion. The equation of motion is primarily a second-order nonlinear time-dependent ordinary differential equation. Our interest is to find the orbit of the particle defined by the radial and angular coordinates. Our goal is to find a relation which connects the radial coordinate with the angular coordinate; that is, we are looking for a relation $r = r(\theta)$. In a first step, we represent the angular momentum of the particle on the orbit by θ'. The total energy then becomes

$$kE = \text{KeplersEquation} \, /. \, l \rightarrow \mu \, r(t)^2 \, \frac{\partial \theta(t)}{\partial t}$$

$$\mu \, r''(t) == \mu \, r(t) \, \theta'(t)^2 - \frac{k}{r(t)^2}$$

This equation is the starting point of our examinations. We get a parameterization of the orbit by θ if we introduce the following transformation:

$$\mathbf{trafo1} = u(\theta(t)) == \frac{1}{r(t)}$$

$$u(\theta(t)) == \frac{1}{r(t)}$$

Differentiation of this transformation with respect to time and a solution for $du/d\theta$, we get

$$\mathbf{sol1} = \mathbf{Flatten}\left[\mathbf{Solve}\left[\frac{\partial\,\mathbf{trafo1}}{\partial t}, \frac{\partial u(\theta(t))}{\partial\theta(t)}\right]\right] \, /. \, \mathbf{Rule} \to \mathbf{Equal}$$

$$\left\{ u'(\theta(t)) == -\frac{r'(t)}{r(t)^2\,\theta'(t)} \right\}$$

On the other hand, we know that the angular momentum is given by the relation $r^2\,\theta' = l/\mu$. This provides the substitution

$$\mathbf{substitution2} = \frac{\partial\theta(t)}{\partial t} \to \frac{l}{\mu\,r(t)^2}$$

$$\theta'(t) \to \frac{l}{\mu\,r(t)^2}$$

With this relation, $du/d\theta$ is represented by

$$\mathbf{sol2} = \mathbf{sol1} \, /. \, \mathbf{substitution2}$$

$$\left\{ u'(\theta(t)) == -\frac{\mu\,r'(t)}{l} \right\}$$

Differentiating a second time with respect to time and solving for r'' delivers

$$\text{sol3} = \text{Flatten}\left[\text{Solve}\left[\frac{\partial \, \text{sol2}}{\partial t}, \frac{\partial^2 r(t)}{\partial t \, \partial t}\right]\right]$$

$$\left\{r''(t) \rightarrow -\frac{l \, \theta'(t) \, u''(\theta(t))}{\mu}\right\}$$

Now, replacing θ' and r by the above derived relations, we finally get

$$\text{substitution3} = \text{sol3} \, /. \, \text{substitution2} \, /. \, r(t) \rightarrow \frac{1}{u(\theta(t))}$$

$$\left\{r''(t) \rightarrow -\frac{l^2 \, u(\theta(t))^2 \, u''(\theta(t))}{\mu^2}\right\}$$

This relation can be simplified by applying the found substitutions for r'', r, and θ'. The equation of motion now reads

$$\text{kEu} = \text{kE} \, /. \, \text{substitution3} \, /. \, \text{substitution2} \, /. \, r(t) \rightarrow \frac{1}{u(\theta(t))}$$

$$-\frac{l^2 \, u(\theta(t))^2 \, u''(\theta(t))}{\mu} \; == \; \frac{l^2 \, u(\theta(t))^3}{\mu} - k \, u(\theta(t))^2$$

A solution with respect to u'' gives

$$\text{kEU} = \text{Solve}\left[\text{kEu}, \frac{\partial^2 u(\theta(t))}{\partial \theta(t) \, \partial \theta(t)}\right] \, /. \, \{\text{Rule} \rightarrow \text{Equal}, \, \theta(t) \rightarrow \theta\}$$

$$\left(u''(\theta) == -\frac{\mu \left(\frac{l^2 \, u(\theta)^3}{\mu} - k \, u(\theta)^2\right)}{l^2 \, u(\theta)^2}\right)$$

This equation can be simplified again if we introduce a translation in u by an amount of $k \, \mu / (l^2)$ providing the new dependent variable $y = u - k \, \mu / (l^2)$. Applying this transformation to the equation of motion gives the simple equation

$$\text{kEUe} = \text{Simplify}\left[\text{kEU} /. u \rightarrow \text{Function}\left[\theta, \frac{k\,\mu}{l^2} + y(\theta)\right]\right]$$

$$(y(\theta) + y''(\theta) == 0)$$

However, this equation is identical with the equation of motion for a harmonic oscillator. We already know the solutions of this equation which are given by harmonic functions with θ as an independent variable. The solution of this equation follows using

$$\text{solution} = \text{Flatten}[\text{DSolve}[\text{kEUe}[[1]], y, \theta]]$$

$$\{y \rightarrow \text{Function}[\{\theta\}, c_1 \cos(\theta) + c_2 \sin(\theta)]\}$$

Here, c_1 and c_2 are constants of integration. c_1 and c_2 are determined by the initial conditions (i.e., the total energy). To fix c_1 and c_2, we multiply the radial equation of motion by r':

$$\text{ke} = \text{KeplersEquation}\,\frac{\partial r(t)}{\partial t}$$

$$\mu\, r'(t)\, r''(t) == \left(\frac{l^2}{\mu\, r(t)^3} - \frac{k}{r(t)^2}\right) r'(t)$$

Integrating with respect to time delivers

$$\text{totalEnergy} = \int \text{ke}\,dt \;/.\; \text{RHSToLHS} \;/.\; 0 \rightarrow H$$

$$\frac{l^2}{2\,\mu\, r(t)^2} + \frac{1}{2}\,\mu\, r'(t)^2 - \frac{k}{r(t)} == H$$

Applying to this relation the transformations for r', r, u, and θ' we gain

$$\textbf{tEnergy} =$$

$$\textbf{totalEnergy} /. \, r(t) \rightarrow \frac{1}{u(\theta(t))} \, /. \, \frac{\partial r(t)}{\partial t} \rightarrow - \frac{l \, \frac{\partial u(\theta(t))}{\partial \theta(t)}}{\mu} \, /. \, \theta(t) \rightarrow \theta \, /.$$

$$u \rightarrow \textbf{Function}\Big[\theta, \frac{k \, \mu}{l^2} + y(\theta)\Big]$$

$$\frac{l^2 \left(\frac{k \mu}{l^2} + y(\theta)\right)^2}{2 \mu} - k \left(\frac{k \mu}{l^2} + y(\theta)\right) + \frac{l^2 \, y'(\theta)^2}{2 \mu} == H$$

Inserting the solution $y = y(\theta)$ into this relation and choosing $c_1 = 0$, we find

$$\textbf{Energy} = \textbf{tEnergy} /. \, \textbf{solution} /. \, c_2 \rightarrow 0 \, // \, \textbf{Simplify}$$

$$\frac{l^2 \, c_1^2}{2 \mu} == \frac{\mu \, k^2}{2 \, l^2} + H$$

This expression relates c_1 with l and H. Solving with respect to c_2 delivers

$$\textbf{const} = \textbf{Simplify[Solve[Energy, } c_1\textbf{]]}$$

$$\Big\{\Big\{c_1 \rightarrow - \frac{\sqrt{\mu} \, \sqrt{\frac{\mu k^2}{l^2} + 2 H}}{l}\Big\}, \Big\{c_1 \rightarrow \frac{\sqrt{\mu} \, \sqrt{\frac{\mu k^2}{l^2} + 2 H}}{l}\Big\}\Big\}$$

Inverting all transformations so far used, we find the final solution

$$\textbf{qh} = u(\theta) == \frac{k \, \mu}{l^2} + y(\theta) /. \, \textbf{solution} /. \, c_2 \rightarrow 0 /. \, u(\theta) \rightarrow \frac{1}{r(\theta)} \, /. \, \textbf{const[2]}$$

$$\frac{1}{r(\theta)} == \frac{k \, \mu}{l^2} + \frac{\sqrt{\frac{\mu k^2}{l^2} + 2 H} \, \cos(\theta) \, \sqrt{\mu}}{l}$$

The representation of the solution can be improved by introducing the following expressions:

$$\texttt{qh = qh}\left(\frac{\texttt{l}^2}{\texttt{k}\,\mu}\right)\ \texttt{// Simplify}$$

$$\frac{l^2}{k\,\mu\,r(\theta)} == \frac{l\,\sqrt{\frac{\mu k^2}{l^2} + 2H}\ \cos(\theta)}{k\,\sqrt{\mu}} + 1$$

Coefficient[qh[[2]], cos(θ)] == ϵ

$$\frac{l\,\sqrt{\frac{\mu k^2}{l^2} + 2H}}{k\,\sqrt{\mu}} == \epsilon$$

and

Coefficient[qh[[2]], cos(θ)]

$$\frac{l\,\sqrt{\frac{\mu k^2}{l^2} + 2H}}{k\,\sqrt{\mu}}$$

sh = Flatten[Solve[Coefficient[qh[[2]], cos(θ)] == ϵ, H]]

$$\left\{ H \to \frac{k^2\,(\epsilon^2 - 1)\,\mu}{2\,l^2} \right\}$$

Applying these substitutions to the original form of the solution, we obtain

shh = SimplifyAll[qh /. sh /. $l^2 \to k\,\mu\,\alpha$]

$$\frac{\alpha}{r(\theta)} == \epsilon\cos(\theta) + 1$$

This relation is known as the standard representation for conical sections. Johann Bernoulli (1667–1748) was the first to demonstrate that orbits in a $1/r$ potential are identical with conic sections (1710). ϵ in the above expression is the eccentricity of the orbit and 2α determines the latus rectum of the orbit.

The above equation approaches a minimum in r if $\cos(\theta)$ reaches a maximum (i.e., $\theta = 0$). Closely related to this behavior is the determination of the integration constant c_2; that is we measure θ starting at r_{min}.

Since the eccentricity is closely related to the energy, the type of the orbit can be determined by this parameter. The following table collects the different types of orbit and connects them with the energy and eccentricity:

$\epsilon > 1$	$H > 0$	hyperbolas
$\epsilon = 1$	$H = 0$	parabolas
$0 < \epsilon < 1$	$U_{min} < H < 0$	ellipses
$\epsilon = 0$	$H = U_{min}$	circle
$\epsilon < 0$	$H < U_{min}$	not allowed

Table 2.5.1.   Different motions in a central force field.

The equation for conical sections is also graphically accessible if we represent r and θ in cartesian coordinates. The equation in cartesian coordinates reads

```
ck = shh /. r[θ] → r /. {r → √x² + y² , θ → ArcTan[x, y]}
```

$$\frac{\alpha}{\sqrt{x^2 + y^2}} == \epsilon \cos(\tan^{-1}(x, y)) + 1$$

The Figure 2.5.9 contains the different types of orbit:

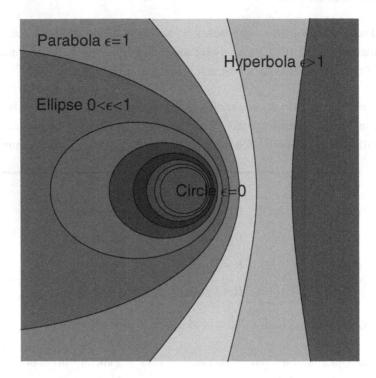

Figure 2.5.9.   The classification of the orbits by means of the eccentricity ϵ is equivalent to the classification due to the energy values in an effective potential.

The following considerations discuss the connections among energy, angular momentum, and the parameters of the orbit (i.e., eccentricity, mean distances of the ellipse from the center, etc.). The geometrical notions are given in the Figure 2.5.10.

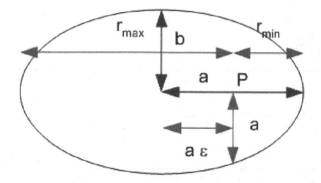

Figure 2.5.10. Geometric relations for the two-body prolem. P denotes the focus of the track, a and b are the principal axis of the ellipse. ϵ and α denote the eccentricity and the latum rectum.

Figure 2.5.10 shows that the larger principal axis can be expressed by minimal and maximal radius of the ellipse:

$$2\,a = r_{\max} + r_{\min}. \tag{2.5.10}$$

By definition, the velocity vanishes in the aphelion and in the perihelion. This behavior guaranties that $r_{\min}$ and $r_{\max}$ are solutions of the following relation:

---

turningPoints = totalEnergy /. $\left\{ \dfrac{\partial r(t)}{\partial t} \to 0,\, r(t) \to r \right\}$

$$\frac{l^2}{2\,r^2\,\mu} - \frac{k}{r} == H$$

The two solutions are

---

st = Solve[turningPoints, r]

$$\left\{ \left\{ r \to \frac{-\sqrt{\mu}\,k - \sqrt{\mu\,k^2 + 2\,H\,l^2}}{2\,H\,\sqrt{\mu}} \right\},\, \left\{ r \to \frac{\sqrt{\mu\,k^2 + 2\,H\,l^2} - k\,\sqrt{\mu}}{2\,H\,\sqrt{\mu}} \right\} \right\}$$

On the other hand, the two extremas in r are represented by ϵ and α with the help of

$$r_{max} = \frac{\alpha}{1 - \epsilon}$$

$$\frac{\alpha}{1 - \epsilon}$$

and

$$r_{min} = \frac{\alpha}{\epsilon + 1}$$

$$\frac{\alpha}{\epsilon + 1}$$

Both relations follow directly from the orbit geometry. The sum of both relations connects ϵ with α to

eq1 = Simplify[$r_{max} + r_{min}$ == 2 a]

$$-\frac{2\alpha}{\epsilon^2 - 1} == 2a$$

which provides the representations for the eccentricity ϵ:

sh = Solve[eq1, ϵ]

$$\left\{\left\{\epsilon \to -\frac{\sqrt{a - \alpha}}{\sqrt{a}}\right\}, \left\{\epsilon \to \frac{\sqrt{a - \alpha}}{\sqrt{a}}\right\}\right\}$$

Using the root with the plus sign and substituting $\alpha = l^2 / (k\mu)$, we obtain

$$\epsilon \text{ /. sh[[2]] /. } \alpha \rightarrow \frac{l^2}{k\,\mu} \text{ // Simplify}$$

$$\frac{\sqrt{a - \frac{l^2}{k\,\mu}}}{\sqrt{a}}$$

The major principal axis is thus represented by

$$\textbf{majorAxis} = \textbf{Simplify}\Big[a == \frac{1}{2} \text{ [Fold[Plus, 0, } r \text{ /. st]]}\Big]$$

$$a == \frac{|\frac{k}{H}|}{2}$$

The smaller principal axis follows:

$$b = \textbf{SimplifyAll}\Big(\frac{\alpha}{\sqrt{1 - \epsilon^2}} \text{ //. Flatten}\Big[\{\alpha \rightarrow \frac{l^2}{k\,\mu}, \text{ sh[[2]], } a \rightarrow \Big|-\frac{k}{2H}\Big|\}\Big]\Big)$$

$$\frac{l\sqrt{|\frac{k}{H}|}}{\sqrt{2}\,\sqrt{k}\,\sqrt{\mu}}$$

At this stage, we know the relations among energy, angular momentum, and principal axes.

In the following, we will derive Kepler's laws from the orbit data. First, let us consider the temporal change of the area swept by the particle. We know that this law is independent of the interacting forces. The temporal change of the area is

$$\frac{dA(t)}{dt} == \frac{r(t)^2\,d\theta(t)}{2\,dt} == \frac{l}{2\mu}$$

$$A'(t) == \frac{1}{2}\,r(t)^2\,\theta'(t) == \frac{l}{2\mu}$$

Because the total area of the ellipse is swept out by the particle in the period τ, we can write

$$\int_0^\tau 1\,dt == \int_0^A \frac{2\mu}{l}\,dA$$

$$\tau == \frac{2A\mu}{l}$$

On the other hand, we know the relation for the total area of an ellipse:

$$A = \pi a b$$

$$\frac{a l \pi \sqrt{|\frac{k}{H}|}}{\sqrt{2}\,\sqrt{k}\,\sqrt{\mu}}$$

Thus, the period is given by

$$\text{period} = \tau == \frac{2(A\mu)}{l}$$

$$\tau == \frac{\sqrt{2}\,a\pi\sqrt{\mu}\,\sqrt{|\frac{k}{H}|}}{\sqrt{k}}$$

The total energy is related to the major principal axis by

---

en = Solve[majorAxis, H]

Solve::ifun : Inverse functions are being used by Solve, so some solutions may not be found.

$$\left\{\left\{H \rightarrow -\frac{k}{2a}\right\}, \left\{H \rightarrow \frac{k}{2a}\right\}\right\}$$

which allows a simplification of the period to

---

prd = period /. en[[2]]

$$\tau == \frac{2a\pi\sqrt{\mu}\sqrt{|a|}}{\sqrt{k}}$$

The fact that the period is proportional to $a^{3/2}$ is known as the third law by Kepler. This result is valid for the fictitious one-particle problem. For this simplification, the reduced mass μ is a combination of two parts. Kepler's original formulation of this law was that the square of the period of a single planet is proportional to the third power of the major principal axis of this planet. A major assumption by Kepler was that the constant relating the square period with the third power of a is a universal constant for all planets. Taking the mass dependence of the planet into account, Kepler's original formulation is valid within this correction. Especially for gravitational forces, we have

$$F(r) = -G\frac{m_1 m_2}{r^2} = \frac{-k}{r^2}; \tag{2.5.11}$$

thus, the constant k is given by

---

r1 = $k \rightarrow G m_1 m_2$

$k \rightarrow G m_1 m_2$

The period can now be expressed as

$$\text{Simplify}\left[\text{prd} \,/.\, \text{r1} \,/.\, \mu \to \frac{m_1 \, m_2}{m_1 + m_2}\right]$$

$$\tau == \frac{2 \, a \, \pi \, \sqrt{|a|} \, \sqrt{\frac{m_1 \, m_2}{m_1 + m_2}}}{\sqrt{G \, m_1 \, m_2}}$$

If we assume $m_2 \gg m_1$, we find

$$\tau == \text{Series}\left[\frac{2 \, a \, \pi \, \sqrt{|a|} \, \sqrt{\frac{m_1 \, m_2}{m_1 + m_2}}}{\sqrt{G \, m_1 \, m_2}}, \{m_1, 0, 1\}\right]$$

$$\tau == \frac{2 \, a \, \pi \, \sqrt{|a|}}{\sqrt{G \, m_2}} - \frac{a \, \pi \, \sqrt{|a|} \, m_1}{m_2 \, \sqrt{G \, m_2}} + O(m_1^2)$$

Thus, the original Kepler formulation is valid if m_1 is much smaller than the mass m_2 of the central star.

2.5.4 Two-Particle Collisions and Scattering

One of the most important methods to gain information on the internal structure of materials is the application of scattering. Scattering is tightly connected to the two-body problem and Kepler's law. The result of a particle bombardment is the scattering of many particles in different directions. The distribution of the particles in space depends on the inner structure and the internal forces of the target particle. To understand the experimental results and how the scattered particles are deflected, we must examine the internal interaction of the target particles and the interaction of the incoming particles. Our main goal here is to understand how the internal structure influences the distribution of these particles.

If two particles interact, the relative motion of these particles are influenced by the interaction force. This interaction can be direct as with two billiard balls or indirect via an interaction potential. For example, a comet is scattered at the Sun due to the existence of the gravitational potential. α-Particles are scattered due to electromagnetic forces near the core of the atom. We demonstrated earlier that in case of known interaction laws, the movement for a two-particle system is completely determined. On the other hand, knowing the conservation laws such as conservation of energy and angular momentum, we are able to derive valuable knowledge in lack of information on the interaction process. The knowledge of conservation laws allows us to determine the final state of the motion from the initial state.

Figures 2.5.11 and 2.5.12 show characteristic scattering processes on a microscopic and macroscopic scale.

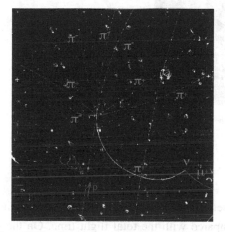

Figure 2.5.11. Proton–proton scattering in a bubble chamber.

Figure 2.5.12.   Orbit of a falling star in the gravitation field of the Earth.

2.5.4.1 Elastic Collisions

A collision usually occurs between two interacting particles. The time of interaction is very short compared with the total flight time. On the other hand, during the collision, the external forces are very small compared with the internal interaction forces. If the interaction time is very short, then we can assume that the forces are central and in the opposite direction. This property guarantees the conservation of the total momentum of the two-particle system. The interaction time usually is very short so that we can assume that the center of mass is at rest.

If the total energy before and after the collision is the same amount, we call this collision elastic. When energy conservation is not satisfied, then an inelastic collision occurred. A completely inelastic collision has occurs if the two particles stick together and all of the kinetic energy is converted to thermal or interaction energy.

An example of an inelastic collision is shown in the Figure 2.5.13. A bullet with initial velocity $v = 850$ m/s hits an apple which destroyed within a few milliseconds.

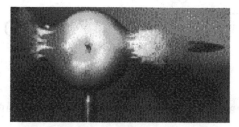

Figure 2.5.13.   Inelastic collision of a bullet with an apple.

Our interest here is in a completely elastic collision. We restrict our discussion to this kind of scattering because elastic collisions can be examined by the use of conservation laws. We also know from the discussions in the previous subsections that the examination simplifies if we assume that the center of mass is at rest. The standard situation of any collision is that a moving particle hits a second particle at rest. A real collision does not have a resting center of mass. Contrary to this simplifying mathematical assumption, one of the two particles will move in the laboratory and the other will be at rest. After the collision, both particles will move in the same direction. It is essential for our discussions that we distinguish between descriptions in the laboratory system and the center of mass system.

Figure 2.5.14 shows the geometry of a two particle collision for masses m_1 and m_2. Mass m_1 is moving with velocity $\vec{u}_1$ toward mass m_2. The movement of particle 1 is along the x-axis. The separation between the two particles in a perpendicular direction to the movement is called the impact parameter.

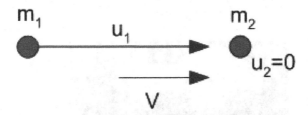

Figure 2.5.14. Two particles in a central collision at the initial stage represented in the laboratory system.

After the collision, the masses m_1 and m_2 travel with velocities $\vec{v}_1$ and $\vec{v}_2$, respectively. The angles ψ and ζ measured with respect to the x-axis determine the directions of the particles (see Figure 2.5.15).

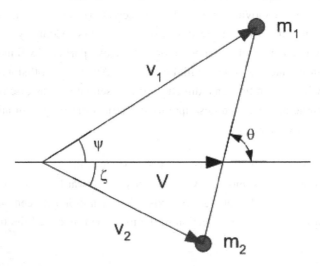

Figure 2.5.15. Two particles in a central collision at the final stage represented in the laboratory system.

The velocity $\vec{V}$ denotes the center of mass velocity in the laboratory system. The following illustration shows the collision in the laboratory system:

In the center of mass system, a collision is represented by two particles moving in the opposite directions (see also Figures 2.5.16 and 2.5.17).

Figure 2.5.16.   Two particles in a central collision at the initial stage represented in the center of mass system.

Primed symbols denote velocities in the center of mass system. After the collision, we find the representation in Figure 2.5.17.

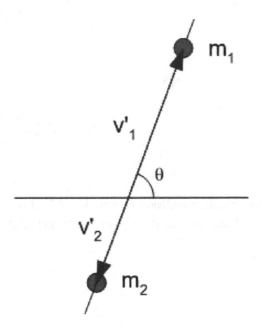

Figure 2.5.17.   Two particles in a central collision at the final stage represented in the center of mass system.

The following illustration shows the collision in the center of mass system:

In figure 2.5.18, θ denotes the scattering angle in the center of mass system. Up to now, we have distinguished four different situations: before collision and after collision and the two reference systems center of mass and laboratory system. These four situations can be combined in a common figure. We combine the end velocities of the laboratory system and the center of mass system as a single vector.

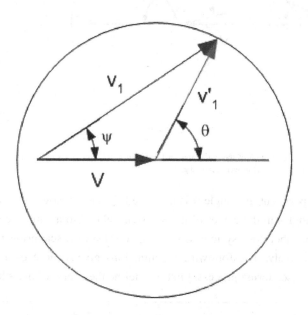

Figure 2.5.18.   Representation of the initial and final states of a collision in a single diagram.

The interpretation of this diagram is the following: If we add to the center of mass velocity $\vec{V}$ the end velocity $\vec{v}_1$ of the particle with mass m_1, then we get end velocity in the laboratory system $\vec{v}_1$. Dependent on the scattering angle θ, $\vec{v}_1$ terminates on a circle with radius v_1'. The center of this circle is the terminal point of the center of mass velocity $\vec{V}$. We find the scattering angle in the laboratory system by connecting the termination point of $\vec{v}_1$ with the origin of $\vec{V}$. If the center of mass velocity $V \leqslant v_1'$, then there exists a unique relation among the velocities $\vec{V}$, $\vec{v}_1$, and $\vec{v}_1$ and the angle θ. However, if $V > v_1'$, the relation is note unique. In this case, there exist two scattering angles in the laboratory system (θ_b, θ_f), a backward and a forward scattering angle, but only one angle ψ in the laboratory system (see Figure 2.5.19).

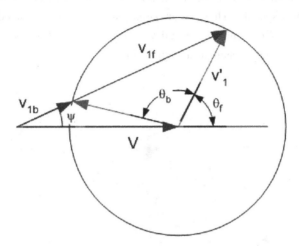

Figure 2.5.19.   Representation of the initial and final states of a collision in a single diagram with backward and forward scattering.

In a real experiment, the angle ψ is measured. In such a case, there are two scattering angles in the center of mass system related to a single scattering angle in the laboratory system. So far, we discussed a scattering process more qualitatively. The following examinations give a more quantitative approach to a scattering process. First, we define the center of mass by

ceneterOfMass = $M\,R(t) == m_1\,r_1(t) + m_2\,r_2(t)$

$M\,R(t) == m_1\,r_1(t) + m_2\,r_2(t)$

Differentiation with respect to time gives

velocity = ∂_t ceneterOfMass

$M\,R'(t) == m_1\,r_1'(t) + m_2\,r_2'(t)$

If we introduce the substitutions

$$velNam = \left\{ \frac{\partial R(t)}{\partial t} \rightarrow V(t), \frac{\partial r_1(t)}{\partial t} \rightarrow u_1(t), \frac{\partial r_2(t)}{\partial t} \rightarrow u_2(t) \right\};$$

velNam // TableForm

$R'(t) \rightarrow V(t)$
$r_1'(t) \rightarrow u_1(t)$
$r_2'(t) \rightarrow u_2(t)$

and consider the second particle at rest in the laboratory system

$$rule = \left\{ M \rightarrow m_1 + m_2, \frac{\partial r_2(t)}{\partial t} \rightarrow 0 \right\};$$

then we can represent the center of mass velocity by

$$\textbf{Flatten}\left[\textbf{Solve}\left[velocity \,/.\, rule, \frac{\partial R(t)}{\partial t}\right]\right] /.\, velNam$$

$$\left\{ V(t) \rightarrow \frac{m_1\, u_1(t)}{m_1 + m_2} \right\}$$

Because mass m_2 is at rest in the initial state, the center of mass velocity of this mass is V (i.e., $v_2' = V$ because the two masses are approaching each other).

A potential advantage of the center of mass system is that the total momentum is zero. Consequently, the two particles are approaching each other in a straight line. After the collision, they depart from each other on opposite directions. In an elastic collision, mass, momentum, and energy are conserved quantities. These conservation laws have the following consequences in the center of mass system for the velocities:

$$u_1' = v_1' \quad \text{and} \quad u_2' = v_2'. \tag{2.5.12}$$

Since u_1 describes the relative velocity of both particles in the center of mass or laboratory system, we have

$$u_1 = u_1'. \tag{2.5.13}$$

Thus, the final velocity for m_2 in the center of mass system is

$$v_2' = \frac{m_1 u_1}{m_1 + m_2}, \tag{2.5.14}$$

$$v_1' = u_1 - u_2' = \frac{m_2 u_1}{m_1 + m_2}. \tag{2.5.15}$$

Referring to Figure 2.5.18 we find

$$v_1' \sin(\theta) = v_1 \sin(\psi) \tag{2.5.16}$$

and

$$v_1' \cos(\theta) + V = v_1 \cos(\psi). \tag{2.5.17}$$

Division of both equations by each other gives

$$\mathbf{rel1} = \frac{v_1 \sin(\psi)}{v_1 \cos(\psi)} == \frac{\tilde{v}_1 \sin(\theta)}{V + \cos(\theta)\, \tilde{v}_1}$$

$$\tan(\psi) == \frac{\sin(\theta)\, \tilde{v}_1}{V + \cos(\theta)\, \tilde{v}_1}$$

Thus, we get for V and v_i,

$$\mathbf{rule2} = \left\{ V \rightarrow \frac{m_1 u_1}{m_1 + m_2}, \tilde{v}_1 \rightarrow \frac{m_2 u_1}{m_1 + m_2} \right\}; \mathbf{rule2\ //\ TableForm}$$

$$V \rightarrow \frac{m_1 u_1}{m_1 + m_2}$$

$$\tilde{v}_1 \rightarrow \frac{m_2 u_1}{m_1 + m_2}$$

Inserting this relation into the angle relation, we find

$$\mathbf{rel2} = \mathbf{Simplify[rel1\ /.\ rule2]}$$

$$\tan(\psi) == \frac{\sin(\theta)\, m_2}{m_1 + \cos(\theta)\, m_2}$$

We observe that the mass ration m_1 / m_2 determines which of the two cases is realized in a collision. We also observe that for $m_1 \ll m_2$, the center of mass system is nearly identical with the laboratory system:

---

rel2[[1]] == Series[rel2[[2]], {m_1, 0, 0}]

$\tan(\psi) == \tan(\theta) + O(m_1^1)$

This property means that the scattered particles do not influence the target particle and, thus, we have

$$\psi \approx \theta \text{ for } m_1 \ll m_2. \tag{2.5.18}$$

On the other hand, for $m_1 = m_2$, we get

---

rel2 /. $m_2 \to m_1$ // Simplify

$\tan(\psi) == \tan\left(\dfrac{\theta}{2}\right)$

---

and, thus,

$$\psi = \frac{\theta}{2} \text{ for } m_1 = m_2. \tag{2.5.19}$$

The scattering angle in the laboratory system is twice as large as the angle in the center of mass system. Since the maximal scattering angle in the lab system is $\psi = \pi$, the scattering angle in the center of mass system can be, at the utmost, $\pi/2$.

Relations relating to the kinetic energy in a scattering process are

---

labEnergy $= T_0 == \dfrac{1}{2} m_1 u_1^2$

$T_0 == \dfrac{1}{2} m_1 u_1^2$

The same in the center of mass system

$$\tilde{T}_0 = \frac{1}{2}\left(m_1\,\tilde{u}_1^2 + m_2\,\tilde{u}_2^2\right)$$

$$\frac{1}{2}\left(m_1\,\tilde{u}_1^2 + m_2\,\tilde{u}_2^2\right)$$

simplifies by applying the relations

rule1 $= \left\{\tilde{u}_2 \to \dfrac{m_1\,u_1}{m_1+m_2},\ \tilde{u}_1 \to \dfrac{m_2\,u_1}{m_1+m_2}\right\}$; **rule1 // TableForm**

$\tilde{u}_2 \to \frac{m_1\,u_1}{m_1+m_2}$

$\tilde{u}_1 \to \frac{m_2\,u_1}{m_1+m_2}$

to the kinetic energy

Simplify$\left[\tilde{T}_0\ /.\ \text{rule1}\right]\ /.\ \text{Flatten}[\text{Solve}[\text{labEnergy},\ u_1]]$

$$\frac{m_2\,T_0}{m_1+m_2}$$

The result demonstrates that the kinetic energy $\tilde{T}_0$ in the center of mass system is always a fraction $m_2/(m_1+m_2) < 1$ of the initial energy in the lab system. The kinetic energy of the final stage in the center of mass system is

$\tilde{T}_1 = \dfrac{1}{2}\,m_1\,\tilde{u}_1^2\ /.\ \text{rule1}\ /.\ \text{Flatten}[\text{Solve}[\text{labEnergy},\ u_1]]$

$$\frac{m_2^2\,T_0}{(m_1+m_2)^2}$$

and

$$\tilde{T}_2 = \frac{1}{2}\, m_2\, \tilde{u}_2^2 \; /. \; \text{rule1} \; /. \; \text{Flatten[Solve[labEnergy, } u_1]]$$

$$\frac{m_1\, m_2\, T_0}{(m_1 + m_2)^2}$$

To express T_1 by T_0, let us consider the ratio

$$\text{ratio} = \frac{T_1}{T_0} \; == \; \frac{m_1\, v_1^2}{\frac{2}{2}\,(m_1\, u_1^2)}$$

$$\frac{T_1}{T_0} \; == \; \frac{v_1^2}{u_1^2}$$

v_1 is connected with $\tilde{v}_1$ and V via the law of cosines:

$$\text{cosineLaw} = \tilde{v}_1^2 \; == \; V^2 - 2\, v_1\, \cos(\psi)\, V + v_1^2$$

$$\tilde{v}_1^2 \; == \; V^2 - 2\cos(\psi)\, v_1\, V + v_1^2$$

Introducing this relation into the energy ratio, we get

$$\frac{T_1}{T_0} = \frac{-V^2\, 2\, v_1\, V \cos\psi + \tilde{v}_1^2}{U_1^2}. \tag{2.5.20}$$

On the other hand, we know the relations

$$\mathbf{rule2} = \left\{ \tilde{v}_1 \rightarrow \frac{m_2\,u_1}{m_1 + m_2}, V \rightarrow \frac{m_1\,u_1}{m_1 + m_2}, v_1 \rightarrow \frac{\tilde{v}_1\,\sin(\theta)}{\sin(\psi)}, \right.$$

$$\left. \psi \rightarrow \tan^{-1}\left(\frac{\sin(\theta)}{\cos(\theta) + \frac{m_1}{m_2}} \right) \right\}; \mathbf{rule2\ //\ TableForm}$$

$\tilde{v}_1 \rightarrow \frac{m_2\,u_1}{m_1 + m_2}$

$V \rightarrow \frac{m_1\,u_1}{m_1 + m_2}$

$v_1 \rightarrow \csc(\psi)\,\sin(\theta)\,\tilde{v}_1$

$\psi \rightarrow \tan^{-1}\left(\frac{\sin(\theta)}{\cos(\theta) + \frac{m_1}{m_2}} \right)$

Inserting all of these relations into the energy ratio, we find

$\mathbf{enrat = ratio\ //.\ rule2\ //\ Simplify}$

$$\frac{T_1}{T_0} == \frac{m_1^2 + 2\cos(\theta)\,m_2\,m_1 + m_2^2}{(m_1 + m_2)^2}$$

This relation allows us to express T_1 by T_0. For identical particles, this simplifies to

$\mathbf{Simplify[enrat\ /.\ m_2 \rightarrow m_1]}$

$$\frac{T_1}{T_0} == \cos^2\!\left(\frac{\theta}{2} \right)$$

More relations for the second particle follow by similar considerations.

2.5.4.2 Scattering Cross Section

From a historical point of view, the theoretical background of the two-body problem was solved by discussing the planet's motion around the Sun. However, the two-body problem is also of great importance if we consider scattering problems in the atomic region. Scattering by atoms is governed by electric central forces determining the behavior of the scattering process.

The scattering problem of particles is governed by different influences. The main influence is defined by the central force acting on the particles. We assume in the following that all scattered particles are of the same kind of material (homogenous beam). All scattered particles have the same mass and the same energy.

In addition, we assume that the central force declines very fast for large distances. We characterize the incoming beam by his intensity I. The beam intensity is a measure for the number of particles transmitted through a normal unique area per second. If a particle approaches the center of force, it either is attracted or repelled. In either case the particle is deflected from his straight way toward the force center. If the particle has passed the center of force, the interaction becomes smaller and smaller and the particle gets on a straight track again. Thus, the scattering process is characterized by three regions: two asymptotics where the particle moves nearly on a straight track and the interaction where the particle is deflected from one direction to another one. The scattering cross section for a certain direction in space is defined by

$$d\sigma(\vec{\Omega}) = \frac{d\sigma}{d\Omega} d\Omega =$$
(Number of scattered particles per time into d Ω) /
(Number of incomming particles per time and per area)
$$= \frac{dN}{I}.$$
(2.5.21)

Here, $d\Omega$ denotes the solid angle in the direction $\vec{\Omega}$. $d\sigma$ is also called the differential scattering cross section. In case of central forces, there exists rotation symmetry along the incoming beam. Taking this symmetry into account, the solid angle $d\Omega$ is given by

$$d\vec{\Omega} = 2\pi\sin(\theta)\,d\theta,$$
(2.5.22)

where θ denotes the angle between the initial and final direction. θ is also called the scattering angle. For an elastic scattering process, we assume conservation of energy and momentum. The angular momentum is also a conserved quantity. It is of great advantage to express the angular momentum by the impact parameter b and the initial energy T_0. The impact parameter b is defined as the perpendicular distance between the initial direction and the scatterer. Let u_1 be the initial velocity of the incoming particle. The angular momentum of the particle with respect to the scattering center is then defined by

$$L = m\,u_1\,b.\qquad\qquad(2.5.23)$$

The initial velocity u_1 is given by the initial energy $T_0 = m\,u_1^2/2$ and, thus, the angular momentum reads

$$L = b\,\sqrt{2\,m\,T_0}.\qquad\qquad(2.5.24)$$

As soon as b and T_0 are fixed, the scattering angle θ is uniquely fixed. Let us for the moment assume that different b values will not result in a single scattering angle. The number of scattered particles into a solid angle $d\,\vec{\Omega}$ in the range θ and $\theta + d\theta$ is given by the incoming particles. The impact parameter is the in the range b to $b + db$. The mathematical relation is

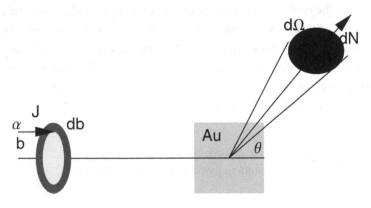

Figure 2.5.20.   The parameters in a scattering process.

$$2\,\pi\mathcal{I}\mathrm{b}\,|\,db\,| = 2\,\pi\,\sigma(\vec{\Omega})\,\mathcal{I}\,\sin(\theta)\,\bigl|\,d\theta\,\bigr|.\qquad\qquad(2.5.25)$$

Since the differentials can change sign, we introduced the amount of db and $d\theta$. The number of particles and all other quantities are positive. Let us assume that the impact parameter is a function of the scattering angle θ and the energy H:

$$b = b(\theta, H). \tag{2.5.26}$$

The scattering cross section becomes a function of the scattering angle

$$\frac{d\,\sigma(\theta)}{d\,\Omega} = \frac{b}{\sin(\theta)} \left| \frac{db}{d\theta} \right|. \tag{2.5.27}$$

The derivation of this formula follows from the assumption of particle conservation; that is, elastic scattering satisfying

$$(dN)_v = (dN)_n \tag{2.5.28}$$

The initial number of particles are given by

```
dNi = I dA /. dA → b db dφ
```

b db dϕ I

The number of particles after the scattering is

```
dNf = −I δσ dΩ /. dΩ → sin(θ) dθ dφ
```

$-$dθ dϕ I $\delta\sigma$ sin(θ)

Conservation of particles implies

```
particleConservation = dNi == dNf
```

b db dϕ I $==$ $-$dθ dϕ I $\delta\sigma$ sin(θ)

Thus, the scattering cross section follows as

$$\text{Flatten}\left[\text{Solve}[\text{particleConservation}, \delta\sigma] \; /. \; \delta\sigma \rightarrow \frac{d\sigma}{d\Omega}\right]$$

$$\left\{\frac{d\sigma}{d\Omega} \rightarrow -\frac{b \, db \, \csc(\theta)}{d\theta}\right\}$$

A formal expression for the scattering angle θ can be derived by symmetry considerations. Since the particle's track is symmetric with respect to the line focus scattering center, we can find from the geometry of the track the relation

$$\theta = \pi - 2\psi \tag{2.5.29}$$

This relation follows from the geometry given in Figure 2.5.21

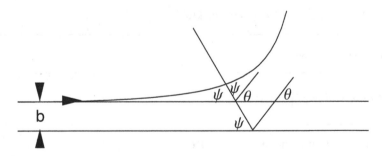

Figure 2.5.21.   Angular relations between the scattering angles and the impact parameter.

We already demonstrated that the change in ψ for a particle with reduced mass μ is given by

$$\Delta\psi = \int_{r_{\min}}^{r_{\max}} \frac{l/r^2}{\sqrt{2\,\mu\left(H - U - \frac{l^2}{2\,\mu\,r^2}\right)}} \, dr. \tag{2.5.30}$$

In case of $r_{\max} \rightarrow \infty$ we gain for ψ

$$\psi = \int_{r_{min}}^{\infty} \frac{b/r^2}{\sqrt{1 - U/\tilde{T}_0 - b^2/r^2}} \, dr \qquad (2.5.31)$$

with $\tilde{T}_0 = \frac{\mu}{2} \tilde{u}_1^2$. Concerning the energy, we find for $r \to \infty$ that $H = T_0$ since $U(r = \infty) = 0$. r_{min} denotes a root of the radicand and measures the shortest distance to the force center.

Since ψ depends on θ and the above integral is a function of b only for given $U(r)$ and T_0, we find that the impact parameter b is a function of θ:

$b = b(\theta)$.

This discussion delivers the scattering cross section in the center of mass system if we assume m_2, the target, at rest. If $m_2 \gg m_1$, the scattering cross section in the center of mass system is nearly the same as in the laboratory system. If this mass relation is not satisfied, a transformation between the center of mass and the laboratory system must be used to convert ψ into θ. Because the number of scattered particles are equal in the center of mass and laboratory system, we find

$$\frac{d\sigma(\theta)}{d\Omega} \, d\Omega' = \frac{d\sigma(\psi)}{d\Omega} \, d\Omega,$$

$$\qquad (2.5.32)$$

$$\frac{d\sigma(\theta)}{d\Omega} \, 2\pi \sin(\theta) \, d\theta = \frac{d\sigma(\psi)}{d\Omega} \, 2\pi \sin(\psi) \, d\psi,$$

with θ and ψ the scattering angles of the center of mass and laboratory system. Thus the scattering cross section in the laboratory system is given by

$$\frac{d\sigma(\psi)}{d\Omega} = \frac{d\sigma(\theta)}{d\Omega} \frac{\sin(\theta)}{\sin(\psi)} \frac{d\theta}{d\psi}. \qquad (2.5.33)$$

The derivation $d\theta/d\psi$ is determined by the transformation

$$eh = \tan(\psi) == \frac{\sin(\theta(\psi)) \, m_2}{m_1 + \cos(\theta(\psi)) \, m_2}$$

$$\tan(\psi) == \frac{\sin(\theta(\psi)) \, m_2}{m_1 + \cos(\theta(\psi)) \, m_2}$$

Differentiating this relation with respect to ψ and solving for $d\theta/d\psi$, we obtain

$$\mathbf{sb = Flatten\left[Simplify\left[Solve\left[\frac{\partial\,eh}{\partial\psi}, \frac{\partial\,\theta(\psi)}{\partial\psi}\right]\right]\right]}$$

$$\left\{\theta'(\psi) \rightarrow \frac{\sec^2(\psi)\,(m_1 + \cos(\theta(\psi))\,m_2)^2}{m_2\,(\cos(\theta(\psi))\,m_1 + m_2)}\right\}$$

With this relation, the scattering angle $\theta(\psi)$ in the laboratory system is given by

$$\mathbf{sh = Simplify[Solve[eh,\ \theta(\psi)]]}$$

Solve::ifun : Inverse functions are being used by Solve, so some solutions
  may not be found; use Reduce for complete solution information. More...

$$\left\{\left\{\theta(\psi) \rightarrow -\cos^{-1}\left(-\frac{1}{m_2^2}\left(\cos^2(\psi)\left(m_1\,m_2\tan^2(\psi) + \right.\right.\right.\right.\right.$$
$$\left.\left.\left.\left.\left.\cot(\psi)\sqrt{\sec^2(\psi)\,m_2^4\tan^2(\psi) - m_1^2\,m_2^2\tan^4(\psi)}\right)\right)\right)\right\},$$
$$\left\{\theta(\psi) \rightarrow \cos^{-1}\left(-\frac{1}{m_2^2}\left(\cos^2(\psi)\left(m_1\,m_2\tan^2(\psi) + \right.\right.\right.\right.$$
$$\left.\left.\left.\left.\cot(\psi)\sqrt{\sec^2(\psi)\,m_2^4\tan^2(\psi) - m_1^2\,m_2^2\tan^4(\psi)}\right)\right)\right)\right\},$$
$$\left\{\theta(\psi) \rightarrow -\cos^{-1}\left(\frac{1}{m_2^2}\left(\cos^2(\psi)\left(\cot(\psi)\sqrt{\sec^2(\psi)\,m_2^4\tan^2(\psi) - m_1^2\,m_2^2\tan^4(\psi)} - \right.\right.\right.\right.$$
$$\left.\left.\left.\left.m_1\,m_2\tan^2(\psi)\right)\right)\right)\right\},$$
$$\left\{\theta(\psi) \rightarrow \cos^{-1}\left(\frac{1}{m_2^2}\left(\cos^2(\psi)\left(\cot(\psi)\sqrt{\sec^2(\psi)\,m_2^4\tan^2(\psi) - m_1^2\,m_2^2\tan^4(\psi)} - \right.\right.\right.\right.$$
$$\left.\left.\left.\left.\left.m_1\,m_2\tan^2(\psi)\right)\right)\right)\right\}\right\}$$

Inserting this expression into the factor $(\sin\theta/\sin\psi)\,d\theta/d\psi$, we get the transformation

$$vh = \text{Simplify}\Big[(\text{PowerExpand} \, /\!/@ \, \#1 \, \&)\Big[\frac{\sin(\theta(\psi)) \, \frac{\partial \theta(\psi)}{\partial \psi}}{\sin(\psi)} \, /. \, sb \, /. \, sh\Big]\Big]$$

$$\Bigg\{ -\Bigg(\cos(\psi)\cot(\psi)\Big(m_1\,m_2 - \cot(\psi)\sqrt{\sec^2(\psi)\,m_2^4\tan^2(\psi) - m_1^2\,m_2^2\tan^4(\psi)}\,\Big)$$

$$\sqrt{1 - \frac{1}{m_2^4}\Big(\cos^4(\psi)\Big(m_1\,m_2\tan^2(\psi) + \cot(\psi)}$$

$$\overline{\sqrt{\sec^2(\psi)\,m_2^4\tan^2(\psi) - m_1^2\,m_2^2\tan^4(\psi)}\,\Big)^2\Big)\Big)\Bigg) \Bigg/$$

$$\Big(m_2\Big(m_2^3 - \sin^2(\psi)\,m_1^2\,m_2 - \cos^2(\psi)\cot(\psi)\,m_1$$

$$\sqrt{\sec^2(\psi)\,m_2^4\tan^2(\psi) - m_1^2\,m_2^2\tan^4(\psi)}\,\Big)\Big),$$

$$\Big(\cos(\psi)\cot(\psi)\Big(m_1\,m_2 - \cot(\psi)\sqrt{\sec^2(\psi)\,m_2^4\tan^2(\psi) - m_1^2\,m_2^2\tan^4(\psi)}\,\Big)^2$$

$$\sqrt{1 - \frac{1}{m_2^4}\Big(\cos^4(\psi)\Big(m_1\,m_2\tan^2(\psi) +}$$

$$\cot(\psi)\sqrt{\sec^2(\psi)\,m_2^4\tan^2(\psi) - m_1^2\,m_2^2\tan^4(\psi)}\,\Big)^2\Big)\Big)\Bigg) \Bigg/$$

$$\Big(m_2\Big(m_2^3 - \sin^2(\psi)\,m_1^2\,m_2 - \cos^2(\psi)\cot(\psi)\,m_1$$

$$\sqrt{\sec^2(\psi)\,m_2^4\tan^2(\psi) - m_1^2\,m_2^2\tan^4(\psi)}\,\Big)\Big),$$

$$-\Big(\cos(\psi)\cot(\psi)\Big(\sqrt{\sec^2(\psi)\,m_2^4\tan^2(\psi) - m_1^2\,m_2^2\tan^4(\psi)}\,\cot(\psi) + m_1\,m_2\Big)^2$$

$$\sqrt{1 - \frac{1}{m_2^4}\Big(\cos^4(\psi)\Big(m_1\,m_2\tan^2(\psi) - \cot(\psi)}$$

$$\overline{\sqrt{\sec^2(\psi)\,m_2^4\tan^2(\psi) - m_1^2\,m_2^2\tan^4(\psi)}\,\Big)^2\Big)\Big)\Bigg) \Bigg/$$

$$\Big(m_2\Big(m_2^3 - \sin^2(\psi)\,m_1^2\,m_2 + \cos^2(\psi)\cot(\psi)\,m_1$$

$$\sqrt{\sec^2(\psi)\,m_2^4\tan^2(\psi) - m_1^2\,m_2^2\tan^4(\psi)}\,\Big)\Big),$$

$$\Big(\cos(\psi)\cot(\psi)\Big(\sqrt{\sec^2(\psi)\,m_2^4\tan^2(\psi) - m_1^2\,m_2^2\tan^4(\psi)}\,\cot(\psi) + m_1\,m_2\Big)^2$$

$$\sqrt{1 - \frac{1}{m_2^4}\Big(\cos^4(\psi)\Big(m_1\,m_2\tan^2(\psi) -}$$

$$\cot(\psi)\sqrt{\sec^2(\psi)\,m_2^4\tan^2(\psi) - m_1^2\,m_2^2\tan^4(\psi)}\,\Big)^2\Big)\Big)\Bigg) \Bigg/$$

$$\Big(m_2\Big(m_2^3 - \sin^2(\psi)\,m_1^2\,m_2 + \cos^2(\psi)\cot(\psi)\,m_1$$

$$\sqrt{\sec^2(\psi)\,m_2^4\tan^2(\psi) - m_1^2\,m_2^2\tan^4(\psi)}\,\Big)\Big)\Bigg\}$$

The two solutions carry out the transformation from the center of mass to the laboratory system.

Let us consider the limiting case of equal masses $m_1 = m_2$; then, this formula reduces to

```
vt = Simplify[(PowerExpand //@ #1 &)[
        Simplify[(PowerExpand //@ #1 &)[Simplify[vh /. m₁ → m₂]]]]]
```

$\{0, 0, -4 \cos(\psi), 4 \cos(\psi)\}$

Thus, the scattering cross section for equal masses is transformed by

$$\frac{d\sigma(\psi)}{d\Omega} = 4 \cos(\psi) \frac{d\sigma(\theta)}{d\Omega} \Big|_{\theta=2\psi}. \tag{2.5.34}$$

The general transformation between the center of mass and the laboratory system is thus given by

```
σ(ψ) == vh[[2]] σ(θ)
```

$$\sigma(\psi) == \left(\cos(\psi) \cot(\psi) \left(m_1 m_2 - \cot(\psi) \sqrt{\sec^2(\psi) m_2^4 \tan^2(\psi) - m_1^2 m_2^2 \tan^4(\psi)} \right)^2 \right.$$

$$\sqrt{\left(1 - \frac{1}{m_2^4} \left(\cos^4(\psi) \left(m_1 m_2 \tan^2(\psi) + \right. \right. \right.}$$

$$\left. \left. \left. \cot(\psi) \sqrt{\sec^2(\psi) m_2^4 \tan^2(\psi) - m_1^2 m_2^2 \tan^4(\psi)} \right)^2 \right) \right)$$

$$\left. \sigma(\theta) \right) \Big/ \left(m_2 \left(m_2^3 - \sin^2(\psi) m_1^2 m_2 - \cos^2(\psi) \cot(\psi) m_1 \right. \right.$$

$$\left. \left. \sqrt{\sec^2(\psi) m_2^4 \tan^2(\psi) - m_1^2 m_2^2 \tan^4(\psi)} \right) \right)$$

For some experimental setups, it is more convenient to know the scattering information on the total space. In such cases, we can calculate the so-called total scattering cross section by integrating over the total space:

$$\sigma_t = \int_{4\pi} \sigma(\theta) \, d\vec{\Omega}$$

$$= 2\pi \int_o^\pi \sigma(\theta) \sin(\theta) \, d\theta. \tag{2.5.35}$$

Contrary to the scattering cross section the total cross section is independent of the scattering system.

Example 1: Hard Sphere Scattering

Let us consider the example of hard spheres scattered on each other. The geometry of the scattering process is represented in Figure 2.5.22.

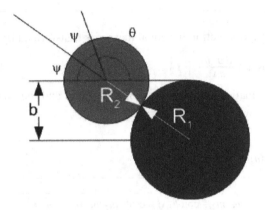

Figure 2.5.22.   Geometry of a hard-sphere scattering process.

Taking the geometric relations for the radii into account, we find

$$1 = R_1 + R_2$$

$$R_1 + R_2$$

For the angles, we have

$$\psi = \frac{\pi}{2} - \frac{\theta}{2}$$

$$\frac{\pi}{2} - \frac{\theta}{2}$$

The impact parametr is given by

$$b = 1 \text{ Sin}[\psi]$$

$$\cos\left(\frac{\theta}{2}\right)(R_1 + R_2)$$

Differentiating the impact parameter b with respect to the scattering angle θ gives

$$db = \partial_\theta b$$

$$-\frac{1}{2} \sin\left(\frac{\theta}{2}\right)(R_1 + R_2)$$

The scattering cross section thus becomes

$$d\sigma = \frac{-b}{\text{Sin}[\theta]} \, db \text{ // Simplify}$$

$$\frac{1}{4}(R_1 + R_2)^2$$

We observe that the scattering cross section for hard spheres is independent of the scattering angle θ. The total cross section of this example is

$$\sigma t = 2\pi \int_0^\pi d\sigma \text{ Sin}[\theta] \, d\theta$$

$$\pi(R_1 + R_2)^2$$

2.5.4.3 Rutherford Scattering

One of the most important applications of the two-body problem is the scattering of charged particles in an electric field. The electric field for a Coulomb scattering is $U(r) = k/r$, where k is a constant determined by the charges q_1 and q_2: $k = q_1 q_2$. k may be a constant of both signs. $k > 0$ resembles a repulsion and $k < 0$ an attraction. The determining equation for the track and scattering angle θ of a particle is determined by

$$\psi = \int_{r_{min}}^{\infty} \frac{b/r^2}{\sqrt{1 - U/\tilde{T}_0 - b^2/r^2}} \, dr = .$$

$$\int_{r_{min}}^{\infty} \frac{b/r^2}{\sqrt{1 - k/(\tilde{T}_0 \, r) - b^2/r^2}} \, dr \tag{2.5.36}$$

This integral is solved by the well-known substitution $u = 1/r$. The result of the integration is

$$\cos(\psi) = \frac{(\kappa/b)}{\sqrt{1 + (\kappa/b)^2}}, \tag{2.5.37}$$

where $\kappa = k/(2\,\tilde{T}_0)$. Solving this equation with respect to the impact parameter b, we find

```
Remove[b]
```

$$\mathbf{impact = PowerExpand\left[Simplify\left[Solve\left[\cos(\psi) == \frac{\kappa}{b\sqrt{\left(\frac{\kappa}{b}\right)^2 + 1}}, b\right]\right]\right]}$$

$$\left\{\left\{b \to -\kappa \cot\left(\frac{\theta}{2}\right)\right\}, \left\{b \to \kappa \cot\left(\frac{\theta}{2}\right)\right\}\right\}$$

The angle ψ is $\psi = \pi/2 - \theta/2$, so we get

$$\text{imp} = \text{impact} \;/.\; \psi \to \frac{\pi}{2} - \frac{\theta}{2}$$

$$\left\{ \left\{ b \to -\kappa \cot\!\left(\frac{\theta}{2}\right) \right\}, \; \left\{ b \to \kappa \cot\!\left(\frac{\theta}{2}\right) \right\} \right\}$$

The derivation of the impact parameter with respect to the scattering angle is given by

$$\text{db} = \frac{\partial\,(\text{imp}\;/.\; b \to b(\theta))}{\partial\theta}$$

$$\left\{ \left\{ b'(\theta) \to \frac{1}{2}\,\kappa\,\csc^2\!\left(\frac{\theta}{2}\right) \right\}, \; \left\{ b'(\theta) \to -\frac{1}{2}\,\kappa\,\csc^2\!\left(\frac{\theta}{2}\right) \right\} \right\}$$

Now, the scattering cross section follows from

$$\text{scatSection} = \text{Simplify}\!\left[\sigma == -\,\frac{b\,\frac{\partial b(\theta)}{\partial\theta}}{\sin(\theta)} \;/.\; \text{Flatten}[\text{Join}[\text{db}, \text{imp}]] \right]$$

$$\sigma == \frac{1}{4}\,\kappa^2\,\csc^4\!\left(\frac{\theta}{2}\right)$$

or

$$\text{sc} = \text{scatSection} \;/.\; \kappa \to \frac{k}{2\,\tilde{T}_0}$$

$$\sigma == \frac{k^2\,\csc^4\!\left(\frac{\theta}{2}\right)}{16\,\tilde{T}_0^2}$$

This formula is known as Rutherford's scattering formula. This relation was experimentally verified for α-particles by Geiger and Marsden in 1913 [2.10].

If the masses m_1 and m_2 or charges q_1 and q_2 are equal, we know that the kinetic energy reduces to $\tilde{T}_0 = T_0/2$. In this case, the scattering section reduces to

scatSection $/. \kappa \to \dfrac{k}{2\, T_0}$ $/. \tilde{T}_0 \to \dfrac{T_0}{2}$

$$\sigma == \frac{k^2 \csc^4(\frac{\theta}{2})}{16\, T_0^2}$$

The transformation of the scattering section for equal masses (charges) from the center of mass to the laboratory system follows from the following formulas:

$$\kappa = \frac{k}{2\,\tilde{T}_0}, \tag{2.5.38}$$

$$\frac{d\sigma(\psi)}{d\Omega} = \frac{1}{4\cos\psi} \left. \frac{d\sigma(\theta)}{d\Omega} \right|_{\theta=2\psi}, \tag{2.5.39}$$

and

$$\tilde{T}_0 = \frac{T_0}{2}. \tag{2.5.40}$$

The result of this transformation is

sc =

Solve$\left[\text{scatSection} /. \left\{\kappa \to \dfrac{k}{2\,T_0}, \sigma \to \dfrac{\sigma}{4\cos(\psi)}, \theta \to 2\,\psi\right\} /. T_0 \to \dfrac{T_0}{2}, \sigma\right]$

$$\left\{\left\{\sigma \to \frac{k^2 \sec^3(\frac{\theta}{2})\tan(\frac{\theta}{2})}{T_0^2}\right\}\right\}$$

The characteristic of Rutherford's scattering is the $1/\sin^4$ dependence of the scattering cross section. This dependence is valid in the laboratory as well as in the center of mass system. The experimental verification of the Rutherford relation was carried out by scattering α-particles on gold atoms. Since the gold particles are much heavier than α-particles, $m_{Au} \gg m_\alpha$, there is no difference between the center of mass and the laboratory system. The scattering cross section is given by

$$sc = scatSection \mathbin{/.} \kappa \rightarrow \frac{k}{2\,T_0}$$

$$\sigma == \frac{k^2 \csc^4(\frac{\theta}{2})}{16\,T_0^2}$$

Plotting this result in a log-log scale we get a nearly straight line for small scattering angles. Figure 2.5.23 shows this relation

```
<< "Graphics`Graphics`";
LogLogPlot(σ /. (sc /. Equal → Rule) /. {k → 4, T₀ → 1},
    {θ, 0.2, π}, AxesLabel → {"θ", "σ16T₀²/k²"});
```

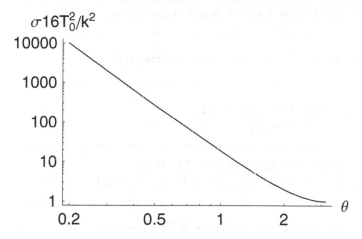

Figure 2.5.23.   Rutherford's scattering cross section in a log-log plot.

From the derivation of the scattering cross section, we know the total number of particles conserved in the scattering process:

$$(dN)_n = -\,\mathcal{I}\,\frac{d\sigma}{d\Omega}\,d\Omega = -\,\mathcal{I}\,\frac{d\sigma}{d\Omega}\sin\theta\,d\theta\,d\varphi, \qquad (2.5.41)$$

$$\frac{dN}{d\Omega} = -\,\mathcal{I}\,\frac{d\sigma}{d\Omega} = \text{const.} \qquad (2.5.42)$$

Measuring the particle number in a certain solid angle and multiplying this quantity by $d\sigma/d\Omega$, we get a constant. This kind of check was applied by Geiger and Marsden to their experimental data. Geiger and Marsden determined for the (Au, α) system scattering angles in the laboratory system, the number of α-particles and the product of the scattering cross section and the number of particles. In the following lines, we collect these data in different lists:

$$10 = \left\{\theta, \ \frac{1}{\mathrm{Sin}\left[\frac{\theta}{2}\right]^4}, \ J, \ \frac{J}{\mathrm{Sin}\left[\frac{\theta}{2}\right]^4}\right\};$$

The scattering angles are

$$11 = \{15, 22.5, 30, 37.5, 45, \\ 60, 75, 105, 120, 135, 150\} * 2 \ \frac{\pi}{360.}$$

{0.261799, 0.392699, 0.523599, 0.654498, 0.785398,
  1.0472, 1.309, 1.8326, 2.0944, 2.35619, 2.61799}

The cross section depends on the scattering angle as

$$12 = \mathrm{Map}\left[\frac{1}{\mathrm{Sin}\left[\frac{\#}{2}\right]^4} \ \&, \ 11\right]$$

{3445.16, 690.331, 222.851, 93.6706, 46.6274,
  16., 7.28134, 2.52426, 1.77778, 1.37258, 1.14875}

The total number of scintillations N for a given angle are

$$13 = \{132000, 27300, 7800, 3300, \\ 1435, 477, 211, 69.5, 51.9, 43, 33.1\}$$

{132000, 27300, 7800, 3300, 1435, 477, 211, 69.5, 51.9, 43, 33.1}

The ratio of N and the Rutherford characteristic is

```
14 = 13 / 12
```

```
{38.3146, 39.5463, 35.0009, 35.2298, 30.7759,
    29.8125, 28.9782, 27.5329, 29.1937, 31.3278, 28.814}
```

The following table collects all of these data:

```
lh = Prepend[Transpose[{11, 12, 13, 14}], 10];
lh // TableForm
```

θ	$\csc^4(\frac{\theta}{2})$	J	$J \csc^4(\frac{\theta}{2})$
0.261799	3445.16	132000	38.3146
0.392699	690.331	27300	39.5463
0.523599	222.851	7800	35.0009
0.654498	93.6706	3300	35.2298
0.785398	46.6274	1435	30.7759
1.0472	16.	477	29.8125
1.309	7.28134	211	28.9782
1.8326	2.52426	69.5	27.5329
2.0944	1.77778	51.9	29.1937
2.35619	1.37258	43	31.3278
2.61799	1.14875	33.1	28.814

If we plot $(\theta, N/\sin(\theta/2)^4)$, we observe that the experiment is in accordance with the theoretical prediction.

```
ListPlot[Transpose[{11, 14}],
  PlotRange → {{0, 2.7}, {0, 39}},
  AxesLabel → {"θ", "N/sin(θ/2)⁴"},
  PlotStyle → RGBColor[0.996109, 0, 0],
  Prolog → {PointSize[0.02]}];
```

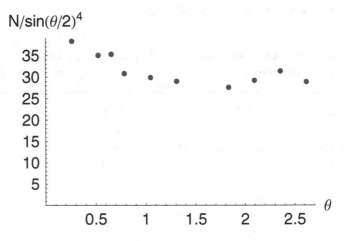

2.5.5 Exercises

1. Show that the relative motion of two particles is not affected by a uniform gravitational field.

2. Two particles connected by an elastic string of stiffness k and equilibrium length a rotate about their center of mass with angular momentum L. Show that their distance r_1 of closest approach and their maximum sparation r_2 are related by

$$\frac{r_1^2 r_2^2 (r_1 + r_2 - 2a)}{r_1 + r_2} = \frac{L^2}{k\mu}$$

where μ is their reduced mass and $r_1 > a$, $r_2 > a$.

3. Find the force law for a central force which allows a particle to move in a logarithmic spiral orbit given by $r = k\theta^2$, where k is a constant.

4. A particle moves in a circular orbit in a force field given by

$$F(r) = -k/r^2.$$

If suddenly k decreases to half its original alue, show that the particle's orbit becomes parabolic.

5. Discuss the motion of a particle in a central inverse square law force field for the case in which there is a superimposed force whose magnitude is inversely proportional to the cube of the distance from the particle to the force center; that is

$$F(r) = \frac{-k}{r^2} - \frac{\lambda}{r^3}, \quad k, \lambda > 0.$$

Show that the motion is described by a precessing ellipse. Consider the cases $\lambda < L^2/\mu$, $\lambda = L^2/\mu$, and $\lambda > L^2/\mu$.

2.5.6 Packages and Programs

Programs

The following lines are used to load special commands used in this notebook. The commands and definitions are contained in the file NewtonsLaws.m. Before you can use this file, you should set the path where it is located. Change the following line in such a way that the file is found.

```
SetDirectory["C:\Mma\Book\ThPh1"];
```

This line loads the contents of the file NewtonsLaws.m.

```
<< NewtonsLaws.m;
```

This line defines a function which maps **PowerExpand[]** to each level of an expression and simlifies the result by **Simplify[]**.

```
SimplifyAll[x_] :=
  MapAll[Simplify[PowerExpand[#]] &, x]
```

2.6 Calculus of Variations

2.6.1 Introduction

The term *calculus of variations* was first coined by Leonhard Euler (see Figure 2.6.1) in 1756. This kind of calculus introduces a special derivative, the variational derivative. We call this derivative the Euler derivative in honor of Euler's achievements in this field. He used it to describe a new method in mechanics which Lagrange had developed a year earlier. Thus, the original application of the Euler derivative originates from mechanics. In this context, Euler and Lagrange used this derivative to write down their famous equations, the Euler–Lagrange equations. Up to now, the main application of this derivative in physics has been the formulation of dynamical equations. Before we discuss the Euler derivative and its implementation, we briefly recall the basic properties of the origin in the calculus of variations.

Figure 2.6.1.   Leonhard Euler (born April 15, 1707, died September 18, 1783) was Switzerland's foremost scientist. He was perhaps the most prolific author of all time in any field. From 1727 to 1783, his writings poured out in a seemingly endless flood, constantly adding knowledge to every known branch of pure and applied mathematics, and also to many that were not known until he created them. Euler was a native of Basel and a student of Johann Bernoulli.

The calculus of variations was first used by Johann Bernoulli in July 1696 when he presented the *brachystochrone problem*. The problem can be formulated as follows: A point mass is moving frictionless in a homogenous force field along a path joining two points. The question is, Which curve connects the two points for the shortest travel? Johann Bernoulli announced the solution of the problem, but did not present his findings in public. He preferred to first challenge his contemporaries to also examine the problem. This challenge was particularly aimed at his brother and teacher Jakob Bernoulli, who was his bitter enemy. Jakob found one solution but did not present it to Johann. It was only upon the intervention of Leibniz, with whom Jakob had a lifelong friendship and a scientific correspondence, that he sent the solution to his brother in May 1697. The most fascinating event was that this solution was a cycloid, a curve also discovered at this time.

2.6.2 The Problem of Variations

As mentioned, the main idea in the calculus of variations arose from the work of Euler and Lagrange. Later, Hamilton contributed the term *minimum principle* to the theory, which is still in use today. The main idea of all these considerations of Euler, Lagrange, and Hamilton is the assumption that there exists a generating functional F. This functional F is responsible for the dynamical development of the motion. The key point in the calculus of variations is to find a function which makes the functional F an extremum. The solution of this issue is to vary the function by introducing a test function. Thus, the variation of F is actually carried out by replacing the function u by a slightly changed new function $u + \epsilon w$, where ϵ is a small parameter and w denotes an arbitrary test function. After replacing u and all of its higher derivatives in the functional F, we have to determine the extreme values of F. The functional in this representation can be considered as a function of the parameter ϵ. The extreme values of F are found if we use the standard procedure of calculus for finding extremums. In mathematical terms, we need to calculate the derivative of F with respect to ϵ under the condition that ϵ vanishes:

$$\frac{dF(\epsilon)}{d\epsilon}\bigg|_{\epsilon=0} = 0. \tag{2.6.1}$$

The basic problem of the calculus of variations is to determine a function $u(x)$ such that the integral

$$\begin{aligned} F[u] &= \int_{x_1}^{x_2} f(x, u, u_x, \dots)\, dx \\ &= \int_{x_1}^{x_2} f(x, u_{(k)})\, dx, \qquad k = 1, 2, \dots, \end{aligned} \tag{2.6.2}$$

assumes an extreme. $f(x, u, u_x, \dots)$ is known as the density of the functional F. An extremum here is either a maximum or a minimum. In Equation (2.6.2), $u_x = \partial u / \partial x$ denotes the partial derivative of u with respect to the independent variables x, where x is a vector of coordinates. Let us assume first that we have only one independent variable x. This assumption will make it easier to represent and discuss the theory. A generalization to more independent variables will be given next.

The expression $F[u]$ given in Equation (2.6.2) is called a functional defined by an integral over a density f which depends on the independent variable x and the unknown function u. In general, this density may also depend on derivatives of u up to a certain order k, denoted by $u_{(k)}$. The limits in the integral (2.6.2) are assumed to be fixed. We note that fixed limits are not necessary. If they are allowed to vary, the problem increases in such a way that not only $u(x)$ but also x_1 and x_2 are needed to bring F to an extreme value. The question is one of how to manage the functional F in becoming an extremum. Let us assume that an extremum of F exists if a function $u = u(x)$ makes the functional F a minimum. Then, any neighboring function, no matter how close it approaches $u(x)$, must make F increase. The definition of a neighboring or test function may be as follows. We introduce a parametric representation of $u = u(x; \epsilon)$ in such a way that for $\epsilon = 0$ and $u = u(x; \epsilon = 0) = u(x)$, we get the identity and the functional yields an extremum. We write the small perturbation of u as

$$u(x; \epsilon) = u(x; 0) + \epsilon w(x), \tag{2.6.3}$$

where $w(x)$ is the test function which has continuous derivatives and vanishes at the endpoints x_1 and x_2. We note that the vanishing of $w(x)$ at x_1 and x_2 $w(x_1) = w(x_2) = 0$ is one of the basic assumptions of the calculus of variations. The above considerations are graphically represented in Figure 2.6.2.

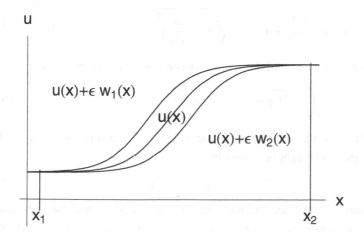

If functions of the type given in Equation (2.6.3) are considered as
variations of u, the functional F becomes a function of ϵ:

$$F[u;\epsilon] = \int_{x_1}^{x_2} f(x, u(x; \epsilon), u_x(x, \epsilon), \ldots)\, dx. \tag{2.6.4}$$

The condition that the integral has a stationary value (in other words, an
extremum) is that F be independent of ϵ in first order. This means that

$$\left. \frac{\partial F}{\partial \epsilon} \right|_{\epsilon=0} = 0 \tag{2.6.5}$$

for all functions $w(x)$. This is a necessary condition but not a sufficient
one. We will not pursue the details of the sufficient conditions here. They
were extensively discussed by Blanchard and Brüning [2.11]. To
demonstrate how these formulas work in detail, let us consider the simple
example of the shortest connection between two points in an Euclidean
plane.

Example 1: Shortest Connection

Let us consider the equation of a curve in an Euclidean space which yields
the shortest distance between two points in the plane. The geometrical
increment of distance ds in the (u, x)-plane is given by

$$ds = \sqrt{dx^2 + du^2} = \sqrt{1 + \left(\frac{du}{dx}\right)^2}\, dx. \tag{2.6.6}$$

The total length s of the curve between two points x_1 and x_2 is

$$s = \int_{x_1}^{x_2} \sqrt{1 + u_x^2}\, dx \equiv F[u]. \tag{2.6.7}$$

We know that the shortest connection between two points in the Euclidean
plane is a straight line given by

$$u(x) = \alpha\, x + \beta, \tag{2.6.8}$$

where α and β are constants determining the slope and the intersection of
the line with the ordinate. Now, let us consider the line in the range $x \in [0,
2\pi]$. To demonstrate the numerical behavior of the functional F, we choose

a special test function $w(x) = \sin(4\,x)$. Using our representation of u given by Equation (2.6.8) with $\alpha=1$ and $\beta=0$ for example, we get for the derivative of u,

$$u_x = 1 + 4\,\epsilon\cos(4\,x). \tag{2.6.9}$$

Inserting this representation into Equation (2.6.7), we find

$$F[\epsilon] = \int_0^{2\pi} \sqrt{1 + 4\,\epsilon\cos(4\,x)}\; dx. \tag{2.6.10}$$

This relation represents our specific functional, now a function solely of ϵ. We are looking for the minimum of this function to get the extremum of the functional. Considered as a function of ϵ, this relation cannot be explicitly solved for ϵ. However, to get an idea of the dependence on the parameter ϵ, we can use *Mathematica*. If we define Equation (2.6.10) as a function depending on ϵ, we can use the numerical capabilities of *Mathematica* to graphically represent the dependence of F on ϵ. First, let us define Equation (2.6.10) by

```
F[ε_] := NIntegrate[
        √(1 + (1 + 4 ε Cos[4 x])²), {x, 0, 2 π}]
```

We then use the defined function **F[]** in connection with **Plot[]** to represent the value of the functional for certain values of ϵ:

```
Plot[Evaluate[F[e]], {e, -1, 1}, AxesLabel → {"e", "F"},
    PlotStyle → RGBColor[1, 0, 0]];
```

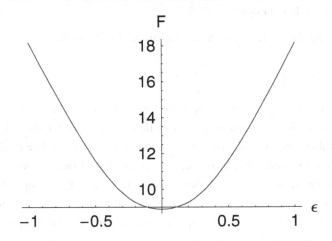

The result of our calculation shows that the value of the functional is minimal for ε=0 and increases for all other values of ε. Thus, we demonstrated numerically that the minimum of the functional exists. In a second plot, we demonstrate the influence of ε on the function $u(x) = x$ for different values of ε. This shows us that the value of $F[u; ε]$ is always greater than $F[u; 0]$, no matter which value (positive or negative) is chosen for ε.

```
Plot[Evaluate[
        {y[x, 0], y[x, 1], y[x, -1/2]} /.
            y → Function[{x, ε}, x + ε Sin[4 x]]],
    {x, 0, 2 π},
    AxesLabel → {"x", "y"},
    PlotRange → All,
    PlotStyle → {RGBColor[0, 0, 0.996109],
            RGBColor[1.000, 0.000, 0.000],
            RGBColor[0.000, 0.251, 0.251]}];
```

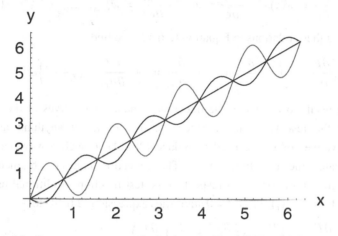

From this figure, we can conclude that the line $u(x) = x$ is one realization of the shortest connection between two points in the Euclidean plane.

2.6.3 Euler's Equation

In this section, we derive the analytical representation of the Euler derivative. The construction of this sort of derivative is based on condition (2.6.5). If we carry out the differentiation with respect to ϵ, Equation (2.6.4) will provide

$$\frac{\partial F}{\partial \epsilon} = \frac{\partial}{\partial \epsilon} \int_{x_1}^{x_2} f(x, u, u_x, \ldots) \, dx. \tag{2.6.11}$$

Since the limits of the integral are fixed, the differentiation affects only the density of the functional F. Hence,

$$\frac{\partial F}{\partial \epsilon} = \int_{x_1}^{x_2} \left(\frac{\partial f}{\partial u} \frac{\partial u}{\partial \epsilon} + \frac{\partial f}{\partial u_x} \frac{\partial u_x}{\partial \epsilon} + \frac{\partial f}{\partial u_{x,x}} \frac{\partial u_{x,x}}{\partial \epsilon} + \cdots \right) dx. \tag{2.6.12}$$

If we now use the representation of $u = u(x; \epsilon)$ as given in Equation (2.6.3) to introduce the ϵ dependence for the variable u and the derivatives $u_{(k)}$, we get

$$\frac{\partial u}{\partial e} = w(x), \qquad \frac{\partial u_x}{\partial \epsilon} = w_x, \qquad \frac{\partial u_{x,x}}{\partial \epsilon} = w_{x,x,x}, \quad \dots \tag{2.6.13}$$

Using these relations in Equation (2.6.12), we find

$$\frac{\partial F}{\partial \epsilon} = \int_{x_1}^{x_2} \left(\frac{\partial f}{\partial u} w(x) + \frac{\partial f}{\partial u_x} w_x + \frac{\partial f}{\partial u_{x,x}} w_{x,x} + \cdots \right) dx. \tag{2.6.14}$$

The result so far is that the integrand contains derivatives of the density f and the test function w. Since we do not know anything about the derivatives of w, we need to reduce (2.6.14) in such a way that it only contains the test function w. The reduction can be obtained by an integration of parts with respect to the test function. Additional use of the conditions $w(x_1) = w(x_2) = 0$ simplifies expression (2.6.14) to

$$\frac{\partial F}{\partial \epsilon} = \int_{x_1}^{x_2} w(x) \left(\frac{\partial f}{\partial u} - \frac{d}{dx} \left(\frac{\partial f}{\partial u_x} \right) \right.$$
$$\left. + \frac{d^2}{dx^2} \left(\frac{\partial f}{\partial u_{x,x}} \right) \mp \cdots \right) dx. \tag{2.6.15}$$

The integral in Equation (2.6.15) seems to be independent of ϵ. However, the function $u = u(x; \epsilon)$ and all derivatives of u are still functions of ϵ. We know from the representation of $u(x; \epsilon)$ that this dependency disappears if we set $\epsilon = 0$. Before we start this calculation, we generalize Equation (2.6.15) to arbitrary orders in the derivatives:

$$\frac{\partial F}{\partial \epsilon} = \int_{x_{12}} w(x) \left(\sum_{n=0}^{\infty} (-1)^n \frac{d^n}{dx^n} \left(\frac{\partial f}{\partial u_{(n)}} \right) \right) dx, \tag{2.6.16}$$

where $u_{(n)} = \partial^n u / \partial x^n$ denotes the nth derivative of u with respect to x. Our aim was to find the extremum of F. A necessary condition for the existence of an extremum is the vanishing of the derivative $\partial F / \partial \epsilon |_{\epsilon=0} = 0$. In our calculations, we assumed that w is an arbitrary function. Thus, the derivative of F can only vanish if the integrand vanishes and so we end up with the result

$$\sum_{n=0}^{\infty} (-1)^n \frac{d^n}{d x^n} \left(\frac{\partial f}{\partial u_{(n)}} \right) = 0, \tag{2.6.17}$$

where u and all the derivatives of u are now independent of ϵ. This result is known as Euler's equation and it is a necessary condition for the functional F to allow an extremum. The Euler equation is reduced to the well-known Euler–Lagrange equation if we restrict the order of the derivatives to 2. Since the Euler equation is needed in the derivation of equations of motion, we define a special symbol for this operation and call it the Euler operator.

2.6.4 Euler Operator

The Euler operator is also known as a variational derivative in the field of dynamical formulations or statistical mechanics. In this subsection, we define this operator as a special type of derivative.

Definition: Euler Operator

Let $f = f(x, u, u_x, \ldots)$ be the density of a functional $F[u]$. Then we call

$$\frac{\delta F}{\delta u} := \sum_{n=0}^{\infty} (-1)^n \frac{d^n}{dx^n} \left(\frac{\partial f}{\partial u_{(n)}} \right)$$

the functional derivative of F and

$$\mathcal{E} := \sum_{n=0}^{\infty} (-1)^n D_n \frac{\partial}{\partial u_{(n)}}$$

an Euler operator. $D_n = d^n / dx^n$ denotes the nth-order total derivative.

The actual information of this definition is that the functional derivative $\delta F / \delta u$ can be replaced by ordinary and partial derivatives if we know the density of the functional F. Consequently, we can introduce a general derivative, the Euler operator, which is based on known operations. The essential content of the above definition is that knowing the density f of a functional F is sufficient to calculate the corresponding functional derivative. The functional derivative follows just by differentiation of the density f. An additional merit is the knowledge of the Euler equation for this functional F. The above definition is a result of the calculus of variations. Thus, the Euler derivative can be calculated by an algorithmic procedure.

2.6.5 Algorithm Used in the Calculus of Variations

Our next goal is to define a *Mathematica* function allowing the calculation of the Euler derivative. Before we present the function, we briefly repeat the main steps of the calculus of variations. These steps are intimately related to the definition of the Euler derivative and are thus the basis of the calculation. The four main steps of the algorithm are as follows:

1. Replacement of the dependent function u by its variation
$$u = u + \epsilon w.$$

2. Differentiation of the functional density with respect to the parameter ϵ and replacement of ϵ by zero after the differentiation.

3. Use the boundary conditions for the test function to eliminate the derivatives in w.

4. The coefficient of the test function w delivers the Euler equation.

These four steps define the calculation of the Euler derivative algorithmically. The function defined in *Mathematica* is based on these four steps. When looking at the definition of the Euler derivative $\mathcal{E}$, we realize that we need at least three pieces of information to carry out the calculation. First, we should know the density of the functional F, second the dependent variable, and third the name of the independent variable. From our discussions of the algorithm, we expect that the highest order of differentiation should be determined by the function itself. Thus, we define the function **EulerLagrange[]** with three necessary arguments. A fourth optional argument allows influencing the representation of the result of the function. The following lines contain the definitions for **EulerLagrange[]**:

```
(* --- Euler derivative for ---*)
(* --- one dependent
    and one independent variable ---*)
Clear[EulerLagrange];
Options[EulerLagrange] = {eXpand → False};


EulerLagrange[density_, depend_, independ_,
    options___] :=
    Block[{f0, rule, fh, ε, w, y, expand},
        (*--- check options ---*)
        {expand} = {eXpand} /. {options} /.
                    Options[EulerD];
        (*--- rule for the variation of u---*)
        f0 = Function[x, y[x] + ε w[x]];
        (*--- rule for the replacement of
                    derivatives of w --- *)
        rule = b_. w^(n_) [independ] :→
                    (-1)^n HoldForm[∂_{independ,n} b];
        (*--- step of variation ---*)
        fh = density /. depend → f0 /.
                    {x → independ, y → depend};
        (*--- differentiation
    with respect to ε ---*)
        fh = Expand[∂_ε fh /. ε → 0];
        (*--- transformation to w ---*)
        fh = fh /. rule /. w[independ] → 1;
        (*---- Euler equations --- *)
        If[expand, fh = ReleaseHold[fh], fh]]
```

This function is part of the *Mathematica* package EulerLagrange. The package also contains functions for larger numbers of independent and dependent variables. To make the use of the Euler operator more convenient, we also defined a single symbol for the Euler operator. This symbol looks like $\mathcal{E}_u^x[f]$, where u denotes the dependent variables, x the independent variables, and f the density of the functional. The symbol is available by a function button which can be generated by the following pattern by using the menu command File+Generate Palette from Selection.

$$\mathcal{E}_{\square}^{\square}[\square]$$

Using the function **EulerLagrange[]** or its equivalent operator $\mathcal{E}$, it is straightforward to calculate the functional derivative of any density containing one dependent and one independent variable. We demonstrate the application of this function by discussing the famous brachystochrone problem already mentioned earlier.

Example 1: Brachystochrone

Let us discuss the classical problem of the brachystochrone solved by Johann Bernoulli in 1696. The physical content of this famous problem is the following: Consider a particle moving in a constant force field. The particle with mass m starts at rest from some higher point in the force field and moves to some lower point. The question is, Which path is selected by the particle to finish the transit in the least possible time? Let us reduce the problem to the point of deriving the Euler equation. The dimensionless functional density governing the movement of the particle can be derived from the integral $t = \int_{p_1}^{p_2} 1/v\,ds$, where t is time, ds is the line element, and v is the velocity. Expressing the line element and the velocity in cartesian coordinates, we can express the density of the functional by

$$f(x, u, u_x) = \left(\frac{1 + u_x^2}{2\,g\,x} \right)^{1/2}, \tag{2.6.18}$$

where u describes the horizontal coordinate and x the vertical one. The application of our function **EulerLagrane[]** to this functional density

$$f = \sqrt{\frac{1 + (\partial_x u[x])^2}{2\,g\,x}}$$

$$\frac{\sqrt{\frac{u'(x)^2 + 1}{g\,x}}}{\sqrt{2}}$$

gives us by applying the Euler operator to the density f a second-order nonlinear ordinary differential equation for the variable u.

---

brachystochroneEquation = Simplify[PowerExpand[$\mathcal{E}_u^x[f]$]]

$$\frac{u'(x)^3 + u'(x) - 2 x u''(x)}{2 \sqrt{2} \sqrt{g} \ x^{3/2} (u'(x)^2 + 1)^{3/2}} == 0$$

This equation of motion determines the movement of the particle. To understand how this equation is generated, we recall Euler's equation for the specific density f Equation (2.6.18) by

$$\frac{\partial f}{\partial u} - \frac{d}{dx} \left(\frac{\partial f}{\partial u_x} \right) == 0. \tag{2.6.19}$$

Since the density f does not depend on u but only on u_x, Euler's equation reduces simply to

$$\frac{d}{dx} \left(\frac{\partial f}{\partial u_x} \right) == 0. \tag{2.6.20}$$

However, relation (2.6.20) indicates that the expression $(\partial f / \partial u_x)$ is a constant with respect to x. On the other hand, this means that our derived second-order nonlinear ordinary differential equation brachystochroneEquation can be integrated once. If we start the integration we fail to get a satisfying result

---

Integrate[brEquation[[1]], x]

$$\frac{\int \frac{u'(x)^3 + u'(x) - 2 x u''(x)}{x^{3/2} (u'(x)^2 + 1)^{3/2}} \, dx}{2 \sqrt{2} \sqrt{g}}$$

---

meaning *Mathematica* is, at the moment, unable to find first integrals of a given second-order ordinary differential equation. However, we know that the a first integral exists which, we denote by $(4 a g)^{-1/2}$ and represent as

```
brachystochroneEquation2 =
Simplify[PowerExpand[∂_∂ₓu[x] (f)]] ==  1 / √(4 a g)
```

$$\frac{u'(x)}{\sqrt{2}\,\sqrt{g}\,\sqrt{x}\,\sqrt{u'(x)^2+1}} == \frac{1}{2\sqrt{a\,g}}$$

Squaring both sides of this equation, we can derive a differential equation which can be solved by integration:

```
dth = Thread[brachystochroneEquation2², Equal]
```

$$\frac{u'(x)^2}{2\,g\,x\,(u'(x)^2+1)} == \frac{1}{4\,a\,g}$$

Solution with respect to first-order derivatives gives an first-order ordinary differential equation which can be solved by separation of variables.

```
dthh = Solve[dth,u'[x]];dthh/.Rule->Equal
```

$$\left(\begin{array}{c} u'(x) == -\dfrac{\sqrt{x}}{\sqrt{2\,a-x}} \\[2ex] u'(x) == \dfrac{\sqrt{x}}{\sqrt{2\,a-x}} \end{array}\right)$$

In the following calculation, we use the second equation, which can be formally integrated to

$$u = \int \frac{\sqrt{x}}{\sqrt{2\,a-x}}\,dx. \tag{2.6.21}$$

The integrand of this relation is represented by

```
int = u'[x] /. dthh[[2]]
```

$$\frac{\sqrt{x}}{\sqrt{2a-x}}$$

The derived expression represents the integrand of the action integral. We simplify the integrand to a more manageable form for *Mathematica* by substituting

```
subst1 = x→a(1-Cos[θ]);
```

The differential dx is replaced by the new differential θ multiplied by a factor.

```
dx = ∂_θ (x /. subst1)
```

$a \sin(\theta)$

The integrand in the updated variables is given by

```
ints = dx int/.subst1
```

$$\frac{a \sqrt{a(1-\cos(\theta))} \sin(\theta)}{\sqrt{2a-a(1-\cos(\theta))}}$$

This expression is simplified by the following chain of functions to

```
ints = √ints² // PowerExpand // Simplify // PowerExpand
```

$$2a \sin^2\left(\frac{\theta}{2}\right)$$

which can easily be integrated with the result

```
u = ∫ ints dθ // Simplify
```

$a\,(\theta - \sin(\theta))$

We now know that the path between p_1 and p_2 in a parametric form given by the coordinates x and y depends on θ. x and y describe the fastest connection between two points in a homogeneous force field. Parameter a contained in the above representation has to be adjusted so that the path of the particle passes point p_2. The curve derived is known as a cycloid.

```
curve = {u,-x/.subst1}
```

$\{a\,(\theta - \sin(\theta)),\ -a\,(1 - \cos(\theta))\}$

A parametric representation of the solution for different parameters a is created by the function **ParametericPlot[]** and given as follows:

```
k1 = curve /. a→1;
k2 = curve /. a→2;
k3 = curve /. a→1.25;
k4 = curve /. a→1.5;

ParametricPlot[{k1,k2,k3,k4},{θ,0,2π},AxesLabel->{"u"
,"x"}];
```

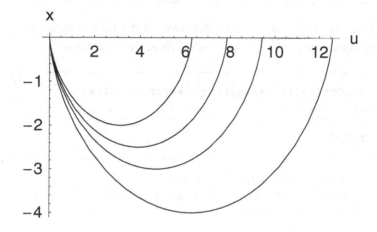

Example 2: Mechanical System

Another example of the application of the function **EulerLagrange[]** is the derivation of the Euler–Lagrange equation for a mechanical system with one degree of freedom. For a detailed discussion of the Euler–Lagrange equation, see Section 2.7. The functional density for such a problem is generally given by the Lagrange function $\mathcal{L}$:

$\mathcal{L} = \text{l}[\text{t, q[t], q'[t]}]$
$l(t, q(t), q'(t))$

where q denotes the generalized coordinate of the particle and t denotes the time. The Euler–Lagrange equation for the general Lagrangian then follows by

```
SetOptions[EulerLagrange, eXpand → True];
```

$$\mathcal{E}_q^t[\mathcal{L}]$$

$$l^{(0,1,0)}(t, q(t), q'(t)) - \frac{\partial l^{(0,0,1)}(t, q(t), q'(t))}{\partial t} == 0$$

If we are interested in the explicit form of the Euler–Lagrange equation, we can set the option *eXpand→False*. Then, the result reads

```
SetOptions[EulerLagrange, eXpand → False];
```

$$\mathcal{E}_q^t[\mathcal{L}]$$

$$-q''(t)\, l^{(0,0,2)}(t, q(t), q'(t)) + l^{(0,1,0)}(t, q(t), q'(t)) - q'(t)\, l^{(0,1,1)}(t, q(t), q'(t)) - l^{(1,0,1)}(t, q(t), q'(t)) == 0$$

This equation is the general representation of the Euler–Lagrange equation.

The Euler operator defined earlier was the result of the variation of a functional. We demonstrated the calculation for a single dependent variable $u = u(x)$ which was a function of one independent variable x. The generic case in applications is more complex. We rarely find systems with only one dependent variable. Thus, we need a generalization of the formulation considering more than one dependent variable in the functional F. In the following exposition, we assume that a set of q dependent variables u^α exists. The functional F for such a case is represented by

$$F[u^1, u^2, u^3, \ldots] = \int_{x_1}^{x_2} f(x, u^1, \ldots, u_x^1, \ldots)\, dx. \tag{2.6.22}$$

The variation of the dependent variables is now performed by introducing a set of test functions w^α. Using this set of auxiliary functions, we can represent the variation by

$$u^\alpha (x; \epsilon) = u^\alpha (x; 0) + \epsilon w^\alpha (x), \quad \alpha = 1, 2, 3, \ldots, q. \tag{2.6.23}$$

The derivation of the Euler operator proceeds in exactly the same way as presented earlier. We skip the detailed calculations and present only the result:

$$\frac{\partial F}{\partial \epsilon} = \int_{x_1}^{x_2} \sum_{\alpha=1}^{q} \{ \sum_{n=1}^{\infty} (-1)^n D_{(n)} \frac{\partial f}{\partial u_{(n)}^\alpha} \} w^\alpha (x) \, dx. \tag{2.6.24}$$

Since the individual variations $w^\alpha (x)$ are all independent of each other, the vanishing of Equation (2.6.24) when evaluated at $\epsilon = 0$ requires the separate vanishing of each expression in curly brackets. Thus, we again can define an Euler operator for each of the q dependent variables u^α.

2.6.6 Euler Operator for q Dependent Variables

In this subsection, we extend the definition of the Euler derivative to a set of q dependent variables. Let $f = f(x, u^1, u^2, \ldots, u_x^1, u_x^2, \ldots)$ be the density of the functional $F[u^1, u^2, \ldots]$. Then, we define the Euler operator $\mathcal{E}_\alpha$ as

$$\mathcal{E}_\alpha := \sum_{n=0}^{\infty} (-1)^n D_{(n)} \frac{\partial}{\partial u_{(n)}^\alpha}, \quad \alpha = 1, 2, \ldots, q, \tag{2.6.25}$$

which will give us the αth Euler equation when applied to the density f:

$$\mathcal{E}_\alpha f = 0. \tag{2.6.26}$$

The only difference between this definition and the definition for a single variable is the number of equations contained in Equation (2.6.26). The occurrence of the q equations in the theoretical formulas must now be incorporated in our *Mathematica* definition for the Euler derivative **EulerLagrange[]**. The theoretical definition (2.6.25) only alters our *Mathematica* function in a way that, for several dependent variables, a set of Euler equations results. Thus, we change our *Mathematica* function in such a way that all dependent variables are taken into account in the application of the $\mathcal{E}_\alpha$ operator. We realize this by including a loop scanning the input list of the dependent variables. The code of this generalized Euler operator is

```
EulerLagrange[density_, depend_List,
    independ_, options___] :=
  Block[{f0, fh, e, w, y, expand,
         euler = {}, wtable},
      {expand} = {eXpand} /. {options} /.
              Options[EulerD];
      wtable = Table[w[i],
                {i, 1, Length[depend]}];
      f0 = Function[x, y[x] + e w[x]];
      rules[i_] :=
          b_. wtable[[i]]^(n_) [independ] :>
              (-1)^n HoldForm[∂_{independ,n} b];
      Do[
          fh = density /. depend[[j]] -> f0 /.
                  {x -> independ, y -> depend[[j]],
                      w -> wtable[[j]]};
          fh = Expand[∂_e fh /. e -> 0];
          fh = fh /. rules[j] /.
                  wtable[[j]][independ] -> 1;
          AppendTo[euler, fh],
          {j, 1, Length[depend]}];
      If[expand,
          euler = ReleaseHold[euler],
          euler]]
```

Let us demonstrate the application of this function by two examples.

Example 1: Two-Dimensional Oscillator System

Assume that we know the functional density of a two-dimensional oscillator system. Let us further assume that the two coordinates of the oscillators are coupled by a product. We expect that the two equations of motion follow by applying the Euler derivative. The Lagrange density of the system reads

```
Clear[u]
```

$$1 = u[t] \, v[t] + (\partial_t u[t])^2 + (\partial_t v[t])^2 - u[t]^2 - v[t]^2$$

$$-u(t)^2 + v(t)\,u(t) - v(t)^2 + u'(t)^2 + v'(t)^2$$

The corresponding system of second-order equations follows by

$$\mathcal{E}^t_{\{u,v\}}[l]$$

$$\{-2\,u(t) + v(t) - 2\,u''(t) == 0, \; u(t) - 2\,v(t) - 2\,v''(t) == 0\}$$

Note that we used the same name, **EulerLagrange[]**, for the operators $\mathcal{E}$ and $\mathcal{E}_\alpha$. This sort of definition is possible and provides a great flexibility in the application of a single symbol for different operations. *Mathematica* is able to distinguish the two different functions by the different arguments.

Example 2: Two-Dimensional Lagrangian

Another example for a two-dimensional Lagrangian is given by the function

$$f = u[t] \, v[t] + (\partial_t u[t])^2 + (\partial_t v[t])^2 + 2\,\partial_t u[t] \, \partial_t v[t]$$

$$u'(t)^2 + 2\,v'(t)\,u'(t) + v'(t)^2 + u(t)\,v(t)$$

This density is a special model of a Dirac Lagrangian containing the derivatives with respect to time as a binomial. The corresponding Euler–Lagrange equations read

$$\mathcal{E}^t_{\{u,v\}}[f]$$

$$\{v(t) - 2\,u''(t) - 2\,v''(t) == 0, \; u(t) - 2\,u''(t) - 2\,v''(t) == 0\}$$

representing a coupled system of second-order ordinary differential equations.

So far, we are able to handle point systems depending on one independent variable. However, equations occurring in real situations depend on more than one independent variable. Thus, we need a generalization of our Euler derivative to more than one independent variable. In fact, the definitions of an Euler operator can be extended from the $q+1$-dimensional case to the $q + p$-dimensional case. We define this operator in the following section.

2.6.7 Euler Operator for $q + p$ Dimensions

Here, we will discuss the general definition of an Euler operator. This sort of operator, for example, is used to write down field equations such as Maxwell's equations, Schrödinger's equation, Euler's equation in hydrodynamics, and many others.

Definition: (q, p)-Dimensional Euler Operator

Let $f = f(x, u_{(n)})$ be the density of the functional $F[u]$ with $x = (x^1, x^2, ..., x^p)$, and $u = (u^1, u^2, ..., u^q)$ be the p- and q-dimensional vectors of the independent and dependent variables, respectively. By $u_{(n)}$ we denote all the derivatives with respect to the independent variables. We call

$$\mathcal{E}_\alpha = \sum_J (-D)_J \frac{\partial}{\partial u_J^\alpha} \tag{2.6.27}$$

the general Euler operator in q dependent and p independent variables. J is a multi-index $J = (j_1, ..., j_k)$ with $1 \le j_k \le p, k \ge 0$. ∎

Since the functional densities f depend on a finite number of derivatives u_J^α, the infinite sum in Equation (2.6.27) is terminated at this upper limit. Again, the Euler equations for a given functional $F[u]$ follow from the application of $\mathcal{E}_\alpha$ to F:

$$\mathcal{E}_\alpha F = 0 , \quad \alpha = 1, 2, ..., q. \tag{2.6.28}$$

From a theoretical point of view, we know the general Euler operator. Our next step is to make this operation available in *Mathematica*. We define the generalized Euler operator by taking into account the different independent variables. The corresponding definition of **EulerLagrange[]** for $q + p$ dimensions is given by

```
EulerLagrange[density_, depend_List,
    independ_List, options___] :=
Block[{f0, fh, e, w, y, x$m, expand,
        euler = {}, wtable},
    {expand} = {eXpand} /. {options} /.
            Options[EulerD];
    wtable = Table[w[i],
            {i, 1, Length[depend]}];
    f0 = Function[x$m, y + e w];
    ruleg[i_] :=
        b_ . wtable[[i]]^(n__) @@ independ :>
            (-1)^Plus@@{n}
HoldForm[∂_Delete[Thread[{independ,{n}}],0] b];
    Do[
        fh = density /. depend[[j]] -> f0 /.
                {x$m -> independ,
                    y ->
        depend[[j]] @@ independ,
                    w -> wtable[[j]] @@ independ};
        fh = Expand[∂_e fh /. e -> 0];
        fh = fh /. ruleg[j] /.
                wtable[[j]] @@ independ -> 1;
        AppendTo[euler, fh],
        {j, 1, Length[depend]}];
    If[Not[expand],
        euler = ReleaseHold[euler],
        euler]]
```

We demonstrate the application of the function **EulerLagrange[]** to the wave equation in 2+1 dimensions and to a system of coupled nonlinear diffusion equations.

Example 1: Quadratic Density

Let us consider a functional in $q = 1$ and $p = 3$ variables and assume that the density is quadratic in the derivatives given by

$$F[u] = \frac{1}{2} \int (u_{x_1}^2(x_1, x_2, x_3) - u_{x_2}^2 - u_{x_3}^2)\, dx_1\, dx_2\, dx_3. \qquad (2.6.29)$$

Calculating the variational derivative, we immediately find that the Euler equations are given by the Laplace equation

$$-u_{x_1 x_1} + u_{x_2, x_2} + u_{x_3, x_3} = 0. \qquad (2.6.30)$$

Using the generalized definition of **EulerLagrange[]**, we can reconstruct the result of our pencil calculation. First, let us define the density by

$$f = \frac{1}{2}\, ((\partial_{x1} u[x1,\, x2,\, x3])^2 -$$
$$(\partial_{x2} u[x1,\, x2,\, x3])^2 - (\partial_{x3} u[x1,\, x2,\, x3])^2)$$

$$\frac{1}{2}\left(-u^{(0,0,1)}(x1,\, x2,\, x3)^2 - u^{(0,1,0)}(x1,\, x2,\, x3)^2 + u^{(1,0,0)}(x1,\, x2,\, x3)^2\right)$$

The application of the Euler operator to f gives

$$\mathbf{wave} = \mathcal{E}_{\{u\}}^{\{x1,x2,x3\}}[f]$$

$$\{u^{(0,0,2)}(x1,\, x2,\, x3) + u^{(0,2,0)}(x1,\, x2,\, x3) - u^{(2,0,0)}(x1,\, x2,\, x3) == 0\}$$

The resulting equation is known as the wave equation in $2 + 1$ dimensions.

Example 2: Diffusion of Two Components

In this example, we will consider a system in two field variables ($q = 2$) and two independent variables ($p = 2$). The physical background of this model is the diffusion of two components in a nonlinear medium. The Lagrange density of this field model has the representation

$$1 = v[x, t] \, \partial_t \, u[x, t] + \partial_x \, u[x, t] \, \partial_x \, v[x, t] + u[x, t]^2 \, v[x, t]^2$$

$$u(x, t)^2 \, v(x, t)^2 + u^{(0,1)}(x, t) \, v(x, t) + u^{(1,0)}(x, t) \, v^{(1,0)}(x, t)$$

The related equations of motion follow by

$$\text{cnondiffu} = \text{TableForm}\big[\mathcal{E}^{\{x,t\}}_{\{u,v\}}[l]\big]$$

$$2 \, u(x, t) \, v(x, t)^2 - v^{(0,1)}(x, t) - v^{(2,0)}(x, t) == 0$$
$$2 \, v(x, t) \, u(x, t)^2 + u^{(0,1)}(x, t) - u^{(2,0)}(x, t) == 0$$

representing two coupled nonlinear diffusion equations for the variables u and v. The same equations of motion can be derived from the functional l_1 given by

$$l1 = -u[x, t] \, \partial_t \, v[x, t] + \partial_x \, u[x, t] \, \partial_x \, v[x, t] + u[x, t]^2 \, v[x, t]^2$$

$$u(x, t)^2 \, v(x, t)^2 - u(x, t) \, v^{(0,1)}(x, t) + u^{(1,0)}(x, t) \, v^{(1,0)}(x, t)$$

The equations of motion follow then from

$$\text{TableForm}\big[\mathcal{E}^{\{x,t\}}_{\{u,v\}}[l1]\big]$$

$$2 \, u(x, t) \, v(x, t)^2 - v^{(0,1)}(x, t) - v^{(2,0)}(x, t) == 0$$
$$2 \, v(x, t) \, u(x, t)^2 + u^{(0,1)}(x, t) - u^{(2,0)}(x, t) == 0$$

This behavior demonstrates that field equations can be derived from different functionals.

2.6.8 Variations with Constraints

This section deals with the problem of having a standard setup for a problem in the calculus of variations and, in addition, some constraints on the function for which we are looking. For example, we are looking for the shortest connection on a curved surface. The fact that the solution we are looking for is part of the surface can be formulated in a condition such as

$$g(q_i -, t) = 0 \tag{2.6.31}$$

defining the surface itself. For a sphere, the condition g is given by

$$g = q_1^2 + q_2^2 + q_3^2 - r^2 = 0, \tag{2.6.32}$$

where r is the radius of the sphere. We call the functional relation g also a boundary condition for the problem of variation.

For the first approach of boundary conditions involved in a variational problem, let us assume that there exist two coordinates $q_1 = y$ and $q_2 = z$ depending on each other. The functional density depends, in addition to the coordinates q_1 and q_2, on the derivatives of the coordinates with respect to t. The density of the functional then is

$$f(t, q_i, q'_i) = f(t, y, y', z, z'). \tag{2.6.33}$$

The corresponding functional reads

$$F[y, z] = \int_{t_1}^{t_2} f(t, y, y', z, z') \, dt. \tag{2.6.34}$$

If we carry out the variation of the two unknown function y and z, we get

$$
\begin{aligned}
&\frac{\partial F}{\partial \varepsilon}\Big|_{\varepsilon=0} = \\
&\int_{t_1}^{t_2} \Big\{ \Big(\frac{\partial f}{\partial (y + \varepsilon W_1)} - \frac{d}{dt} \Big(\frac{\partial f}{\partial (y' + \varepsilon W'_1)} \Big) \Big) \frac{\partial (y + \varepsilon W_1)}{\partial \varepsilon} + \\
&\Big(\frac{\partial f}{\partial (z + \varepsilon W_2)} - \frac{d}{dt} \Big(\frac{\partial f}{\partial (z' + \varepsilon W'_2)} \Big) \Big) \frac{\partial (z + \varepsilon W_2)}{\partial \varepsilon} \Big\} \, dt \Big|_{\varepsilon=0}.
\end{aligned}
\tag{2.6.35}
$$

In addition, we have the boundary condition in the form

$$g(t, y, z) = 0. \tag{2.6.36}$$

Applying the variations also to this condition, we find

$$g(t, y + \varepsilon w_1, z + \varepsilon w_2) = 0. \tag{2.6.37}$$

This condition shows that the two independent variations (test functions w_1 and w_2) become dependent on each other. Differentiation g with respect to the parameter ϵ, we find

$$\frac{dg}{d\varepsilon} = \frac{\partial g}{\partial (y + \varepsilon w_1)} \frac{\partial (y + \varepsilon w_1)}{\partial \varepsilon}$$
$$+ \frac{\partial g}{\partial (z + \varepsilon w_2)} \frac{\partial (z + \varepsilon w_2)}{\partial \varepsilon} = 0 \qquad (2.6.38)$$

$$\Longleftrightarrow \frac{\partial g}{\partial (y + \varepsilon w_1)} w_1 + \frac{\partial g}{\partial (z + \varepsilon w_2)} w_2 = 0 \qquad (2.6.39)$$

$$\Longleftrightarrow w_2 = -\left(\frac{\partial g}{\partial \bar{y}}\right) w_1 \frac{1}{(\partial g/\partial \bar{z})} . \qquad (2.6.40)$$

Inserting this result into the functional F we get

$$\frac{\partial F}{\partial \varepsilon}\Big|_{\varepsilon=0} = \int_{t_1}^{t_2} \left\{\left(\frac{\partial f}{\partial y} - \frac{d}{dt}\left(\frac{\partial f}{\partial y'}\right)\right) w_1 \right.$$
$$+ \left(\frac{\partial f}{\partial z} - \frac{d}{dt}\left(\frac{\partial f}{\partial z'}\right)\right) w_2 \Big\} dt$$
$$= \int_{t_1}^{t_2} \left\{\frac{\partial f}{\partial y} - \frac{d}{dt}\left(\frac{\partial f}{\partial y'}\right)\right. \qquad (2.6.41)$$
$$\left. - \left(\frac{\partial f}{\partial z} - \frac{d}{dt}\left(\frac{\partial f}{\partial z'}\right)\right)\left(\frac{\partial g/\partial y}{\partial g/\partial z}\right)\right\} w_i \, dt = 0.$$

Since the w_j are arbitrary, we find

$$\frac{\partial f}{\partial y} - \frac{d}{dt}\left(\frac{\partial f}{\partial y'}\right) = \left(\frac{\partial f}{\partial z} - \frac{d}{dt}\left(\frac{\partial f}{\partial z'}\right)\right) \frac{\partial g/\partial y}{\partial g/\partial z} \qquad (2.6.42)$$

$$\Longleftrightarrow \left(\frac{\partial f}{\partial y} - \frac{d}{dt}\left(\frac{\partial f}{\partial y'}\right)\right) \frac{1}{\frac{\partial g}{\partial y}} = \left(\frac{\partial f}{\partial z} - \frac{d}{dt}\left(\frac{\partial f}{\partial z'}\right)\right) \frac{1}{\partial g/\partial z} . \qquad (2.6.43)$$

Since the left-hand side contains only derivatives of f and g with respect to y and y' and the right-hand side contains only derivatives with respect to z and z', we can separate the relation by introducing a common function λ depending only on the independent variable t. Thus, the resulting determining equations for f and g are

$$\frac{\partial f}{\partial y} - \frac{d}{dt}\left(\frac{\partial f}{\partial y'}\right) + \lambda(t) \frac{\partial g}{\partial y} = 0, \qquad (2.6.44)$$

$$\frac{\partial f}{\partial z} - \frac{d}{dt}\left(\frac{\partial f}{\partial z'}\right) + \lambda(t) \frac{\partial g}{\partial z} = 0. \qquad (2.6.45)$$

The problem is solved if we can determine the three unknown functions $y = y(t)$, $z = z(t)$, and $\lambda = \lambda(t)$. For these three unknowns, we know three equations first the two Euler equations resulting from the functional F (2.6.44) and (2.6.45), second the boundary condition $g = 0$. Thus, we have a sufficient number of equations to determine the unknowns y, z, and λ. λ, the additional unknown, is called a Lagrange multiplier, which Lagrange in 1788 originally introduced in his *Mechanique Analytique*. The generalization from two variable to many variables and many boundary conditions is now obvious. The procedure demonstrated above can be applied to a more complicated problem. The resulting determining equations are

$$\frac{\partial f}{\partial q_i} - \frac{d}{dt}\left(\frac{\partial f}{\partial q'_i}\right) + \sum_{j=1}^{M} \lambda_j(t)\,\frac{\partial g_j}{\partial q} = \; = 0, \tag{2.6.46}$$

$$g_j(q_i, t) = 0, \tag{2.6.47}$$

with $i = 1, 2, ..., N$ and $j = 1, 2, ..., M$. The first equation represents a system of equations consisting of N equations for $N + M$ unknowns. In addition, there exist M boundary conditions which allow a consistent solution of the problem. For $N + M$ unknown functions, there exist $N + M$ equations.

In practical applications, the system of equations $g_j(q_i, t) = 0$ is equivalent to a system of M differential equations

$$\sum_i \frac{\partial g_j}{\partial q_i}\, g\, q_i = 0, \quad i = 1, 2, ..., N, \tag{2.6.48}$$
$$j = 1, 2, ..., M.$$

Mechanical problems are usually formulated in such a way that the M boundary conditions are represented by differential equations.

Example 1: Rolling Wheel on an Inclined Plane

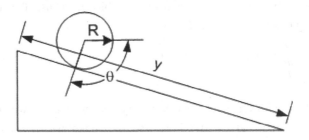

Figure 2.6.3.   On a inclined plane, a wheel is rolling downward without any slip.

Let us consider a rolling wheel on an inclined plane (see Figure 2.6.3). The y coordinate is then given by

$$y = R\Theta, \tag{2.6.49}$$

where R is the radius of the wheel. The boundary condition for the movement is thus

$$g(y, Q) = y - R\Theta = 0$$

and

$$\frac{\partial g}{\partial y} = 1,$$

$$\frac{\partial g}{\partial \Theta} = R$$

are the quantities related to the Lagrange multiplier.

2.6.9 Exercises

1. Show that the shortest distance between two points in three-dimensional space is a straight line.

2. Show that the geodesic on the surface of a right circular cylinder is a helix.

3. Find the dimensions of the parallelepiped of maximum volume that is circumscribed by 1) a sphere of radius R and 2) an ellipsoid with semiaxes a, b, and c.

4. Find the ratio of the radius R to the height h of a right circular cylinder of fixed volume V that will minimize the surface area A.

5. A disk of radius R rolls without slipping inside the parabola $y = a x^2$. Find the equaion of constraint. Express the condition which allows the disk to roll so that it contacts the parabola at one and only one point, independent of its position.

2.6.10 Packages and Programs

EulerLagrange Package

The EulerLagrange package serves to derive the Euler–Lagrange equations from a given Lagrangian.

```
If[$MachineType == "PC",
  $EulerLagrangePath = $TopDirectory <>
    "/AddOns/Applications/EulerLagrange/";
  AppendTo[$Path, $EulerLagrangePath],
  $EulerLagrangePath =
   StringJoin[$HomeDirectory, "/.Mathematica/3.0/
      AddOns/Applications/EulerLagrange", "/"];
  AppendTo[$Path, $EulerLagrangePath]];
```

The next line loads the package.

```
<< EulerLagrange.m
```

```
Notation[$\mathcal{E}_{u\_}^{x\_}$[den_] $\Leftrightarrow$ EulerLagrange[den_, u_, x_]]
```

$$\mathcal{E}_{\square}^{\square}[\square]$$

2.7 Lagrange Dynamics

2.7.1 Introduction

In this chapter, we discuss one of the fundamental principles of classical mechanics – the Lagrangian formulation (see Figure 2.7.1). This formulation also provides the necessary background to learn about the Hamiltonian formulation, which, in turn, provides the natural framework in which to investigate the ideas of integrability and nonintegrability in a wide class of mechanical systems. Many of the differential equations so far discussed describe the motion of a particle moving in some force field and, as such, they are examples of Newtonian equations of motion. Since Newton's work, the Laws of Mechanics have been the subject of ever more general and elegant formulations.

Figure 2.7.1.   Joseph Louis Lagrange born January 25, 1736; died April 10, 1813.

In order to circumvent some of the practical difficulties which arise in attempts to apply Newton's equations to particular problems, alternative procedures can be developed. All such approaches are, in essence, *a posteriori* because it is known beforehand that the result equivalent to the Newtonian equations must be obtained. Thus, in order to effect a simplification, it is not required to formulate a new theory of mechanics —

the Newtonian theory is quite correct — but only to devise an alternative method of dealing with complicated problems in a general manner. Such a method is contained in Hamilton's principle and the equations of motion which result from the application of this principle are called Lagrange's equations.

General equations of motion can be seductively derived by invoking such fundamental principles as the homogeneity of space and time and the use of an almost magical variational principle, Hamilton's principle, to the extent that the resulting laws would appear to have been determined from purely deductive principles. In view of the wide range of applicability that Hamilton's principle has been found to possess, it is not unreasonable to assert that Hamilton's principle is more fundamental than are Newton's equations. Therefore, we will proceed by first postulating Hamilton's principle; we will then obtain Lagrange's equation and show that these are equivalent to Newton's equations.

2.7.2 Hamilton's Principle Historical Remarks

Minimal principles in physics have a long and interesting history. The search for such principles is predicated on the notion that Nature always acts in such a way that certain important quantities are minimized when a physical process takes place. The first such minimum principles were developed in the field of optics.

Hero of Alexandria, in the second century BC, found that the reflection of light is based on the shortest possible path of a ray.

In 1657, Fermat (see Figure 2.7.2) reformulated the principle by postulating that a light ray travels in such a way that on its path it requires the least time. Fermat's principle of least time leads immediately not only to the correct law of reflection but also to Snell's law of refraction.

Figure 2.7.2.   Pierre de Fermat (born August 17, 1601; died January 12, 1665), a French lawyer, linguist, and amateur mathematician.

Newton, Leibniz, and Bernoulli discussed the problem of the brachystochrone and the shape of a hanging chain (a catenary).

In 1747, Maupertius first applied the general minimum principle in mechanics (see Figure 2.7.3). He asserted that dynamical motion takes place with minimum action. His theological reasoning was that action is minimized through the wisdom of God.

Figure 2.7.3.   Pierre de Maupertuis (born September 28, 1698; died August 27, 1759), a French mathematician and astronomer. He is most famous for formulating the principle of least action. The first use to which Maupertius put the principle of least action was to restate Fermat's derivation of the law of refraction (1744).

In 1760, Lagrange put the principle of least action on a firm basis (see Figure 2.7.1). However, the principle of least action is less general than Hamilton's principle.

In 1828, Gauss developed a method of treating mechanics by his principle of least constraint; a modification was later made by Hertz and embodied in his principle of least curvature. These principles, which were formulated 6 years later are closely related to Hamilton's principle. However, Hamilton's more general formulation is still today in use.

Figure 2.7.4.   Carl Friedrich Gauss (born April 30, 1777; died February 23, 1855), worked in a wide variety of fields in both mathematics and physics, including number theory, analysis, differential geometry, geodesy, magnetism, astronomy, and optics. His work has had an immense influence in many areas.

In 1834 and 1835 Hamilton (see Figure 2.7.5) announced the dynamical principle upon which it is possible to base all of mechanics and, indeed, most of classical physics. Hamilton's principle reads:

> Of all the possible paths along which a dynamical system can move from one point to another within a specific time interval, the actual path followed is that which minimizes the time integral of the difference between the kinetic and potential energies.

In terms of the calculus of variations, Hamilton's principle becomes

$$\delta \int_{t_1}^{t_2} (T - V)\, dt = 0.$$

This variational statement of the principle requires only that $T - V$ be an extremum, not necessarily a minimum, but in almost all applications of importance in dynamics, the minimum condition obtains.

Figure 2.7.5.   Sir William Rowan Hamilton (born August 04, 1805; died Septembe 02, 1865), Scottish mathematician and astronomer, and later, Irish Astronomer Royal. In 1843, Hamilton discovered the quaternions, the first noncommutative algebra to be studied. He felt this would revolutionize mathematical physics and he spent the rest of his life working on quaternions.

Let us consider a mechanical system consisting of a collection of particles — interacting among each other according to well-defined force laws; experience has shown that the *state of the system* is completely described by the set of all the positions and velocities of the particles. The coordinate frame need not be cartesian, as was the case in Newton's work, and the description can be effected by means of some set of *generalized coordinates* q_i, $(i = 1, \ldots, n)$ and generalized velocities q'_i, $(i = 1, \ldots, n)$.

If the system moves from a position at some time t_1, labeled by the coordinate set $\vec{q}^{(1)} = (q_1(t_1), \ldots, q_n(t_1))$, to a position $\vec{q}^{(2)} = (q_1(t_2), \ldots, q_n(t_2))$ at another time t_2, then the actual motion can be determined from Hamilton's principle of least action. This requires that the integral of the so-called *Lagrange* function takes the minimum possible value between the initial and final times. For the moment, we treat the Lagrangian as a black box, merely stating that it can only be some function of those variables on which the state of a system can depend, namely

$$L = L(q_1, \ldots, q_n, q'_1, \ldots, q'_n, t). \tag{2.7.1}$$

The famous principle of least action or Hamilton's principle requires that the action integral

$$W = \int_{t_1}^{t_2} L(\vec{q}, \vec{q}\,', t)\, dt \tag{2.7.2}$$

be a minimum. For the moment, we drop the subscript on the q_i's and q'_i's and assume a single degree of freedom. The positions $q^{(1)}$ and $q^{(2)}$ at the initial and final times t_1 and t_2, respectively, are assumed fixed. There can be many different paths $q(t)$ connecting $q^{(1)}$ and $q^{(2)}$, and the aim is to find those that extremize the action (2.7.2). This is done by looking at the effect of a first variation, that is adding a small alteration along the path which vanishes at either end. A remarkable feature of this procedure is that we are considering the effect of these variations about a path which we do not yet know. The first variation of the action is then determined by

---

```
 δ    ⌠t₂
───  ⎮   L[q[t], ∂ₜq[t], t] ⅆ t
δ q   ⌡t₁
```

```
-∂{t,1} L^(0,1,0) [q[t], q'[t], t] +
   L^(1,0,0) [q[t], q'[t], t] == 0
```

This equation is known as Lagrange's equation.

For *n* degrees of freedom q_1, q_2, ..., q_n, the variation must be effected for each variable independently (i.e., $q_i + \epsilon w_i$). The result gained is a set of equations

---

```
      ⎡  δ    ⌠t₂
Map[ ⎢ ───  ⎮   L[q1[t], q2[t], q3[t], ∂ₜq1[t],
      ⎣  δ #  ⌡t₁

           ⎤
∂ₜq2[t], ∂ₜq3[t], t] ⅆt⎥ &, {q1, q2, q3}]
           ⎦
```

```
{-∂{t,1} L^(0,0,0,1,0,0,0) [q1[t], q2[t], q3[t], q1'[t],
        q2'[t], q3'[t], t] + L^(1,0,0,0,0,0,0) [q1[t],
     q2[t], q3[t], q1'[t], q2'[t], q3'[t], t] == 0,
 -∂{t,1} L^(0,0,0,0,1,0,0) [q1[t], q2[t], q3[t], q1'[t],
        q2'[t], q3'[t], t] + L^(0,1,0,0,0,0,0) [q1[t],
     q2[t], q3[t], q1'[t], q2'[t], q3'[t], t] == 0,
 -∂{t,1} L^(0,0,0,0,0,1,0) [q1[t], q2[t], q3[t], q1'[t],
        q2'[t], q3'[t], t] + L^(0,0,1,0,0,0,0) [q1[t],
     q2[t], q3[t], q1'[t], q2'[t], q3'[t], t] == 0}
```

which are the celebrated Lagrange equations. If the explicit form of the Lagrangian is known, then the set of equations of motion are a set of second-order equations. If, in addition, the initial dates $(q_i(0), q'_i(0)$, $i = 1, 2, 3, ...)$ are given, the entire history of the system is determined. For Laplace, this deterministic framework appeared so powerful that he claimed: *We ought then to regard the present state of the universe as the effect of its preceding state and as the cause of its succeeding state.*

At this point of the theory, we know how to derive the Lagrange equations from a given Lagrangian. However, up to now, we did not discuss how we can find the Lagrangian. In determining the correct form for the Lagrangian function, it is interesting to see how far one can go in making this choice by invoking only the most basic principles. Landua and Lifshitz [2.2] argue that for a free particle, the principles of homogeneity of time and isotropy of space determine that the Lagrangian can only be proportional to the square of the velocities. The two mentioned properties ensure that the motion can be considered in the context of an *inertial frame*, (i.e., independent of its absolute position in space and time). If the constant of proportionality is taken to be half the particle mass, then the Lagrangian for a system of noninteracting particles is just their total kinetic energy; that is,

$$L = \sum_{i=1}^{n} \tfrac{1}{2} m q'^2_i = T. \tag{2.7.3}$$

Beyond this, experimental facts have to be invoked in that if the particles interact among each other according to some force law contained in a potential energy function $V(q_1, q_2, ..., q_n)$, then Landau and Lifshitz say *experience has shown* that the correct form of the Lagrangian is

$$L = T - V = \sum_{i=1}^{n} \tfrac{1}{2} m q'^2_i - V(q_1, q_2, ..., q_n). \tag{2.7.4}$$

The potential energy function is such that the force acting on each particle is determined by

$$\vec{F}_i = -\frac{\partial}{\partial q_i} V(q_1, q_2, ..., q_n). \tag{2.7.5}$$

This provides a definition of the potential energy because it ensures that the net work done by a system in traversing a closed path in the configuration space is zero. For velocity-independent potentials, Lagrange's equations become

```
SetOptions[EulerLagrange, eXpand -> False];
```

$$\frac{\delta}{\delta q} \int_{t_1}^{t_2} \left(\frac{1}{2} m \, (\partial_t q[t])^2 - V[q[t]] \right) dt$$

```
-V'[q[t]] - m q''[t] == 0
```

which, in the case of cartesian coordinates, are just Newton's equations.

In general, the Lagrange equations of motion are a set of n ordinary differential equations of second order. A complete solution will contain $2n$ arbitrary constants. These constants are usually taken to specify the state of the system at some initial time. Instead of giving the initial state of the system, one might give the initial configuration and a later configuration. These conditions might not be self-consistent because the second configuration might not result from the first one under the action of the given forces, no matter how the initial velocity components are chosen.

One of the most useful devices for solving the Lagrange equations of motion is to discover the first integrals of the motion. A first integral of a set of differential equations is a function of the unknowns and contains derivatives of one order lower than the order of the differential equations themselves and remains constant by virtue of the differential equations. Examples of such integrals are the energy and the momentum of isolated systems. The advantage of having an integral of the motion is that it reduces the order of the system of equations to be solved.

2.7.3 Hamilton's Principle

To see how Hamilton's principle works, let us consider a mechanical system consisting of N interacting particles. As we noted in Section 2.7.2, it is sufficient to introduce a functional depending on coordinates and velocities. The choice of coordinates and velocities only is a matter of experience. The chosen coordinates must not be cartesian coordinates. However, Newton's mechanics is based on cartesian coordinates. In Hamilton's principle, it is sufficient to choose so-called generalized coordinates q_i $(i = 1, 2, ..., N)$ and generalized velocities q'_i $(i = 1, 2, ..., N)$. These coordinates are chosen in such a way that the mathematical description of the problem is simplified.

The choice of generalized (appropriate) coordinates is motivated by the following arguments. A point system consisting of N points has, in general, to satisfy a number r constraints. These restrictions are given by

$$g_\alpha (x_\beta, t) = 0 \quad \alpha = 1, 2, ... r, \quad \beta = 1, 2, ..., 3N$$

$$\tag{2.7.6}$$

where x_β denotes the $3N$ cartesian coordinates. The degrees of freedom for such a system are determined by $f = 3N - r$. Our aim is to replace the $3N$ cartesian coordinates x_β by f generalized coordinates q_i. These f generalized coordinates q_i are free of any constraints and allow a complete description of the system. The physical meaning of the coordinates can be different from the cartesian coordinates. For example, the generalized coordinates can be distances, angles, line elements, and so forth. It does not matter how one interprets these coordinates, but it is important that the number of the coordinates equal the degrees of freedom of the system. Such coordinates are optimized coordinates for the system.

2.7.3.1 Classes of Constraints

Constraints which are given by algebraic expressions like

$$g_\alpha (x_\beta, t) = 0, \quad \alpha = 1,2,...,r, \tag{2.7.7}$$

with $r < f = 3N - r$ are called holonomic.

Constraints not representable by algebraic relations are called nonholonomic. Another classification of constraints is based on the time dependence or independence. Time-dependent constraints are termed rheonimic. Constraints independent of time are called scleronomic. The following table summarizes the terms used to classify mechanical constraints.

	rheonom with time	skleronom without time
holonom	$g_\alpha(q_i,t)=0$	$g_\alpha(q_i)=0$
nonholonom	$g_\alpha(q_i,t)\leqq 0$	$g_\alpha(q_i)\leqq 0$

Table 2.7.1.   Classification of constraints as rehonom, skleronom, holonon, and nonholonom conditions.

The motion of a particle system with N generalized coordinates from a position $\vec{q}^{(1)} = (q_1(t_1),\ q_2(t_1) \ldots q_N(t_1))$ at $t = t_1$ to a different position $\vec{q}^{(2)} = (q_1(t_2),\ q_2(t_2) \ldots q_N(t_2))$ at $t = t_2$ is governed by Hamilton's principle. Hamilton's principle itself is governed by a functional called the Lagrange functional whose density is a function of generalized coordinates and velocities:

$$\mathcal{L} = \mathcal{L}(q_1,\ q_2,\ \ldots,\ q_N,\ q'_1,\ q'_2,\ \ldots,\ q'_N,\ t). \qquad (2.7.8)$$

This kind of density takes on an extremal value in a time interval t_1 to t_2 if the right path is chosen. At the moment we assume that such a density exists and ask for consequences for the density. If the density exists then we are able to write down the corresponding functional

$$L[q_1] = \int_{t_1}^{t_2} \mathcal{L}(q_1,\ q_2,\ \ldots,\ q_N,\ q'_1,\ q'_2,\ \ldots,\ q'_N,\ t)\,dt. \qquad (2.7.9)$$

Calculus of variations tells us that this functional assumes an extremal value if the Euler equations are satisfied; that is

$$\mathcal{E}_i \mathcal{L} = 0 = \frac{\delta L}{\delta q_i}, \qquad\qquad i = 1,\ 2,\ \ldots,\ N \qquad (2.7.10)$$

or explicitly

$$\frac{\partial \mathcal{L}}{\partial q} - \frac{d}{dt}\left(\frac{\partial \mathcal{L}}{\partial \dot{q}}\right) = 0. \qquad (2.7.11)$$

In *Mathematica,* we get for a system with N coordinates the expression

```
SetOptions[EulerLagrange, eXpand -> True];
```

$\mathcal{E}^t_{q[i]}[\mathcal{L}[q[i][t], \partial_t q[i][t], t]]$

```
-∂{t,1} L^(0,1,0)[q[i][t], q[i]'[t], t] +
   L^(1,0,0)[q[i][t], q[i]'[t], t] == 0
```

which is identical with relation (2.7.11). This kind of equation is also known as Euler–Lagrange equation. The Euler–Lagrange equations are ordinary differential equations of second order. If we carry out the differentiation explicitly, we get a second-order ODE.

```
SetOptions[EulerLagrange, eXpand -> False];
```

$\mathcal{E}^t_{q[i]}[\mathcal{L}[q[i][t], \partial_t q[i][t], t]]$

```
-L^(0,1,1)[q[i][t], q[i]'[t], t] -
   q[i]''[t] L^(0,2,0)[q[i][t], q[i]'[t], t] +
   L^(1,0,0)[q[i][t], q[i]'[t], t] -
   q[i]'[t] L^(1,1,0)[q[i][t], q[i]'[t], t] == 0
```

At this stage of our calculations we note that the order of differentiation of Euler–Lagrange equations is identical with the order of differentiation of Newton's equation.

If Hamilton's principle has a real physical meaning, then the equations of motion must be identical with Newton's equation of motion. To establish this connection, we define a Lagrange density which separates into two parts. The first part contains only velocity-dependent components and the second part contains only information on coordinates. This separation is motivated by the two energies known as kinetic energy and potential energy. Let us first assume that both energies are linearly combined:

$$\mathcal{L} = \alpha T + \beta V. \tag{2.7.12}$$

where T and V denote kinetic and potential energies, respectively. The parameters α and β are, up to now, unknown. The kinetic energy is a function of generalized velocities q'_i given by

$$T = T(q'_1, \ q'_2, \ ..., \ q'_N) = T(q'_i). \tag{2.7.13}$$

This function is defined in *Mathematica* by

```
T = T[∂_t q[i][t]]
```

```
T[q[i]'[t]]
```

The potential energy is a function of the generalized coordinates q_i given by

$$V = V(q_1, \ q_2, \ ..., \ q_N) = V(q_i). \tag{2.7.14}$$

or in *Mathematica* by

```
V = V[q[i][t]]
```

```
V[q[i][t]]
```

The Lagrange density is the given by relation (2.7.12)

```
L = α T + β V
```

```
α T[q[i]'[t]] + β V[q[i][t]]
```

From the Euler–Lagrange equations, we get the following system of equations of motion

$$\frac{\partial \mathcal{L}}{\partial q_i} - \frac{d}{dt}\left(\frac{\partial \mathcal{L}}{\partial \dot{q}_i}\right) =$$

$$\beta \frac{\partial V(q_i)}{\partial q_i} - \alpha \frac{d}{dt}\left(\frac{\partial T}{\partial \dot{q}_i}\right) = \beta \frac{\partial V(q_i)}{\partial q_i} - \alpha \frac{\partial^2 T}{\partial \dot{q}^2_i}\ \ddot{q}_i = 0 \tag{2.7.15}$$

$$\Longleftrightarrow -\frac{\beta}{\alpha}\ \frac{\partial V}{\partial q_i} + \frac{\partial^2 T}{\partial \dot{q}^2_i}\ \ddot{q}_i = 0 \qquad i = 1, \ 2, \ ..., \ N. \tag{2.7.16}$$

in *Mathematica*, it follows that

```
SetOptions[EulerLagrange, eXpand -> False];
```

```
ElerLagrangeEquation = ℰᵗ_q[i] [L]
```

$$\beta \mathcal{V}'[q[i][t]] - \alpha \mathcal{T}''[q[i]'[t]] \, q[i]''[t] == 0$$

Newton's theory provides for an N-particle system the following system of equations

$$m_i q''_i = F_i, \qquad\qquad i = 1, 2, \dots, N. \qquad\qquad (2.7.17)$$

If we, in addition, assume that the forces F_i can be represented by a potential gradient

$$F_i = -\frac{\partial V(q_i)}{\partial q_i}, \qquad\qquad i = 1, 2, \dots, N, \qquad\qquad (2.7.18)$$

then we get Newton's equation in the form

$$\frac{\partial V(q_i)}{\partial q_i} + m_i q''_i = 0, \qquad\qquad (2.7.19)$$

or in *Mathematica*,

```
NwtonsEquations =
   ∂_q[i][t] (𝒱[q[i][t]]) + m[i] ∂_t,t q[i][t] == 0
```

$$\mathcal{V}'[q[i][t]] + m[i] \, q[i]''[t] == 0$$

If both systems of equations are identical, the difference of the two systems must vanish:

```
rel1 = NwtonsEquations[[1]] -
   ElerLagrangeEquation[[1]] // Simplify
```

$$-(-1 + \beta) \, \mathcal{V}'[q[i][t]] + (m[i] + \alpha \mathcal{T}''[q[i]'[t]]) \, q[i]''[t]$$

Because the second-order derivative in the q_i's and the potential gradient are not equal to zero, the coefficients of these terms must vanish. The coefficient with respect to the potential gives

```
r1 =
  Solve[Coefficient[rel1, ∂q[i][t] 𝒱[q[i][t]]] == 0, β] //
    Flatten
```

```
{β → 1}
```

The relation for α is gained by

```
r2 = Solve[
    Coefficient[rel1, ∂t,t q[i][t]] == 0, α] // Flatten
```

$$\left\{\alpha \rightarrow -\frac{m[i]}{\mathcal{T}''[q[i]'[t]]}\right\}$$

If we, in addition, assume that the kinetic energy is a quadratic function in the generalized coordinates, then the masses m_i are the front factors of the quadratic term. Thus, we can set

```
r2 = r2 /. ∂q[i]'[t],q[i]'[t] 𝒯[q[i]'[t]] -> m[i]
```

```
{α → -1}
```

Now, the two unknowns α and β are determined and the Lagrange density becomes

```
ℒ = L /. r1 /. r2
```

```
-𝒯[q[i]'[t]] + 𝒱[q[i][t]]
```

In standard mechanics texts, the Lagrange density is defined by

$$\mathcal{L} = T(q'_i) - V(q_i). \tag{2.7.20}$$

However, the sign does not matter because the resulting system of equations of motion is invariant with respect to a change of all signs. This is demonstrated by the derivation of the equations of motion by

$\mathcal{E}^t_{q[i]} [\mathcal{L}]$

$\mathcal{V}[q[i][t]] + \mathcal{T}''[q[i]'[t]] q[i]''[t] == 0$

and

$\mathcal{E}^t_{q[i]} [-\mathcal{L}]$

$-\mathcal{V}[q[i][t]] - \mathcal{T}''[q[i]'[t]] q[i]''[t] == 0$

The major assumption in the derivation of the Lagrange density was that the kinetic energy is a quadratic function in the generalized velocities q'_i. A simple realization is given by

$$T(q'_i) = \frac{m_i}{2} q'^2_i + c_1 q'_i + c_2. \tag{2.7.21}$$

The simplest form of the kinetic energy for an N-particle system is thus

$$T(q'_i) = \sum_{i=1}^{N} \frac{m_i}{2} q'^2_i. \tag{2.7.22}$$

In general, the kinetic energy is a homogenous quadratic function in the generalized velocities q'_i:

$$T = \sum_{j,k} a_{jk} q'_j q'_k. \tag{2.7.23}$$

Differentiation of this relation with respect to q'_i delivers

$$\frac{dT}{d\dot{q}_i} = \frac{d}{d\dot{q}_i} \left\{ \sum_{k,j} a_{jk} q'_j q'_k \right\} = \sum_{k,j} a_{jk} \frac{d}{d\dot{q}_i} (q'_j q'_k)$$

$$= \sum_{k,j} \delta_{jk} \frac{dq'_j}{dq'_{i\delta_{ji}}} q'_k + q'_j \frac{dq'_k}{dq'_{i\delta_{ki}}} \tag{2.7.24}$$

$$= \sum_{k,j} a_{jk} (\delta_{ji} q'_k + q'_j \delta_{ki}) = \sum_k a_{ik} q'_k + \sum_j a_{ji} q'_j.$$

Multiplying this with q'_i and summing over i gives

$$\sum_i q'_i \frac{dT}{dq'_i} = \sum_{i,k} a_{ik} q'_k q'_l + \sum_{i,j} a_{ji} q'_j q'_i \tag{2.7.25}$$

which is equivalent to

$$\sum_i q'_i \frac{dT}{dq'_i} = 2 \sum_{i,k} a_{ik} q'_k q'_i \qquad (2.7.26)$$

because the indices in the second sum are changeable. Then, it follows that

$$\sum_i q'_i \frac{dT}{dq'_i} = 2 T. \qquad (2.7.27)$$

This result, however, is a special case of the more general Euler theorem on homogenous functions $f(y_k)$ given by

$$\sum_k y_k \frac{\partial f}{\partial y_k} = n f. \qquad (2.7.28)$$

The main result is that the Lagrange density can be chosen as the difference of kinetic and potential energy if we require that Newton's equations be the target of Hamilton's principle. We also realized that the Lagrange density is gauge invariant with respect to a common factor which does not alter the resulting equations of motion. We demonstrated that the variation of

$$\frac{d}{d\varepsilon} \int_{t_1}^{t_2} (T(q'_i) - V(q_i)) \, dt \, |_{\varepsilon=0} = 0 \qquad (2.7.29)$$

delivers the equations of motion, which is just Hamilton's principle.

Example 1: Harmonic Oscillator

As a first example let us examine the harmonic oscillator. This kind of system is central in different fields of physics (e.g., in solid state physics to describe crystals, in quantum physics to examine harmonic interactions). The kinetic energy of a single harmonic oscillator in generalized velocities is given by

```
T = m
    ─ (∂_t q[t])²
    2
```

$$\frac{1}{2} m \, q'[t]^2$$

The potential energy is given by the harmonic function

$$V = \frac{k}{2} q[t]^2$$

$$\frac{1}{2} k q[t]^2$$

where *m* is mass and *k* is a force constant. The Lagrange density follows by

$$L = T - V$$

$$-\frac{1}{2} k q[t]^2 + \frac{1}{2} m q'[t]^2$$

Applying the Euler–Lagrange operator to this density, we find the governing equation of motion

$$\text{harmonicOs} = \mathcal{E}_q^t[L]$$

$$-k q[t] - m q''[t] == 0$$

The solution of this equation demonstrates that the motion is described by harmonic functions:

```
DSolve[harmonicOs, q, t] // Flatten
```

$$\left\{ q \to \text{Function}\left[\{t\}, \text{C[1] Cos}\left[\frac{\sqrt{k}\, t}{\sqrt{m}}\right] + \text{C[2] Sin}\left[\frac{\sqrt{k}\, t}{\sqrt{m}}\right]\right]\right\}$$

Example 2: Rolling Wheel on an Inclined Plane

Let us consider a wheel rolling on a inclined plane. The kinetic energy consists of two parts. The first part is purely translational and the second purely rotational. The total kinetic energy is given by

$$T = \frac{1}{2} m (\partial_t y[t])^2 + \frac{1}{2} I (\partial_t \theta[t])^2 \;/. \; I \; -> \; \frac{1}{2} m R^2$$

$$\frac{1}{2} m y'[t]^2 + \frac{1}{4} m R^2 \theta'[t]^2$$

where m is the mass, $I = m R^2/2$ is the moment of inertia with respect to the center, and R is the radius of the wheel. The potential energy is mainly generated by Earth's gravitation:

$$V = m g (1 - y[t]) \; Sin[\alpha]$$

$$m g \, Sin[\alpha] \; (1 - y[t])$$

where l is the total length of the plane. The generalized coordinates here are y and θ. The origin of the potential is chosen in such a way that at the bottom of the ramp, $V = 0$. The Lagrange density of the system is given by

$$L = T - V$$

$$-m g \, Sin[\alpha] \; (1 - y[t]) + \frac{1}{2} m y'[t]^2 + \frac{1}{4} m R^2 \theta'[t]^2$$

representing a function in y, y', and θ'.

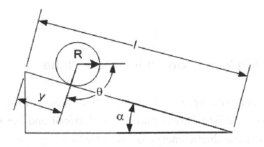

Figure 2.7.6.   Wheel on a ramp. Definition of constraints and coordinates.

In addition to the Lagrange density, the system has to satisfy the additional constraint of nonslip; that is,

```
g = y[t] - R θ[t] == 0
```

```
y[t] - R θ[t] == 0
```

The degrees of freedom f for the system is then determined by

$$f = N - M = 2 - 1 = 1; \qquad (2.7.30)$$

that the system has one degree of freedom if the wheel rolls without slipping. Thus, we can use either y or θ as the generalized coordinate. Let us choose y as the appropriate coordinate. Then, from the constraint g, we get

```
gconst = θ -> Function[t, y[t]/R ]
```

```
θ → Function[t, y[t]/R ]
```

Inserting this relation into the Lagrangian density, we get

```
Ly = L /. gconst
```

$$-m g \, Sin[\alpha] \, (1 - y[t]) + \frac{3}{4} \, m \, y'[t]^2$$

If we prefer to chose θ as the appropriate coordinate we find

```
Lθ = L /. y -> Function[t, R θ[t]]
```

$$-m g \, Sin[\alpha] \, (1 - R θ[t]) + \frac{3}{4} \, m \, R^2 \, θ'[t]^2$$

Both Lagrangians are equivalent for the description of motion. The governing equation of motion follows for each case

```
eqy = ℰ_y^t[Ly]
```

$$m\,g\,Sin[\alpha] - \frac{3}{2}\,m\,y''[t] == 0$$

and

```
eqθ = ℰ_θ^t[Lθ]
```

$$m\,R\,g\,Sin[\alpha] - \frac{3}{2}\,m\,R^2\,\theta''[t] == 0$$

The solutions for each case follows by

```
soly = DSolve[eqy, y, t] // Flatten
```

$$\left\{y \to Function\left[\{t\},\ C[1] + t\,C[2] + \frac{1}{3}\,t^2\,g\,Sin[\alpha]\right]\right\}$$

```
solθ = DSolve[eqθ, θ, t] // Flatten
```

$$\left\{\theta \to Function\left[\{t\},\ C[1] + t\,C[2] + \frac{t^2\,g\,Sin[\alpha]}{3\,R}\right]\right\}$$

The point of view of this problem is to assume that y and θ are independent of each other. In this case, we have to carry out Hamilton's principle under the action of constraints. The constraints are used to determine the Lagrange multiplier. The Lagrange equations now read

```
el1 = ℰ_y^t[L][[1]] + λ[t] ∂_y[t] g[[1]] == 0
```

$$m\,g\,Sin[\alpha] + \lambda[t] - m\,y''[t] == 0$$

```
el2 = 𝓔ᵗₑ[L]〚1〛 + λ[t] ∂ₑ[t] g〚1〛 == 0
```

$$-R \lambda[t] - \frac{1}{2} m R^2 \theta''[t] == 0$$

In addition, the constraint gives

```
gconst
```

$$\theta \to \text{Function}\left[t, \frac{y[t]}{R}\right]$$

These three relations are the basis for the solution of the problem.

Let us first differentiae the constraint relation twice with respect to *t* and solve the resulting relation with respect to θ'':

```
solconst = Solve[∂ₜ,ₜ g〚1〛 == 0, ∂ₜ,ₜ θ[t]] // Flatten
```

$$\cdot \left\{\theta''[t] \to \frac{y''[t]}{R}\right\}$$

Then, we can use the result in the second Euler–Lagrange equation and solve for the Lagrange multiplier:

```
solλ = Solve[el2 /. solconst, λ[t]] // Flatten
```

$$\left\{\lambda[t] \to -\frac{1}{2} m y''[t]\right\}$$

Inserting the result into the first Euler–Lagrange equation, we find

```
eql1 = el1 /. solλ
```

$$m g \sin[\alpha] - \frac{3}{2} m y''[t] == 0$$

which determines the Lagrange multiplier completely:

```
LagrangeMultiplier =
  solλ /. Flatten[Solve[eql1, ∂_{t,t}y[t]]]
```

$$\left\{ \lambda[t] \rightarrow -\frac{1}{3} \, m \, g \, Sin[\alpha] \right\}$$

The Euler–Lagrange equations then follow by inserting the Lagrange multiplier:

```
el1f = el1 /. LagrangeMultiplier
```

$$\frac{2}{3} \, m \, g \, Sin[\alpha] - m \, y''[t] == 0$$

```
el2f = el2 /. LagrangeMultiplier
```

$$\frac{1}{3} \, m \, R \, g \, Sin[\alpha] - \frac{1}{2} \, m \, R^2 \, \theta''[t] == 0$$

The integration of the two equations with initial conditions introduced deliver

```
DSolve[Join[{el1f}, {y[0] == y0, y'[0] == v0}], y, t] //
  Flatten
```

$$\left\{ y \rightarrow Function\left[\{t\}, \, \frac{1}{3} \, (3 \, t \, v0 + 3 \, y0 + t^2 \, g \, Sin[\alpha]) \right] \right\}$$

```
DSolve[Join[{el2f}, {θ[0] == θ0, θ'[0] == ω0}], θ, t] //
  Flatten
```

$$\left\{ \theta \rightarrow Function\left[\{t\}, \, \frac{3 \, R \, \theta0 + 3 \, R \, t \, \omega0 + t^2 \, g \, Sin[\alpha]}{3 \, R} \right] \right\}$$

Example 3: Sliding Mass Connected to a Pendulum

Let us consider two mass points as a coupled pendulum. The first mass m_1 is sliding on a horizontal bar in the x-direction. The second mass is connected with the first one by a stiff rod. At each end of the rod, one mass point is located (see Figure 2.7.7). The second mass m_2 is the pendulum mass.

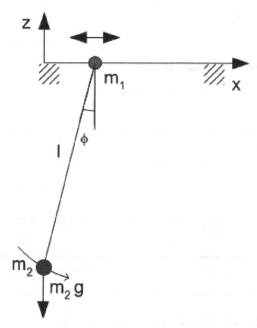

Figure 2.7.7.   Sliding mass pendulum.

The movement of mass m_1 is restricted to the x-direction. The second mass m_2 undergoes translations in x as well as oscillations around its support. The total kinetic energy is generated by two parts:

$$T1 = \frac{m1}{2}\left(\left(\partial_t x1[t]\right)^2 + \left(\partial_t z1[t]\right)^2\right)$$

$$\frac{1}{2}\, m1\,\left(x1'[t]^2 + z1'[t]^2\right)$$

and

$$T2 = \frac{m2}{2}\left(\left(\partial_t x2[t]\right)^2 + \left(\partial_t z2[t]\right)^2\right)$$

$$\frac{1}{2}\, m2\,\left(x2'[t]^2 + z2'[t]^2\right)$$

The potential energies of the two masses are

$$V1 = 0$$

$$0$$

and

$$V2 = m2\, g\, z2[t]$$

$$m2\, g\, z2[t]$$

The total kinetic and potential energies are

$$T = T1 + T2$$

$$\frac{1}{2}\, m1\,\left(x1'[t]^2 + z1'[t]^2\right) + \frac{1}{2}\, m2\,\left(x2'[t]^2 + z2'[t]^2\right)$$

and

```
V = V1 + V2
```

```
m2 g z2[t]
```

To introduce generalized coordinates, we have to take the constraints into account. The following rules define a transformation between original coordinates and generalized coordinates:

```
generalizedCoordinates =
  {x1 -> Function[t, x[t]], z1 -> Function[t, 0],
   x2 -> Function[t, x[t] + l Sin[φ[t]]],
   z2 -> Function[t, -l Cos[φ[t]]]}
```

```
{x1 → Function[t, x[t]], z1 → Function[t, 0],
  x2 → Function[t, x[t] + l Sin[φ[t]]],
  z2 → Function[t, -l Cos[φ[t]]]}
```

The transformed kinetic energy follows with

```
𝒯 = T /. generalizedCoordinates // Simplify
```

$$\frac{1}{2} \left((m1 + m2) \, x'[t]^2 + 2 \, l \, m2 \, Cos[φ[t]] \, x'[t] \, φ'[t] + l^2 \, m2 \, φ'[t]^2 \right)$$

The potential energy is

```
𝒱 = V /. generalizedCoordinates // Simplify
```

```
-l m2 g Cos[φ[t]]
```

The Lagrangien density in x and $φ$ is given by

```
L = 𝒯 - 𝒱
```

```
l m2 g Cos[ϕ[t]] + 1/2 ((m1 + m2) x'[t]² +
    2 l m2 Cos[ϕ[t]] x'[t] ϕ'[t] + l² m2 ϕ'[t]²)
```

From the Lagrange density, the two Euler–Lagrange equations are derived
via the application of the Euler operator:

```
el1 = 𝓔ˣᵗ[L]
```

```
l m2 Sin[ϕ[t]] ϕ'[t]² - m1 x''[t] -
    m2 x''[t] - l m2 Cos[ϕ[t]] ϕ''[t] == 0
```

```
el2 = 𝓔ᵗϕ[L]
```

```
-l m2 g Sin[ϕ[t]] - l m2 Cos[ϕ[t]] x''[t] - l² m2 ϕ''[t] == 0
```

A view at these two equations shows that the second-order derivative in x
can be used to decouple the two equations. Solving for the generalized
acceleration x'', we find

```
sol2 = Solve[el2, ∂_{t,t} x[t]] // Flatten
```

$$\left\{ x''[t] \rightarrow - \frac{Sec[\phi[t]] \, (l \, m2 \, g \, Sin[\phi[t]] + l^2 \, m2 \, \phi''[t])}{l \, m2} \right\}$$

This result is used to eliminate x in the first Euler–Lagrange equation:

```
el1ϕ = el1 /. sol2 // Simplify
```

```
(m1 + m2) g Tan[ϕ[t]] + l m2 Sin[ϕ[t]] ϕ'[t]² +
    l (-m2 Cos[ϕ[t]] + (m1 + m2) Sec[ϕ[t]]) ϕ''[t] == 0
```

The resulting equation is an equation containing only ϕ as the unknown quantity. Because the derived equation is nonlinear, there is no direct method to find an analytic solution. If we assume that ϕ and the first derivatives of ϕ are small quantities, we are able to Taylor expand the equation around the equilibrium point $\phi = 0$. Because ϕ is a small quantity, squares of ϕ are even smaller than ϕ itself. If we use these information in the expansion, we get

```
linell𝜙 = (Series[ell𝜙[[1]], {𝜙[t], 0, 2}] // Normal) /.
    {(∂ₜ𝜙[t])² → 0, 𝜙[t]² → 0}
```

```
(m1 + m2) g 𝜙[t] + l m1 𝜙″[t]
```

a linear harmonic equation. The solution of this equation follows with

```
sol𝜙 = DSolve[linell𝜙 == 0, 𝜙, t] // Flatten
```

$$\left\{\phi \to \text{Function}\left[\{t\},\right.\right.$$
$$\left.\left. \text{C[1] Cos}\left[\frac{t\sqrt{m1\ g + m2\ g}}{\sqrt{l}\ \sqrt{m1}}\right] + \text{C[2] Sin}\left[\frac{t\sqrt{m1\ g + m2\ g}}{\sqrt{l}\ \sqrt{m1}}\right]\right]\right\}$$

Knowing the solution for ϕ, we are able to get an equation for x. At this stage, we also need to approximate the resulting equation under the same assumptions as for ϕ. The solution of the equation is a function linear in time with oscillations around this trend.

```
DSolve[x''[t] ==
    Normal[Series[x''[t] /. sol2, {ϕ[t], 0, 1}]] /.
    solϕ // Simplify, x, t] // Flatten
```

$$\left\{ x \rightarrow \text{Function}\left[\{t\}, \right. \right.$$

$$C[3] + t\,C[4] - \frac{1}{m1} \left(\left(\sqrt{1}\ \sqrt{m1}\ C[1]\ \text{Cos}\left[\frac{t\ \sqrt{(m1+m2)\ g}}{\sqrt{1}\ \sqrt{m1}} \right] \right.\right.$$

$$\left(\frac{\sqrt{1}\ \sqrt{m1}\ m2\ g\ C[2]\ \text{Cos}\left[\frac{t\ \sqrt{(m1+m2)\ g}}{\sqrt{1}\ \sqrt{m1}} \right]}{\sqrt{(m1+m2)\ g}} - \right.$$

$$\left.\left. \frac{\sqrt{1}\ \sqrt{m1}\ m2\ g\ C[1]\ \text{Sin}\left[\frac{t\ \sqrt{(m1+m2)\ g}}{\sqrt{1}\ \sqrt{m1}} \right]}{\sqrt{(m1+m2)\ g}} \right) \right) \Bigg/$$

$$\left(\sqrt{(m1+m2)\ g}\ \left(C[2]\ \text{Cos}\left[\frac{t\ \sqrt{(m1+m2)\ g}}{\sqrt{1}\ \sqrt{m1}} \right] - \right.\right.$$

$$\left.\left. C[1]\ \text{Sin}\left[\frac{t\ \sqrt{(m1+m2)\ g}}{\sqrt{1}\ \sqrt{m1}} \right] \right) \right) +$$

$$\left(\sqrt{1}\ \sqrt{m1}\ C[2]\ \text{Sin}\left[\frac{t\ \sqrt{(m1+m2)\ g}}{\sqrt{1}\ \sqrt{m1}} \right] \right.$$

$$\left(\frac{\sqrt{1}\ \sqrt{m1}\ m2\ g\ C[2]\ \text{Cos}\left[\frac{t\ \sqrt{(m1+m2)\ g}}{\sqrt{1}\ \sqrt{m1}} \right]}{\sqrt{(m1+m2)\ g}} - \right.$$

$$\left.\left. \frac{\sqrt{1}\ \sqrt{m1}\ m2\ g\ C[1]\ \text{Sin}\left[\frac{t\ \sqrt{(m1+m2)\ g}}{\sqrt{1}\ \sqrt{m1}} \right]}{\sqrt{(m1+m2)\ g}} \right) \right) \Bigg/$$

$$\left(\sqrt{(m1+m2)\ g}\ \left(C[2]\ \text{Cos}\left[\frac{t\ \sqrt{(m1+m2)\ g}}{\sqrt{1}\ \sqrt{m1}} \right] - \right.\right.$$

$$\left.\left.\left.\left. C[1]\ \text{Sin}\left[\frac{t\ \sqrt{(m1+m2)\ g}}{\sqrt{1}\ \sqrt{m1}} \right] \right) \right) \right] \right\}$$

Thus, we derived a harmonic solution for ϕ and an increasing solution with oscillations for x. The question arises of whether this kind of solution is also observed for the nonlinear coupled system of x and ϕ. To find an answer to this question, we first have to specify the parameters in this model (i.e., the masses, the length of the pendulum, and the acceleration g). The following list contains one example for these parameters:

```
parameters = {m1 -> 1, m2 -> .5, 1 -> .7, g -> 9.81}
```

```
{m1 → 1, m2 → 0.5, 1 → 0.7, g → 9.81}
```

The numerical solution of the two Euler–Lagrange equations then follows upon specifying the initial conditions for x, x', ϕ, and ϕ'. The following line contains all of these steps:

```
nsol = NDSolve[
    {el1, el2, x[0] == .1, x'[0] == 0.01, ϕ[0] == 0.1,
        ϕ'[0] == 0.01} /. parameters, {x, ϕ}, {t, 0, 13}]
```

```
{{x → InterpolatingFunction[{{0., 13.}}, <>],
    ϕ → InterpolatingFunction[{{0., 13.}}, <>]}}
```

The resulting functions can be represented in a plot showing that both coordinates oscillate with a certain frequency. It is also obvious that the solution for x increases in time as expected from the linear approximation of the Euler–Lagrange equations.

```
Plot[Evaluate[{x[t], ϕ[t]} /. nsol], {t, 0, 13},
  AxesLabel -> {"t", "x,ϕ"}, PlotStyle ->
  {RGBColor[1, 0, 0], RGBColor[0, 0, 1]}];
```

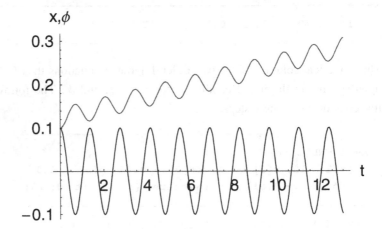

The solutions gained can be used to generate a flip-chart movie showing the movement of the two masses

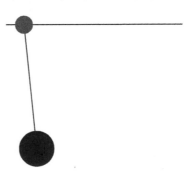

Thus, we get the information on how the two masses move under a specific initial condition.

Example 4: Sliding Mass on a Curve

This example is an extension of the previous example. The change here is the movement of mass m_1. We assume that mass 1 can move in the x-direction and z-direction restricted by a given curve. The second mass is again connected with the first one by a stiff rod. At each end of the rod, one mass point is located (see Figure 2.7.8). The second mass m_2 is the pendulum mass.

The movement of mass m_1 is governed by a function of x. We assume that this curve is given by a polynomial of order 8. A plot of the polynomial is as follows:

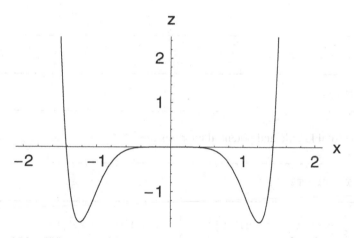

Figure 2.7.8. Sliding mass pendulum on a curve. Here we used the relation $z = x^8 - 2x^6$ as an example.

The second mass m_2 undergoes translations in x as well as oscillations around its support. The total kinetic energy is generated by two parts:

$$T1 = \frac{m1}{2} \left((\partial_t x1[t])^2 + (\partial_t z1[t])^2 \right)$$

$$\frac{1}{2} m1 \left(x1'[t]^2 + z1'[t]^2 \right)$$

and

$$T2 = \frac{m2}{2} \left((\partial_t x2[t])^2 + (\partial_t z2[t])^2 \right)$$

$$\frac{1}{2} m2 \left(x2'[t]^2 + z2'[t]^2 \right)$$

The potential energies of the two masses are

$$V1 = m1 \, g \, z1[t]$$

$$m1 \, g \, z1[t]$$

and

$$V2 = m2 \, g \, z2[t]$$

$$m2 \, g \, z2[t]$$

The total kinetic and potential energies are

$$T = T1 + T2$$

$$\frac{1}{2} m1 \left(x1'[t]^2 + z1'[t]^2 \right) + \frac{1}{2} m2 \left(x2'[t]^2 + z2'[t]^2 \right)$$

and

$$V = V1 + V2$$

$$m1 \, g \, z1[t] + m2 \, g \, z2[t]$$

To introduce generalized coordinates, we have to take the constraints into account. The following rules define a transformation between original coordinates and generalized coordinates:

```
generalizedCoordinates = {x1 -> Function[t, x[t]],
    z1 -> Function[t, x[t]^8 - 2 x[t]^6],
    x2 -> Function[t, x[t] + 1 Sin[ϕ[t]]],
    z2 -> Function[t, x[t]^8 - 2 x[t]^6 - 1 Cos[ϕ[t]]]}
```

```
{x1 → Function[t, x[t]],
  z1 → Function[t, x[t]^8 - 2 x[t]^6],
  x2 → Function[t, x[t] + 1 Sin[ϕ[t]]],
  z2 → Function[t, x[t]^8 - 2 x[t]^6 - 1 Cos[ϕ[t]]]}
```

The transformed kinetic energy follows by

```
𝒯 = T /. generalizedCoordinates // Simplify
```

$$\frac{1}{2} \left(m1 \left(1 + 16\, x[t]^{10}\, (3 - 2\, x[t]^2)^2\right) x'[t]^2 + \right.$$
$$m2 \left((x'[t] + 1\, Cos[ϕ[t]]\, ϕ'[t])^2 + (-12\, x[t]^5\, x'[t] + \right.$$
$$\left.\left. 8\, x[t]^7\, x'[t] + 1\, Sin[ϕ[t]]\, ϕ'[t])^2\right)\right)$$

The potential energy is

```
𝒱 = V /. generalizedCoordinates // Simplify
```

$$g \left(-1\, m2\, Cos[ϕ[t]] - 2\, (m1 + m2)\, x[t]^6 + (m1 + m2)\, x[t]^8\right)$$

The Lagrangian density in x and $ϕ$ is thus given by

```
L = 𝒯 - 𝒱
```

$$-g \left(-1\, m2\, \text{Cos}[\phi[t]] - 2\,(m1+m2)\,x[t]^6 + (m1+m2)\,x[t]^8\right) +$$
$$\frac{1}{2} \left(m1\,\left(1 + 16\,x[t]^{10}\,(3 - 2\,x[t]^2)^2\right)\,x'[t]^2 + \right.$$
$$m2\,\left((x'[t] + 1\,\text{Cos}[\phi[t]]\,\phi'[t])^2 + (-12\,x[t]^5\,x'[t] + \right.$$
$$\left.\left.8\,x[t]^7\,x'[t] + 1\,\text{Sin}[\phi[t]]\,\phi'[t])^2\right)\right)$$

From the Lagrange density, the two Euler–Lagrange equations are derived via the application of the Euler derivative:

```
SetOptions[EulerLagrange, eXpand → False];
```

```
el1 = 𝓔ₓᵗ[L]
```

$$12\,m1\,g\,x[t]^5 + 12\,m2\,g\,x[t]^5 - 8\,m1\,g\,x[t]^7 -$$
$$8\,m2\,g\,x[t]^7 - 720\,m1\,x[t]^9\,x'[t]^2 - 720\,m2\,x[t]^9\,x'[t]^2 +$$
$$1152\,m1\,x[t]^{11}\,x'[t]^2 + 1152\,m2\,x[t]^{11}\,x'[t]^2 -$$
$$448\,m1\,x[t]^{13}\,x'[t]^2 - 448\,m2\,x[t]^{13}\,x'[t]^2 +$$
$$1\,m2\,\text{Sin}[\phi[t]]\,\phi'[t]^2 + 12\,1\,m2\,\text{Cos}[\phi[t]]\,x[t]^5\,\phi'[t]^2 -$$
$$8\,1\,m2\,\text{Cos}[\phi[t]]\,x[t]^7\,\phi'[t]^2 - m1\,x''[t] - m2\,x''[t] -$$
$$144\,m1\,x[t]^{10}\,x''[t] - 144\,m2\,x[t]^{10}\,x''[t] +$$
$$192\,m1\,x[t]^{12}\,x''[t] + 192\,m2\,x[t]^{12}\,x''[t] -$$
$$64\,m1\,x[t]^{14}\,x''[t] - 64\,m2\,x[t]^{14}\,x''[t] -$$
$$1\,m2\,\text{Cos}[\phi[t]]\,\phi''[t] + 12\,1\,m2\,\text{Sin}[\phi[t]]\,x[t]^5\,\phi''[t] -$$
$$8\,1\,m2\,\text{Sin}[\phi[t]]\,x[t]^7\,\phi''[t] == 0$$

```
el2 = 𝓔φᵗ[L]
```

$$-1\,m2\,g\,\text{Sin}[\phi[t]] + 60\,1\,m2\,\text{Sin}[\phi[t]]\,x[t]^4\,x'[t]^2 -$$
$$56\,1\,m2\,\text{Sin}[\phi[t]]\,x[t]^6\,x'[t]^2 -$$
$$1\,m2\,\text{Cos}[\phi[t]]\,x''[t] + 12\,1\,m2\,\text{Sin}[\phi[t]]\,x[t]^5\,x''[t] -$$
$$8\,1\,m2\,\text{Sin}[\phi[t]]\,x[t]^7\,x''[t] -$$
$$1^2\,m2\,\text{Cos}[\phi[t]]^2\,\phi''[t] - 1^2\,m2\,\text{Sin}[\phi[t]]^2\,\phi''[t] == 0$$

The derived Euler–Lagrange equations are a set of coupled nonlinear second-order equations. It is likely that this set of equations does not allow a symbolic solution Thus, we switch to numerical work and specify the values for parameters as well as initial conditions. The following list contains one example for the parameters:

```
parameters = {m1 -> 1, m2 -> 1.5, 1 -> .7, g -> 9.81}
```

$$\{m1 \to 1, \ m2 \to 1.5, \ 1 \to 0.7, \ g \to 9.81\}$$

The numerical solution of the two Euler–Lagrange equations then follows upon specifying the initial conditions for x, x', ϕ, and ϕ'. The following line contains all these steps:

```
nsol = NDSolve[{el1, el2, x[0] == .01, x'[0] == 0.3,
     φ[0] == 0.5, φ'[0] == 0.01} /. parameters,
   {x, φ}, {t, 0, 43}, MaxSteps -> 11000]
```

$$\{\{x \to \text{InterpolatingFunction}[\{\{0., 43.\}\}, <>],$$
$$\phi \to \text{InterpolatingFunction}[\{\{0., 43.\}\}, <>]\}\}$$

The resulting functions can be represented in a plot showing that both coordinates oscillate. It is also obvious that the solution for x increases in time. Thus, the presnt model shows similar behavior as the solution of the original model.

```
Plot[Evaluate[{x[t], φ[t]} /. nsol], {t, 0, 43},
  AxesLabel -> {"t", "x,φ"}, PlotStyle ->
  {RGBColor[1, 0, 0], RGBColor[0, 0, 1]}];
```

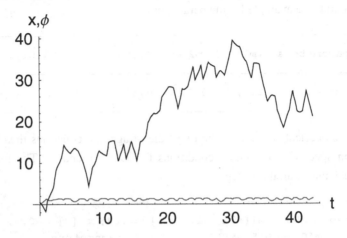

The solutions obtained can be used to generate a flip-chart movie showing the movement of the two masses:

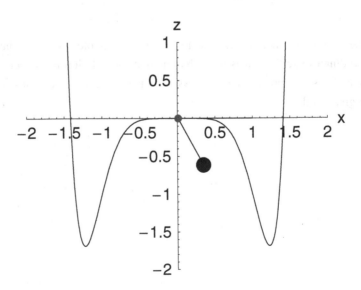

2.7.4 Symmetries and Conservation Laws

The solution of equations of motion are tightly connected with conservation laws. Conservation laws allow to reduce the number of integration steps, iff they are known. A method for determining symmetries of differential equations is given in the author's book [2.9]. We start our examinations with the Euler–Lagrange equations

$$\frac{\partial \mathcal{L}}{\partial q_i} - \frac{d}{dt}\left(\frac{\partial \mathcal{L}}{\partial q'_i}\right) = 0, \qquad i = 1, 2, \ .... \qquad (2.7.31)$$

If the Lagrange density $\mathcal{L} = \mathcal{L}(q_i, q'_i, t)$ is independent of a coordinate q_i, that is

$$\frac{\partial \mathcal{L}}{\partial q_i} = 0, \qquad (2.7.32)$$

we call this coordinate cyclic or ignorable. For the ith Euler–Lagrange equation, it immediately follows that

$$-\frac{d}{dt}\left(\frac{\partial \mathcal{L}}{\partial q'_i}\right) = 0. \qquad (2.7.33)$$

In other words,

$$\frac{\partial \mathcal{L}}{\partial q'_i} = \text{const.} \qquad (2.7.34)$$

representing a conserved quantity. If a Lagrangian contains cyclic variables (i.e., is independent of this variable), then the derivative with respect to the generalized velocity is a conserved quantity. This conserved quantity represents an ordinary differential equation of first order. Thus, the second-order differential equation in the ith component is replaced by a first-order one.

The cyclic behavior of the Lagrangian is mainly dependent on the choice of coordinates for the problem. Thus, it is useful to choose such coordinates that generate a large number of cyclic coordinates. Conservation laws are thus related to cyclic coordinates. On the other hand, conservation laws are related to symmetries allowed by the Lagrangian. For example, conservation of energy is connected with the symmetry of translations with respect to time. Conservation of momentum is a consequence of the translation symmetry in space. Conservation of angular momentum follows from the rotation symmetry of the Lagrangian.

All of the mentioned conservation laws can be represented as balance equations.

2.7.4.1 Conservation of Energy and Translation in Time

To check conservation of energy, we examine the Lagrangian with respect to translations in time. A consistent representation of the formulas is gained by using the Euler–Lagrange equations:

```
Remove[L]
```

```
∂_t L[q_i[t], ∂_t q_i[t], t]
```

```
L^(0,0,1) [q_i[t], q'_i[t], t] + q''_i[t] L^(0,1,0) [q_i[t], q'_i[t], t] +
  q'_i[t] L^(1,0,0) [q_i[t], q'_i[t], t]
```

$$\frac{d}{dt}\mathcal{L}(q_i, q'_i, t) = \sum_i \left\{ \frac{\partial\mathcal{L}}{\partial q_i} q'_i + \frac{\partial\mathcal{L}}{\partial q'_i} q''_i \right\} + \frac{\partial\mathcal{L}}{\partial t}$$

$$= \sum_i \left\{ q'_i \frac{d}{dt}\left(\frac{\partial\mathcal{L}}{\partial q'_i} \right) + \frac{\partial\mathcal{L}}{\partial q'_i} q''_i \right\} + \frac{\partial\mathcal{L}}{\partial t}$$

$$= \sum_i \frac{d}{dt}\left(\frac{\partial\mathcal{L}}{\partial q'_i} q'_i \right) + \frac{\partial\mathcal{L}}{\partial t} \tag{2.7.35}$$

$$= \frac{d}{dt}\left(\sum_i \frac{\partial\mathcal{L}}{\partial q'_i} q'_i \right) + \frac{\partial\mathcal{L}}{\partial t}$$

Collecting the time derivatives to a total time derivative we obtain

$$\frac{d}{dt}\left(\mathcal{L} - \sum_i \frac{\partial\mathcal{L}}{\partial q'_i} q'_i \right) = \frac{\partial\mathcal{L}}{\partial t}, \tag{2.7.36}$$

$$\frac{d}{dt}\left(\sum_i \frac{\partial\mathcal{L}}{\partial q'_i} q'_i - \mathcal{L} \right) = -\frac{\partial\mathcal{L}}{\partial t}. \tag{2.7.37}$$

This relation represents the energy balance in terms of the Lagrangian and the generalized coordinates.

Assuming scleronomic constraints, the cartesian coordinates $x_\beta = x_\beta(q_i)$ are independent of time. Thus, the kinetic and potential energies are also independent of time. Consequently, the Lagrangian is a pure function of the coordinates independent of time (i.e., $\partial\mathcal{L}/\partial t = 0$). Thus, we get

$$\frac{d}{dt}\left(\sum_i \frac{\partial\mathcal{L}}{\partial q'_i} q'_i - \mathcal{L} \right) = 0 \tag{2.7.38}$$

and

$$\sum_i \frac{\partial \mathcal{L}}{\partial q'_i} q'_i - \mathcal{L} = \text{const.} \tag{2.7.39}$$

This expression is a conserved quantity remaining constant in a time evolution. Applying Euler's homogeneity relation on the sum of the left-hand side, we get

$$\sum_i \frac{\partial \mathcal{L}}{\partial q'_i} q'_i = \sum_i \frac{\partial T}{\partial q'_i} q'_i = 2T, \tag{2.7.40}$$

and taking the Lagrangian as $\mathcal{L} = T - V$ that it follows,

$$2T - T + V = T + V = \text{const.} = H. \tag{2.7.41}$$

Conservation of energy is guaranteed if the Lagrangian is invariant with respect to time translations (i.e., independent of time). For such a case, the Lagrangian does not change if we move in time. This behavior also means that the total number of possible tracks starting at a fixed time are independent of the initial time. Consequently, there is no way to determine by observation of the tracks the initial time if the acting forces are known.

The connection between conservation laws and invariants or symmetries are very important in all fields of modern physics.

The derived function H is known as Hamilton's function and is also called the Hamiltonian. The Hamiltonian in terms of the Lagrangian is given by

$$H = \sum_i \frac{\partial \mathcal{L}}{\partial q'_i} q'_i - \mathcal{L}. \tag{2.7.42}$$

Note: The Hamiltonian is identical to the total energy if the following two requirements are satisfied:

i)   The kinetic energy is homogeneous of degree 2.

ii)   The potential energy is independent of the velocity.

2.7.4.2 Conservation of Momentum

Assuming that space is homogenous in an inertial system, we can conclude that the Lagrangian is invariant with respect to spatial translations in the case of a closed system. To prove this conclusion, let us consider an infinitesimal transformation of the coordinates:

```
itrafo = q -> Function[ε, q + ε ξ[q]]
```

```
q → Function[ε, q + ε ξ[q]]
```

$$\tilde{q}_i = q_i + \varepsilon \xi_i(q_i), \qquad\qquad\qquad (2.7.43)$$

with ε an infinitesimal parameter and $\xi_i(q_i)$ as the infinitesimal element of the global transformation:

```
Series[q[q, ε], {ε, 0, 1}]
```

```
q[q, 0] + q^(0,1)[q, 0] ε + O[ε]²
```

$$\begin{aligned}
\tilde{q}_i &= \tilde{q}_i(q_i, \varepsilon) \\
&= \tilde{q}_i(q_i, \varepsilon = 0) + \frac{\partial \tilde{q}_i}{\partial \varepsilon}\Big|_{\varepsilon=0} + O(\varepsilon^2) \qquad (2.7.44) \\
&= q_i + \varepsilon \, \xi_i(q_i) + O(\varepsilon^2)
\end{aligned}$$

with $\xi_i = \partial \tilde{q}_i / \partial \varepsilon \big|_{\varepsilon=0}$.

Consider the Lagrangian as a function of the new coordinates $\tilde{q}_i$, so that $\tilde{\mathcal{L}} = \tilde{\mathcal{L}}(\tilde{q}_i, \tilde{q}'_i)$ represents the transformed Lagrangian. Expanding this new Lagrangian around the identity $\varepsilon = 0$, we find

$$\tilde{\mathcal{L}} = \tilde{\mathcal{L}}\Big|_{\varepsilon=0} + \varepsilon \frac{\partial \tilde{\mathcal{L}}}{\partial \varepsilon}\Big|_{\varepsilon=0} + O(\varepsilon^2) \qquad\qquad (2.7.45)$$

with

$$\tilde{\mathcal{L}}\big|_{\varepsilon=0} = \mathcal{L}(q_i, q'_i) = \mathcal{L}. \qquad\qquad (2.7.46)$$

Now, if we set

$$\tilde{\mathcal{L}} - \mathcal{L} = \delta\mathcal{L} = \frac{\partial \tilde{\mathcal{L}}}{\partial \varepsilon}\Big|_{\varepsilon=0} \varepsilon + O(\varepsilon^2), \qquad\qquad (2.7.47)$$

where terms of order $0(\varepsilon^2)$ vanish. If we assume that $\mathcal{L}$ is invariant with respect to the infinitesimal transformation then, we find

$$\tilde{\mathcal{L}} = \mathcal{L} \tag{2.7.48}$$

and thus we get the sufficient condition

$$\delta \mathcal{L} = 0 = \left. \frac{\partial \tilde{\mathcal{L}}}{\partial \varepsilon} \right|_{\varepsilon=0} \varepsilon + 0(\varepsilon^2). \tag{2.7.49}$$

In first-order ε, we can set

$$\left. \frac{\partial \tilde{\mathcal{L}}}{\partial \varepsilon} \right|_{\varepsilon=0} = 0. \tag{2.7.50}$$

Writing this formula explicitly, we find

$$\sum_i \frac{\partial \tilde{\mathcal{L}}}{\partial \tilde{q}_i} \left. \frac{\partial \tilde{q}_i}{\partial \varepsilon} \right|_{\varepsilon=0} + \sum_i \frac{\partial \tilde{\mathcal{L}}}{\partial \tilde{q}'_i} \left. \frac{\partial \tilde{q}'_i}{\partial \varepsilon} \right|_{\varepsilon=0} = 0 \tag{2.7.51}$$

$$\left. \frac{\partial \tilde{q}_i}{\partial \varepsilon} \right|_{\varepsilon=0} = \xi_i(q_i); \tag{2.7.52}$$

$\partial \tilde{q}'_i / \partial \varepsilon = 0$ since $\tilde{q}'_i = q'_i$ velocities are not due to transformations and thus, we can write

$$\sum_i \frac{\partial \tilde{\mathcal{L}}}{\partial \tilde{q}_i} \left. \frac{\partial \tilde{q}_i}{\partial \varepsilon} \right|_{\varepsilon=0} = 0 \iff \sum_i \frac{\partial \mathcal{L}}{\partial q_i} \xi_i(q_i) = 0 \tag{2.7.53}$$

$$\longrightarrow \frac{\partial \mathcal{L}}{\partial q_i} = 0. \tag{2.7.54}$$

Taking the Euler–Lagrange equations into account, we get

$$- \frac{d}{dt} \left(\frac{\partial \mathcal{L}}{\partial q'_i} \right) = 0 \tag{2.7.55}$$

or

$$\frac{\partial \mathcal{L}}{\partial q'_i} = \text{const.} \tag{2.7.56}$$

The Lagrangian is assumed to be expressed by the difference of kinetic and potential energy. In addition, the kinetic energy is a homogenous function of degree 2. Taking these considerations into account, we get

$$\frac{\partial \mathcal{L}}{\partial q'_i} = \frac{\partial}{\partial q'_i} (T - V) = \frac{\partial T}{\partial q'_i} = m_i q'_i = \text{const.}, \tag{2.7.57}$$

$$\frac{\partial \mathcal{L}}{\partial q'_i} = p_i(t) = p_i(0). \tag{2.7.58}$$

The total linear momentum thus becomes

$$\sum_i \frac{\partial \mathcal{L}}{\partial q'_i} = \sum_i p_i(t) = \sum_i p_i(0) = P(0). \tag{2.7.59}$$

In conclusion, the total momentum is a conserved quantity. This result holds for a spatial homogenous system.

2.7.4.3 Conservation of Angular Momentum

The discussion of inertial systems revealed that the related space is isotropic, meaning that the mechanical properties are independent of the orientation in space. Especially the Lagrangian is invariant with respect to an infinitesimal rotation. We restrict our considerations to infinitesimal rotations because global rotations are generated by many infinitesimal rotations.

Rotation of a system by an infinitesimal angle $\delta\theta$ transforms a position vector $\vec{r}$ to another position vector $\vec{r} + \delta\vec{r}$ (see Figure 2.7.9)

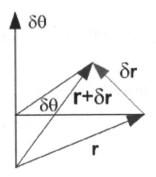

Figure 2.7.9.   Rotation of a position vector $\vec{r}$.

The infinitesimal position vector is determined by

$$\delta\vec{r} = \delta\vec{\theta} \times \vec{r}. \tag{2.7.60}$$

In addition to the change of the position vector, an infinitesimal rotation changes the velocity also. The infinitesimal velocity change is determined by

$$\delta\vec{r}' = \delta\vec{\theta} \times \vec{r}'. \tag{2.7.61}$$

Now, consider a single particle in cartesian coordinates. The infinitesimal change of the Lagrangian in these coordinates is given by

$$\delta \mathcal{L} = \sum_i \frac{\partial \mathcal{L}}{\partial x_i} \delta x_i + \sum_i \frac{\partial \mathcal{L}}{\partial x'_i} \delta x'_i = 0. \tag{2.7.62}$$

On the other hand, the linear momenta are represented by means of the Lagrangian by

$$P_i = \frac{\partial \mathcal{L}}{\partial x'_i}. \tag{2.7.63}$$

The temporal change of the momenta are thus

$$\frac{dp_i}{dt} = \frac{d}{dt}\left(\frac{\partial \mathcal{L}}{\partial x'_i}\right) = \frac{\partial \mathcal{L}}{\partial x_i}. \tag{2.7.64}$$

This relation holds because the Euler–Lagrange equations are satisfied. Thus, the infinitesimal change of the Lagrangian is given by

$$\delta \mathcal{L} = \sum_{i=1}^3 p'_i \delta x_i + \sum_{i=1}^3 p_i \delta x'_i = 0. \tag{2.7.65}$$

The components can be replaced by the vectors and result in the relation

$$\vec{p}'.\delta \vec{r} + \vec{p}\delta \vec{r}' = 0. \tag{2.7.66}$$

Using the infinitesimal representations of the position vector and the velocity, we obtain

$$\vec{p}'\left(\delta\vec{\theta} \times \vec{r}\right) + \vec{p}\left(\delta\vec{\theta} \times \vec{r}'\right) = 0. \tag{2.7.67}$$

A cyclic interchange of infinitesimal vectors and vectors provides the compact relation

$$\delta\vec{\theta}\{(\vec{r} \times \vec{p}') + (\vec{r}' \times \vec{p})\} = 0 \tag{2.7.68}$$
$$\Leftrightarrow \delta\vec{\theta} . \frac{d}{dt}(\vec{r} \times \vec{p}) = 0. \tag{2.7.69}$$

Since the infinitesimal change of the angle was arbitrary, we conclude that the temporal change of the cross-product vanishes:

$$\frac{d}{dt}(\vec{r} \times \vec{p}) = 0, \tag{2.7.70}$$

meaning that the quantity

$$\vec{r} \times \vec{p} = \text{const.} = \vec{l}. \tag{2.7.71}$$

is a conserved quantity. The presented infinitesimal changes of the Lagrangian all result in a conserved quantity. In general, the infinitesimal changes are related to symmetries of the Lagrangian. The symmetries itself are determined by infinitesimal transformations. In modern physics, this relation between symmetries and conserved quantities is very important. Symmetries determine the conserved quantities and vice versa. In the above discussions, we considered the simplest symmetries (translations and rotations) under which a Lagrangian may be invariant. However, there are many more symmetry transformations related to other conserved quantities. The results so far derived are collected in the following table:

Properties of the inertial system	propeties of $\mathcal{L}$	conserved quantity
homogenity in time	independent of time	total energy
homogenity in space	translation invariance	linear omentum
isotropy of space	rotation invariance	angular momentum

Table 2.7.2.   Lagrangian properties and conserved quantities.

The symmetry considerations are far more general then presented above. This generalization was given by Emmy Noether (Figure 2.7.10) in her famous theorem in 1915.

Figure 2.7.10.   Emmi Noether born March 23, 1882; died April 14, 1935.

Theorem: Noether Theorem

Given the time-independent Lagrangian $\mathcal{L}(q_i, q'_i)$ of an holonomic system which is invariant with respect to an invertible transformation around the identity with $\varepsilon = 0$,

$$q_i \to \tilde{q}_i = \tilde{q}_i(q_j, \varepsilon) \qquad (2.7.72)$$

with $\tilde{q}_i(q_j, \varepsilon = 0) = q_i$, that is for all ε, we have

$$\mathcal{L}(q_i, q'_i) = \mathcal{L}\left(q_i, (\tilde{q}_j, \varepsilon), \sum_{j=1}^{N} \frac{\partial q_i}{\partial \tilde{q}_j} \tilde{q}_j\right)$$
$$= \mathcal{L}(\tilde{q}_i, \tilde{q}'_i, \varepsilon) = \mathcal{L}(\tilde{q}_i, \tilde{q}'_i), \qquad (2.7.73)$$

then the quantity

$$I(q_i, q'_i) = \sum_j \frac{\partial \mathcal{L}}{\partial q'_j} \frac{\partial \tilde{q}_j(q_i, \varepsilon)}{\partial \varepsilon} \bigg|_{\varepsilon=0} \qquad (2.7.74)$$

is a conserved quantity of the Euler–Lagrange equations.■

The symmetries under which the Lagrangian is invariant are also known as continuous symmetries. This notion was introduced because ε is a continuous parameter determining the symmetries of the corresponding group.

In the following, we prove the Noether theorem. We start by checking the invariance of the Euler–Lagrange equations under coordinate transformations

$$\frac{\partial \mathcal{L}}{\partial \tilde{q}_i} - \frac{d}{dt} \frac{\partial \mathcal{L}}{\partial \tilde{q}'_i} = 0. \qquad (2.7.75)$$

The check can be carried out by replacing $q \to Q$; this is left as an exercise for the reader. The derivation of the transformed equation with respect to ε gives us

$$\frac{d\mathcal{L}(q_i, q'_i)}{d\varepsilon} = 0 = \sum_i \frac{\partial \mathcal{L}}{\partial \tilde{q}_i} \frac{d\tilde{q}_i}{d\varepsilon} + \frac{\partial \mathcal{L}}{\partial \tilde{q}'_i} \frac{d\tilde{q}'_i}{d\varepsilon}; \qquad (2.7.76)$$

again using the Euler–Lagrange equation, we find

$$0 = \sum_i \frac{d}{dt}\left(\frac{\partial \mathcal{L}}{\partial \tilde{q}'_i}\right) \frac{d\tilde{q}_i}{d\varepsilon} + \frac{\partial \mathcal{L}}{\partial \tilde{q}'_i} \frac{d}{dt} \frac{d\tilde{q}_i}{d\varepsilon}$$
$$= \frac{d}{dt}\left(\sum_i \frac{\partial \mathcal{L}}{\partial \tilde{q}'_i} \frac{d\tilde{q}_i}{d\varepsilon}\right). \qquad (2.7.77)$$

Thus, the expression

$$I(q_i, q'_i, \varepsilon) = \sum_i \frac{\partial \mathcal{L}}{\partial \tilde{q}'_i} \frac{d\tilde{q}_i}{d\varepsilon} \qquad (2.7.78)$$

is a conserved quantity for any ε.

The resulting integrals are linearly dependent on each other for different ε's. Thus, it is sufficient to consider only one value for ε. We chose $\varepsilon = 0$ and get

$$I(q_i, q'_i, \varepsilon = 0) = \sum_i \frac{\partial \mathcal{L}}{\partial \tilde{q}'_i} \frac{d\tilde{q}_i}{d\varepsilon} \Big|_{\varepsilon=0} = \sum_i \frac{\partial \mathcal{L}}{\partial q'_i} \xi_i. \qquad (2.7.79)$$

This is the conserved quantity given in the Noether theorem.

Example 1: Invariant Lagrangian

Let us consider the Lagrangian

$$\mathcal{L} = \frac{m}{2} (x'^2 + y'^2) - V(x) = \mathcal{L}(x, x', y'). \qquad (2.7.80)$$

As a transformation consider

$$X = x, \qquad (2.7.81)$$
$$Y = y + \varepsilon, \qquad (2.7.82)$$

with ε a constant. Applying the transformation to the Lagrangian with ε a constant, we find

$$\begin{aligned}
\mathcal{L}(x, x', y') &= \mathcal{L}(X, X', Y' - \dot{\varepsilon}) \\
&= \mathcal{L}(X, X', Y', \varepsilon) \\
&= \frac{m}{2} (X'^2 + Y'^2) - V(X) \\
&= \mathcal{L}(X, X', Y').
\end{aligned} \qquad (2.7.83)$$

Invariance of the Lagrangian guaranties the assumptions in the Noether theorem. The conserved quantity is thus given by

$$\begin{aligned}
I &= \left(\frac{\partial \mathcal{L}}{\partial X'} \frac{\partial x}{\partial \varepsilon} + \frac{\partial \mathcal{L}}{\partial Y'} \frac{\partial (y+\varepsilon)}{\partial \varepsilon} \right)\Big|_{\varepsilon=0} \\
&= \frac{\partial \mathcal{L}}{\partial Y'} \Big|_{\varepsilon=0} \\
&= m Y' \big|_{\varepsilon=0} \\
&= m y'.
\end{aligned} \qquad (2.7.84)$$

Thus, the y-component of the linear momentum is a conserved quantity.

2.7.5 Exercises

1. Show that the equations of motion derivable from a Lagrangian are unchanged if to the Lagrangian there is added the total time derivative of an arbitrary function of q_m, and t.

2. Write down the expressions for the kinetic energy of the following systems, using the minimum number of coordinates: (i) a free particle; (ii) a particle constrained to remain on a sphere; (iii) a particle constrained to remain on a circular cylinder.

3. Write down the Lagrangian for a particle confined to a horizontal plane in cartesian coordinates. Introduce the additional constraint $x^2 + y^2 = a^2$ by means of a Lagrange multiplier λ and show that λ is proportional to the centripedal force exerted by the constraint upon the particle.

2.7.6 Packages and Programs

Euler–Lagrange Package

The EulerLagrange package serves to derive the Euler–Lagrange equations from a given Lagrangian.

```
If[$MachineType == "PC",
  $EulerLagrangePath = $TopDirectory <>
    "/AddOns/Applications/EulerLagrange/";
  AppendTo[$Path, $EulerLagrangePath],
  $EulerLagrangePath =
   StringJoin[$HomeDirectory, "/.Mathematica/3.0/
     AddOns/Applications/EulerLagrange", "/"];
  AppendTo[$Path, $EulerLagrangePath]];
```

The next line loads the package.

```
<< EulerLagrange.m
```

Get::noopen : Cannot open EulerLagrange.m. More…

```
$Failed
```

```
Options[EulerLagrange]
```

```
{eXpand → False}
```

```
SetOptions[EulerLagrange, eXpand → True]
```

SetOptions::optnf :
  eXpand is not a known option for EulerLagrange. More…

```
SetOptions[EulerLagrange, eXpand → True]
```

Define some notations.

```
<< Utilities`Notation`
```

Define the notation of a variational derivative connected with the Euler–Lagrange function.

$$\texttt{Notation}\left[\frac{\delta}{\delta u\_}\int_{t_1}^{t_2}f\_\,d\,t\_ \iff \texttt{EulerLagrange[f\_, u\_, t\_]}\right]$$

To access the variational derivative, we define an alias variable *var* allowing us to access the symbolic definition by the escape sequence [ESC] *var* [ESC].

$$\texttt{AddInputAlias}\left[\frac{\delta}{\delta \square}\int_{t_1}^{t_2}\square\,d\,\square, \texttt{"var"}\right]$$

Here is an example for an arbitrary Lagrangian:

$$\frac{\delta}{\delta u} \int_{t_1}^{t_2} \texttt{L[u[t], } \partial_t \texttt{ u[t]] d t}$$

$$-\partial_{\{t,1\}} \texttt{L}^{(0,1)} \texttt{[u[t], u'[t]]} + \texttt{L}^{(1,0)} \texttt{[u[t], u'[t]] == 0}$$

We also define an Euler–Lagrange operator allowing us to access the Euler–Lagrange functon as a symbol:

$$\texttt{Notation} \Big[\mathcal{E}_{\texttt{u\_}}^{\texttt{x\_}} \texttt{[den\_]} \Longleftrightarrow \texttt{EulerLagrange[den\_, u\_, x\_]} \Big]$$

Here is the alias notation for the Euler–Lagrange operator:

$$\texttt{AddInputAlias} \Big[\mathcal{E}_{\square}^{\square} \texttt{[}\square\texttt{]}, \texttt{ "ELop"} \Big]$$

2.8 Hamiltonian Dynamics

2.8.1 Introduction

Hamiltonian dynamics is an alternative formulation of the Lagrangian dynamics. In Lagrangian dynamics, we used the generalized coordinates q_i and velocities q'_i as basic variables. Hamilton's dynamic introduces a set of canonical variables which are basically the coordinates q_i and the generalized momenta p_i. We defined the generalized momenta in Lagrange's dynamic by the relation

$$p_i = \frac{\partial \mathcal{L}}{\partial q'_i}, \qquad\qquad i = 1, 2, \ldots \qquad\qquad (2.8.1)$$

In a similar way, the generalized forces F_i were expressed by the relations

$$F_i = \frac{\partial \mathcal{L}}{\partial q_i}, \qquad\qquad i = 1, 2, \ldots \qquad\qquad (2.8.2)$$

If the generalized coordinates q_i are identical with the cartesian coordinates, we can identify the generalized momenta with the linear momenta $p_i = mq'_i$. On the other hand, the Euler–Lagrange equations are reduced to Newton's second law:

$$p'_i = F_i, \qquad\qquad i = 1, 2, \ldots, N. \qquad\qquad (2.8.3)$$

The main advantage of the Hamilton formulation is that different theories such as quantum mechanics, statistical physics, and perturbation theory can be based on this formulation. Hamilton's formulation of classical mechanics also allows a natural approach to chaotic systems and the question of integrability. The concept of a phase space opens the door for an efficient study of integrability and nonintegrability. However, Hamilton's formulation of classical mechanics introduces nothing new in physics but allows an efficient treatment of mechanical systems. The two formulations, Lagrange's and Hamilton's, are equivalent to each other and allow a direct transition between the two theories.

2.8.2 Legendre Transform

We demonstrate here that the Hamilton and Lagrange formulation of classical mechanics can be transformed into each other. Lagrange used for his formulation of mechanics the generalized coordinates and velocities (q_i, q'_i) as basic quantities. Contrary Hamilton decided to introduce the fundamental coordinate set (q_i, p_i) where q_i are the generalized coordinates as in the Lagrange formulation and p_i are the generalized momenta. Already Euler and Leibniz knew that a transformation between such basic quantities exists. The two sets of coordinates can be converted into each other by a so called Legendre transform. This transform uses the property that a function $f = f(x)$ can be either represented by the standard set of coordinates (x, f) or by the coordinate and the functions tangent. To demonstrate these relations let us consider a function

$$y = f(x) \tag{2.8.4}$$

under the restriction that $\partial^2 f / \partial x^2 > 0$; tat is, we consider convex functions. Under this assumption, the Legendre transform of f is a new function g depending on a new variable s. The relations among f, g, and s are defined in Figure 2.81.

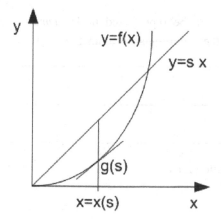

Figure 2.8.1.   Legendre transform of a function $y = f(x)$ to its Legendre representation $g(s)$.

Figure 2.8.1 shows that $g(s)$ counts the maximal distance between the inclined line $y = sx$ and the function $f(x)$; that is,

$$g(s) = \text{s}x - f(x) = G(s, \, x(s)). \tag{2.8.5}$$

Since $x(s)$ is defined as maximum of g, we find from this relation

$$\frac{\partial G}{\partial x} = s - f'(x) = 0. \tag{2.8.6}$$

It is obvious that the new variable s can be identified with the tangent of $f(x)$; that is,

$$s = f'(x). \tag{2.8.7}$$

Since f is convex, $x = x(s)$ is uniquely determined.

Let us consider a mechanical example which allows a Hamilton function of the kind $H = H(p)$. We also state at this moment that one of Hamilton's equations is given by $q' = \partial H / \partial p$. If we carry out the above construction in the $(y, \, p)$-plane and call the new function $L(s)$, we find

$$L(s) = sp - H(p). \tag{2.8.8}$$

The new variable s follows now from the extremal condition as $s = \partial H / \partial p = q'$ so that the Legendre transform becomes

$$L(q') = q'p - H(p), \tag{2.8.9}$$

where p is a function of q' defined by $q' = \partial H / \partial p$.

The above theoretical steps can be represented in *Mathematica* by the following lines. First, define the Hamiltonian as a function of p:

```
H = h[p]
```

$h(p)$

Then, introduce the new function L as

```
L = s p - H
```

$$p\, s - h(p)$$

The extremal condition allows on to establish an equation which provides the new variable s:

```
et1 = ∂_p L == 0
```

$$s - h'(p) == 0$$

If we solve with respect to s and take into account one of Hamilton's equations

```
subst = Flatten[Solve[et1, s]] /. h'[p] -> q'
```

$$\{s \rightarrow q'\}$$

we find that s is just given by q'. Substituting this knowledge into the function L, we obtain

```
L /. subst
```

$$p\, q' - h(p)$$

as a function of q'. The procedure to carry out a Legendre transform is thus algorithmic and can be implemented in a single function. Before we implement the Legendre transform, let us consider the more general case when the Hamiltonian is a function of several independent variables.

The generalization of this result to a Hamiltonian depending on a set of N coordinates $(q_i,\ p_i)$ is given by

$$H(q_i,\ p_i) = \sum_{i=1}^{N} p_i\, q'_i - \mathcal{L}; \qquad (2.8.10)$$

the corresponding Lagrangian is then defined by

$$\mathcal{L} = \sum_{i=1}^{N} p_i q'_i - H, \tag{2.8.11}$$

where the generalized momenta p_i are defined by the standard relation

$$p_i = \frac{\partial \mathcal{L}}{\partial q'_i}. \tag{2.8.12}$$

These relations are valid under the assumption that the Jacobian determinant

$$\Delta = \det\left(\frac{\partial^2 \mathcal{L}}{\partial q'_i \, \partial q'_j}\right) \neq 0 \tag{2.8.13}$$

does not vanish. The property that $\Delta \neq 0$ guarantees that the generalized velocities q'_i can be uniquely solved for p_i and vice versa. This relation is a generalization of the convexity.

As an example, let us consider the following Lagrange density:

$$\mathcal{L} = \sum_{i=1}^{N} \frac{1}{2} m_i q'^2_i - V(q_i). \tag{2.8.14}$$

First, we determine the generalized moment by

$$p_i = \frac{\partial \mathcal{L}}{\partial q'_i} = m_i q'_i. \tag{2.8.15}$$

The check of convexity shows

$$\frac{\partial^2 \mathcal{L}}{\partial q'_i \, \partial q'_j} = m_i \delta_{ij}, \tag{2.8.16}$$

$$\det\left(\frac{\partial^2 \mathcal{L}}{\partial q'_i \, \partial q'_j}\right) = \det(m_i \delta_{ij}) \neq 0. \tag{2.8.17}$$

This relation guarantees that the generalized momenta can be expressed by the generalized velocities; that is,

$$q'_i = \frac{p_i}{m_i}. \tag{2.8.18}$$

Thus, the Hamiltonian in q_i and p_i is given by

$$
\begin{aligned}
H(q_i, \ p_i) \\
&= \sum_{i=1}^{N} p_i \frac{p_i}{m_i} - \left\{ \sum_{i=1}^{N} \frac{1}{2} m_i \left(\frac{p_i}{m_i}\right)^2 - V(q_i) \right\} \\
&= \sum_{i=1}^{N} \frac{p^2_i}{2m} + V(q_i).
\end{aligned}
\tag{2.8.19}
$$

The procedure discussed above is implemented in the following function. The function **LegendreTranform[]** allows one to transform a given density to an alternate representation:

```
LegendreTransform[A_, x_List, momenta_List,
   indep_ : {t}] := Block[{momentaRelations},
   momentaRelations =
   MapThread[∂#1 A == #2 &, {x, momenta}];
   sol = Flatten[Solve[momentaRelations, x]];
            Length[x]
   Simplify[Expand[   ∑     x[[i]] ∂x[[i]] A - A] /. sol]]
            i=1
```

The following Lagrangian density with two degrees of freedom describes two particles interacting by a general potential V:

```
Clear[V]
```

```
l = m1/2 (∂t q1[t])² + m2/2 (∂t q2[t])² - V[q1[t], q2[t]]
```

$$\frac{1}{2} m1\, q1'(t)^2 + \frac{1}{2} m2\, q2'(t)^2 - V(q1(t), q2(t))$$

The transformation to a Hamiltonian needs the Lagrangian and the sets of original and final variables.

```
h = LegendreTransform[l,
   {∂t q1[t], ∂t q2[t]}, {p1[t], p2[t]}]
```

$$\frac{p1(t)^2}{2\,m1} + \frac{p2(t)^2}{2\,m2} + V(q1(t), q2(t))$$

The result is a representation of the Hamiltonian in a new set of coordinates (q_i, p_i). The back transformation to the Lagrangian uses the Hamiltonian as density, the set of momenta as initial coordinates, and the generalized velocities as target coordinates of the transformation:

```
LegendreTransform[h,
  {p1[t], p2[t]}, {∂_t q1[t], ∂_t q2[t]}]
```

$$\frac{1}{2} \left(m1\, q1'(t)^2 + m2\, q2'(t)^2 - 2\, V(q1(t), q2(t)) \right)$$

This simple example can be extended to a more complicated one.

Example 1: Moving Beat on a String

Let us consider a beat (mass point) in a homogenous gravitational field. The beat is restricted to move on a string of the form $y = f(x)$. The functional relation of the string acts as a constraint on the movement of the mass point. Let us first discuss the movement without any constraint. Thus, we have to use two coordinates in the Lagrangian. The kinetic energy for a plane movement is given by

```
T = 1/2 m ((∂_t x[t])² + (∂_t y[t])²)
```

$$\frac{1}{2} m \left(x'(t)^2 + y'(t)^2 \right)$$

The potential energy is

```
V = m g y[t]
```

$$g\, m\, y(t)$$

and, thus, the Lagrangian is

```
L = T - V
```

$$\frac{1}{2} m \left(x'(t)^2 + y'(t)^2 \right) - g\, m\, y(t)$$

If we now introduce the constraint of the movement by $y = f(x)$, we can write the Lagrangian as

```
lconstr = L /. y -> Function[t, f[x[t]]]
```

$$\frac{1}{2} m \left(f'(x(t))^2 \, x'(t)^2 + x'(t)^2 \right) - g \, m \, f(x(t))$$

We observe that the degree of freedom of this problem reduces from two to one if the constraint is introduced in the Lagrangian. The Hamiltonian for this Lagrangian then follows by applying the function **LegendreTransform[]** to the Lagrangian:

```
ham = LegendreTransform[lconstr, {∂ₜx[t]}, {p[t]}]
```

$$\frac{2 g \, f(x(t)) \, (f'(x(t))^2 + 1) \, m^2 + p(t)^2}{2 \, m \, (f'(x(t))^2 + 1)}$$

The result is a nontrivial expression for the Hamiltonian combining the coordinate and momenta by means of the arbitrary function f. To understand how the transformation was carried out, let us calculate the generalized momentum from the Lagrangian by

```
mom = ∂∂ₜx[t] lconstr // Simplify
```

$$m \left(f'(x(t))^2 + 1 \right) x'(t)$$

The result shows that for this case, the generalized momentum is not only a function of the velocity x' but also a function of the coordinate x. The generalized velocity thus is

```
Solve[mom == p[t], ∂ₜx[t]]
```

$$\left\{\left\{x'(t) \to \frac{p(t)}{m\left(f'(x(t))^2 + 1\right)}\right\}\right\}$$

These two relations were applied to the transformation from the Lagrangian to the Hamiltonian. Especially the last relation was necessary to eliminate the velocity by means of the generalized momentum.

2.8.3 Hamilton's Equation of Motion

If we know the Lagrangian of a mechanical system, the equations of motion follow by the application of Hamilton's principle. Another way to derive the equations of motion is by applying Hamilton's formalism to the Hamiltonian. To derive Hamilton's equations, let us consider the Hamiltonian as a function of generalized coordinates. On the other hand, the same Hamiltonian can be derived from the Lagrangian. The equivalence of both approaches delivers Hamilton's equation. First, let us demonstrate this procedure for a mechanical system with one degree of freedom. The Hamiltonian for this case can be derived from the Lagrangian by means of the Legendre transform:

```
h = p v - l[q, v, t]
```

$$p\,v - l(q, v, t)$$

where v represents the generalized velocity of the system. If we calculate the total derivative of this representation of the Hamiltonian and use the Euler–Lagrange equations as well as the definition of the generalized momentum, we get

```
r1 = Dt[h] /. {∂_q l[q, v, t] -> pp, ∂_v l[q, v, t] -> p}
```

$$v\,d\,p - pp\,d\,q - d\,t\,l^{(0,0,1)}(q, v, t)$$

On the other hand, let us consider the Hamiltonian as a function of the two generalized coordinates (q, p). Then, the total derivative of this representation is

```
r2 = Dt[H[q, p, t]]
```

$$dt\,H^{(0,0,1)}(q,\,p,\,t) + dp\,H^{(0,1,0)}(q,\,p,\,t) + dq\,H^{(1,0,0)}(q,\,p,\,t)$$

If both relations describe the same system, we are able to extract the factors of the total differentials. The following line examines the difference of both expressions and extracts the coefficients of the total differentials:

```
Map[# == 0 &,
    Map[Coefficient[r1 - r2, #] &, {Dt[t], Dt[p],
        Dt[q]}] /. {v -> q', pp -> p'}] // TableForm
```

$$-H^{(0,0,1)}(q,\,p,\,t) - l^{(0,0,1)}(q,\,q',\,t) == 0$$
$$q' - H^{(0,1,0)}(q,\,p,\,t) == 0$$
$$-p' - H^{(1,0,0)}(q,\,p,\,t) == 0$$

The result is that the time derivative of the Hamiltonian is equal to the negative time derivative of the Lagrangian. The two other equations represent the time derivative of the generalized coordinate and momentum. The first of these relations state that the time evolution of the coordinate is given by the derivative of the Hamiltonian with respect to the momentum. The evolution of the momentum is determined by the negative derivative of the Hamiltonian with respect to the coordinate. If the Hamiltonian or Lagrangian is independent of time, the first relation does not exist.

The same procedure as demonstrated above can be applied to a mechanical system of more than one degree of freedom. First, let us calculate the total derivative of the Hamiltonian in the Legendre representation. Carrying out the calculation, we find

$$dH = \sum_i p_i \, dq'_i + q'_i \, dp_i - \frac{\partial \mathcal{L}}{\partial q_i} \, dq'_i - \frac{\partial \mathcal{L}}{\partial q_i} \, dq'_i$$

$$- \frac{\partial \mathcal{L}}{\partial t} \, dt \tag{2.8.20}$$

$$= \sum_i q'_i \, dp_i - \frac{\partial \mathcal{L}}{\partial q_i} \, dq'_i - \frac{\partial \mathcal{L}}{\partial t} \, dt.$$

The Euler–Lagrange equations provide

$$\frac{\partial \mathcal{L}}{\partial q_i} = \frac{d}{dt} \left(\frac{\partial \mathcal{L}}{\partial q'_i} \right) = p'_i. \tag{2.8.21}$$

Thus, the total derivative of the Hamiltonian becomes

$$dH = \sum_i q'_i \, dp_i - p'_i \, dq_i - \frac{\partial \mathcal{L}}{\partial t} \, dt. \tag{2.8.22}$$

On the other hand, the Hamiltonian is a function of the q_i and p_i, so that

$$dH = \sum_i \frac{\partial H}{\partial q_i} \, dq_i + \frac{\partial H}{\partial p_i} \, dp_i + \frac{\partial H}{\partial t} \, dt. \tag{2.8.23}$$

Comparing both expressions, we gain the relations

$$p'_i = - \frac{\partial H}{\partial q_i}, \tag{2.8.24}$$

$$q'_i = \frac{\partial H}{\partial p_i}, \tag{2.8.25}$$

$$\frac{\partial \mathcal{L}}{\partial t} = - \frac{\partial H}{\partial t}. \tag{2.8.26}$$

These relations are Hamilton's famous equations. Because of their symmetrical appearance, these equations are also called canonical equations. The set of variables (q_i, p_i) are known as canonical variables. The above system of equations is a first-order ordinary differential system of $2N$ equations. This system of equation is equivalent to the second-order equations resulting from Hamilton's principle.

The above system of equations is called Hamilton's equations although these equations are known since 1809 to be derived by Lagrange and Poisson. However, both did not realize the importance of their derived results in mechanics. Until 1831, when Cauchy pointed out the importance of these equations for mechanical systems, the equations were applied to mechanical problems. Hamilton derived these equations in 1834 from a variational principle. He opened a wide field of applications with his work.

To simplify the derivation of Hamilton's equations of motion, we collect the necessary steps in the function **HamiltonsEquation[]**. This function assumes that the Hamiltonian is a function of the generalized coordinates and momenta. It is also assumed that the coordinates are functions of time by default:

```
HamiltonsEquation[hamiltonian_,
   gcoordinates_List, gmomenta_List, indep_ : t] :=
  Block[{qp, pp}, qp = Map[∂indep # &, gcoordinates];
  pp = Map[∂indep # &, gmomenta]; Flatten[
   {MapThread[#1 == ∂#2 hamiltonian &, {qp, gmomenta}],
    MapThread[#1 == -∂#2 hamiltonian &,
     {pp, gcoordinates}]}]]]
```

To see how this function works, let us examine an example.

Example 1: Hamilton's Equation for a Sliding Beat

We already know that the Hamiltonian of a sliding bead is given by

$$hamB = \frac{p(t)^2}{2m\left(\left(\frac{\partial f(x(t))}{\partial x(t)}\right)^2 + 1\right)} + g\, m\, f(x(t))$$

$$\frac{p(t)^2}{2m\left(f'(x(t))^2 + 1\right)} + g\, m\, f(x(t))$$

Applying the above function to this Hamiltonian, we find

hamEqs = FullSimplify[HamiltonsEquation(hamB, {x(t)}, {p(t)})];
TableForm[hamEqs]

$$x'(t) == \frac{p(t)}{m\, f'(x(t))^2 + m}$$

$$p'(t) == \frac{f'(x(t))\left(\frac{p(t)^2 f''(x(t))}{(f'(x(t))^2 + 1)^2} - g\, m^2\right)}{m}$$

These two equations describe the time evolution of the coordinate x and the momentum p. The constraint of the movement is defined by the arbitrary function f. If we choose this function in a specific way, for example as a parabola, we find the explicit representation of Hamilton's equation:

---

hamEqs /. $f \to$ Function[κ, κ^2] // TableForm

$$x'(t) == \frac{p(t)}{4\,m\,x(t)^2 + m}$$

$$p'(t) == \frac{2\,x(t)\left(\frac{2\,p(t)^2}{(4\,x(t)^2+1)^2} - g\,m^2\right)}{m}$$

2.8.4 Hamilton's Equations and the Calculus of Variation

Hamilton's principle is the basis for the derivation of Euler–Lagrange equations. The mathematical background of this derivation is the variational principle for the Lagrangian

$$\frac{d}{d\varepsilon}\, L[q_i]\, \Big|_{\varepsilon=0} = \frac{d}{d\varepsilon} \int_{t_1}^{t_2} \mathcal{L}(q_i,\, q'_i,\, t)\,dt\, \Big|_{\varepsilon=0} = 0. \qquad (2.8.27)$$

The variation of the coordinates delivered the equation of motion in the representation

$$\frac{\partial \mathcal{L}}{\partial q_i} - \frac{d}{dt}\left(\frac{\partial \mathcal{L}}{\partial q'_i}\right) = 0, \qquad i = 1, 2, \dots. \qquad (2.8.28)$$

We also know that the Hamiltonian can be obtained from the Lagrangian by means of a Legendre transform by

$$H = \textstyle\sum_i p_i\,q'_i - \mathcal{L}. \qquad (2.8.29)$$

Since the Legendre transform is invertible, we find

$$\mathcal{L} = \textstyle\sum_i p_i\,q'_i - H\,(q_i,\, -p_i,\, t). \qquad (2.8.30)$$

The variational principle based on $\mathcal{L}$ from which the equations of motion follow is represented by

$$\left(\frac{dL}{d\varepsilon}\right)\Big|_{\varepsilon=0} = \frac{d}{d\varepsilon} \int_{t_1}^{t_2} \left(\textstyle\sum_i p_i\,q'_i - H\right) dt\, \Big|_{\varepsilon=0} = 0. \qquad (2.8.31)$$

The variation here means that all variables q_i, q'_i and p_i take part in the variation of the functional. This is expressed by the following relations

$$\tilde{q}_i = q_i + \varepsilon\omega_i, \qquad (2.8.32)$$

$$\tilde{p}_i = p_i + \varepsilon\nu_i. \qquad (2.8.33)$$

Inserting this representations of the changed functions q_i and p_i into the functional L, we find

$$\frac{dL}{d\varepsilon}\Big|_{\varepsilon=0} = \frac{d}{d\varepsilon} \int_{t_1}^{t_2} (\textstyle\sum_i \tilde{p}_i \, \tilde{q}'_i - H(\tilde{q}_i, \tilde{p}_i, t)) \, dt \Big|_{\varepsilon=0} = 0$$

$$= \int_{t_1}^{t_2} \sum_i \left(\tilde{q}'_i \, v_i + \tilde{p}_i \, \omega'_i - \left(\frac{\partial H}{\partial \tilde{q}_i} \frac{d\tilde{q}_i}{d\varepsilon} + \frac{\partial H}{\partial \tilde{p}_i} \frac{d\tilde{p}_i}{d\varepsilon} \right) \right) dt \Big|_{\varepsilon=0}$$

$$= \int_{t_1}^{t_2} \left\{ \sum_i (q'_i, \, v_i + p_i \, w'_i) - \sum_i \left(\frac{\partial H}{\partial q_i} w_i + \frac{\partial H}{\partial p_i} v_i \right) \right\} dt$$

$$= \int_{t_1}^{t_2} \left(\sum_i \left(q'_i - \frac{\partial H}{\partial p_i} \right) v_i + \sum_i \left(- p'_i - \frac{\partial H}{\partial q_i} \right) w_i \right) dt.$$

If v_i and w_i are independent of each other, we get

$$q'_i = \frac{\partial H}{\partial p_i}, \tag{2.8.34}$$

$$p'_i = -\frac{\partial H}{\partial q_i}. \tag{2.8.35}$$

This set of equations are Hamilton's equation of motion. Another example demonstrating the application of Hamilton's equations is the motion of a particle on a cylindric surface.

Example 1: Motion on a Cylinder

Let us assume that a mass point moves on the surface of a cylinder which extends in, z-direction to infinity. For the geometry, see Figure 2.8.2.

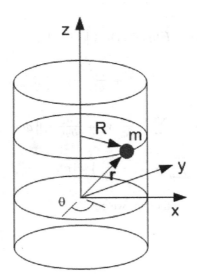

Figure 2.8.2.   A beat on the surface of a cylinder. The cylinder is the surface of movement.

The surface of the cylinder is defined by

$$x^2 + y^2 = R^2. \tag{2.8.36}$$

In addition, we assume that the force on the particle is proportional to the distance measured from the center of the cylinder. We assume

$$\vec{F} = -k\,\vec{r}, \tag{2.8.37}$$

where k is a scalar constant. The potential related to this force is

$$V = \frac{k}{2}\ (R^2 + z[t]^2)$$

$$\frac{1}{2}\,k\,(R^2 + z(t)^2)$$

The squared velocity of the particle in cylindrical coordinates is

$$v2 = (\partial_t r[t])^2 + r[t]^2 (\partial_t \theta[t])^2 + (\partial_t z[t])^2$$

$$r'(t)^2 + z'(t)^2 + r(t)^2 \theta'(t)^2$$

Since the motion of the particle is restricted to the surface with $r = R$ and R is a constant, the kinetic energy becomes

$$T = \frac{m}{2} v2 \; /. \; r \rightarrow \text{Function}[t, R]$$

$$\frac{1}{2} m (z'(t)^2 + R^2 \theta'(t)^2)$$

The Lagrangian then follows as

$$\mathcal{L} = T - V$$

$$\frac{1}{2} m (z'(t)^2 + R^2 \theta'(t)^2) - \frac{1}{2} k (R^2 + z(t)^2)$$

The two generalized coordinates of $\mathcal{L}$ are θ and z. The corresponding generalized momenta are

$$p_\theta = \partial_{\partial_t \theta[t]} \mathcal{L}$$

$$m R^2 \theta'(t)$$

and

$$p_z = \partial_{\partial_t z[t]} \mathcal{L}$$

$$m z'(t)$$

Since we are dealing with a conservative system and time is not explicitly present in the Lagrangian, we can find the Hamiltonian by a Legendre transform. The resulting Hamiltonian is a sum of kinetic and potential energy. The Hamiltonian is calculated by

```
hamCyl = LegendreTransform[ℒ,
    {∂ₜ θ[t], ∂ₜ z[t]}, {pθ[t], pz[t]}] // Expand
```

$$\frac{k\,R^2}{2} + \frac{pz(t)^2}{2\,m} + \frac{1}{2}\,k\,z(t)^2 + \frac{p\theta(t)^2}{2\,m\,R^2}$$

Hamilton's equation of motion follow from

```
hEqs = HamiltonsEquation[hamCyl,
    {θ[t], z[t]}, {pθ[t], pz[t]}]; hEqs // TableForm
```

$$\theta'(t) == \frac{p\theta(t)}{m\,R^2}$$
$$z'(t) == \frac{pz(t)}{m}$$
$$p\theta'(t) == 0$$
$$pz'(t) == -k\,z(t)$$

We observe that the temporal change in the angular momentum vanishes. This property states that p_θ is a conserved quantity which is defined by

```
eqθ = pθ == κ
```

$$m\,R^2\,\theta'(t) == \kappa$$

This relation states that the angular momentum with respect to the z-axis is a conserved quantity. We expect this result because the system is invariant with respect to rotations around the z-axis. If we use the second of these equations and differentiate with respect to time and replace the temporal changes of p_z with the last equation, we find

```
eqz = Map[∂_t # &, hEqs[[2]]] /. (hEqs[[4]] /. Equal -> Rule)
```

$$z''(t) == -\frac{k\,z(t)}{m}$$

This equation is a harmonic equation for the z coordinate. Thus the movement along the z direction is harmonic. The solution is given by

```
solZ = DSolve[eqz, z, t] // Flatten
```

$$\left\{z \to \text{Function}\left[\{t\},\ c_1 \cos\left(\frac{\sqrt{k}\,t}{\sqrt{m}}\right) + c_2 \sin\left(\frac{\sqrt{k}\,t}{\sqrt{m}}\right)\right]\right\}$$

The solution for the angular coordinate follows from

```
solθ = DSolve[eqθ, θ, t] // Flatten
```

$$\left\{\theta \to \text{Function}\left[\{t\},\ \frac{l\,\kappa}{m\,R^2} + c_1\right]\right\}$$

The track of the beat is generated by using the symbolic solutions.

```
gra1 = ParametricPlot3D[
    {R Sin[θ[t]], R Cos[θ[t]], z[t]} /. solZ /. solθ /.
    {R -> 1, m -> 1, k -> 0.1, κ -> 2, C[1] -> 0, C[2] -> 1},
    {t, 0, 6 π}, PlotPoints -> 120];
```

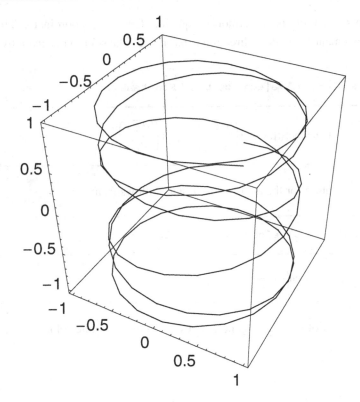

To see how the beat is moving along the track, we generate some points and collect them in a table.

```
points =
    Table[{RGBColor[0, 0, 1], PointSize[0.1], Point[
        {R Sin[θ[t]], R Cos[θ[t]], z[t]} /. solZ /. solθ /.
        {R -> 1, m -> 1, k -> 0.1, κ -> 2, C[1] -> 0,
            C[2] -> 1}]}, {t, 0, 6 π, 0.2}];
```

These points are used in sequenz of plots generating an illustration of the motion.

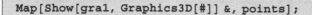

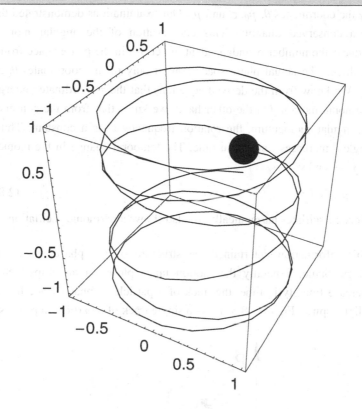

2.8.5 Liouville's Theorem

A mechanical system in Hamiltonian dynamics is represented by two sets of canonical coordinates: the generalized coordinates q_i and the generalized momenta p_i. Each set of coordinates is the basis of a space. Both spaces are completely independent of each other. The two spaces are called the configuration space and the momentum space, respectively. The union of both spaces allows one to collect the total information on the mechanical system in a single space, the so, called Hamiltonian phase space.

Let us consider the example discussed in the last subsection. We examined the motion of particle on an infinite cylinder. The phase space is generated by the coordinates θ, p_θ, z, and p_z. Our examinations demonstrated that p_θ is a conserved quantity. This conservation of the angular momentum reduces the number of independent directions in the phase space from four to three. The actual phase space consist only of the coordinates θ, z, and p_z. We know from the derived equations that the z-coordinate undergoes a harmonic motion. On the other hand, we know that from the conservation of angular momentum, the rotation frequency θ' is a constant. Thus, the angle θ increases linearly in time. The temporal change in the momentum p_z is given by

$$p'_z = k\,z. \tag{2.8.38}$$

Since z oscillates harmonically, p_z also shows a harmonic oscillation.

This information determines the structure of the phase space. In the (z, p_z)-plane, generally the motion takes place on an ellipse. Since θ increase linearly in time, the track of a particle in phase space lies on an elliptic spiral. Figure 2.8.3 shows a single track of a particle in phase space.

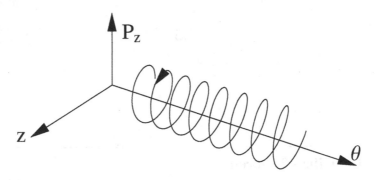

Figure 2.8.3.   Motion on a cylinder represented in phase space coordinates.

An orbit in phase space at constant energy $H =$ const. is an elliptic spiral. If we know the initial conditions of a mechanical system (i.e., $q_i\,(t = 0)$, $p_i\,(t = 0)$), and if the system is conservative, then a unique orbit in phase space is defined. The initial conditions determine the total energy. This kind of description is not restricted on a single particle but can be extended on an arbitrary number of particles. From a practical point of view, we face the problem that each additional particle in the phases space extends its dimension by six coordinates. For example, if we want to treat an ensemble of 10^{23} particles, we have a phase space of the same number of freedoms. From a practical point of view, such an approach is not efficient.

We need to introduce a method allowing us an appropriate description of a large number of particles. One such method is to measure the density in phase space. The number of particles in phase space dv is

$$N = \rho\, dv, \tag{2.8.39}$$

with

$$dv = dq_1\, dq_2 \dots dq_\kappa\, dp_1\, dp_2 \dots dp_\kappa, \tag{2.8.40}$$

where κ is the dimensionality of the configuration space.

Let us consider an infinitesimal volume element in phase space. Because the underlying dynamical system generates continuous changes in phase space, we observe that a certain amount in the coordinates q_i and p_i will flow into the volume and another part will flow out of the volume.

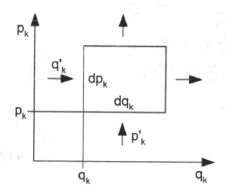

Figure 2.8.4.   Infinitesimal phase space volume. There is flow into and out of the volume.

For example, the inflow on the left surface of the volume is determined by the density of particles in phase space at this location and by the temporal change of the coordinate:

$$\rho\, \frac{dq'_k}{dt}\, dp_k = \rho\, q'_k\, dp_k . \tag{2.8.41}$$

The inflow from the bottom is

$$\rho \, \frac{dp'_k}{dt} \, dq_k = \rho \, p'_k \, dq_k. \tag{2.8.42}$$

Thus, the total number of incoming particles are

$$j_{\text{in}} = \rho \, (q'_k \, dp_k + p'_k \, dq_k). \tag{2.8.43}$$

The drain of particles from the volume can be approximated by the gradient in the coordinates:

$$j_{\text{out}} = \left(\rho \, q'_k + \frac{\partial}{\partial q_k} \, (\rho \, q'_k) \, dq_k \right) dp_k + \left(\rho \, p'_k + \frac{\partial}{\partial p_k} \, (\rho \, p'_k) \, dp_k \right) dq_k. \tag{2.8.44}$$

The particle balance is thus given by

$$j_{\text{in}} - j_{\text{out}} = \left\{ - \frac{\partial}{\partial q_k} \, (\rho \, q'_k) - \frac{\partial}{\partial p_k} \, (\rho \, p'_k) \right\} dq_k \, dp_k. \tag{2.8.45}$$

The sum balance currant must be equal the temporal changes in the density for all possible configurations of the volumes

$$
\begin{aligned}
& \frac{\partial \rho}{\partial t} + \sum_{k=1}^{\rho} \frac{\partial}{\partial q_k} \, (\rho \, q'_k) + \frac{\partial}{\partial p_k} \, (\rho \, p'_k) = 0 \\
\Leftrightarrow \quad & \frac{\partial \rho}{\partial t} + \sum_{k=1}^{\rho} \frac{\partial \rho}{\partial q_k} \, q'_k + \frac{\partial \rho}{\partial p_k} \, p'_k \\
& + \sum_{k=1}^{\rho} \rho \, \frac{\partial q'_k}{\partial q_k} + \rho \, \frac{\partial p'_k}{\partial p_k} \, \frac{\partial \rho}{\partial t} \\
& + \sum_{k=1}^{\rho} \frac{\partial \rho}{\partial q_k} \, q'_k + \frac{\partial \rho}{\partial p_k} \, p'_k \\
& + \rho \sum_{k=1}^{\rho} \left(\frac{\partial q'_k}{\partial q_k} + \frac{\partial p'_k}{\partial p_k} \right).
\end{aligned}
\tag{2.8.46}
$$

Hamilton's equations provide

$$p'_k = - \frac{\partial H}{\partial q_k}, \tag{2.8.47}$$

$$q'_k = \frac{\partial H}{\partial p_k}, \tag{2.8.48}$$

or

$$\frac{\partial p'_k}{\partial p_k} = - \frac{\partial^2 H}{\partial q_k \, \partial p_k} \tag{2.8.49}$$

and

$$\frac{\partial q'_k}{\partial q_k} = \frac{\partial^2 H}{\partial p_k \, \partial q_k} \tag{2.8.50}$$

$$\Longrightarrow - \frac{\partial p'_k}{\partial p_k} = \frac{\partial q'_k}{\partial q_k}. \tag{2.8.51}$$

Thus, the equation for ρ reduces to

$$\frac{\partial \rho}{\partial t} + \sum_{k=1}^{K} \frac{\partial \rho}{\partial q_k} q'_k + \frac{\partial \rho}{\partial p_k} p'_k = 0. \tag{2.8.52}$$

This formula is equivalent to a total temporal change of ρ, meaning the density ρ in phase space is a conserved quantity. This result is equivalent with Liouville's theorem that the density of the phase space is conserved while the system develops dynamically. This result was published by Liouville in 1838. The theorem by Liouville is a special case of a more general theory based on Poisson brackets.

2.8.6 Poisson Brackets

Let us consider a function similar to the phase space density which depends on phase space coordinates q_k, p_k, and t:

$$f = f(q_k, p_k, t). \tag{2.8.53}$$

The structure of the phase space is governed by Hamilton's equations

$$q'_k = \frac{\partial H}{\partial p_k}, \tag{2.8.54}$$
$$p'_k = -\frac{\partial H}{\partial q_k}. \tag{2.8.55}$$

The total temporal change of f is given by

$$\frac{df}{dt} = \frac{\partial f}{\partial t} + \sum_{k=1}^{P} \frac{\partial f}{\partial q_k} q'_k + \frac{\partial f}{\partial p_k} p'_k. \tag{2.8.56}$$

Inserting Hamilton's equation of motion into this expression gives us

$$\frac{df}{dt} = \frac{\partial f}{\partial t} + \sum_{k=1}^{P} \frac{\partial f}{\partial q_k} \frac{\partial H}{\partial p_k} - \frac{\partial f}{\partial p_k} \frac{\partial H}{\partial q_k}. \tag{2.8.57}$$

On the phase space spanned by the coordinates q_k and p_k, let us define an abbreviation for the following expression:

$$\sum_{k=1}^{P} \frac{\partial f}{\partial q_k} \frac{\partial H}{\partial p_k} - \frac{\partial f}{\partial p_k} \frac{\partial H}{\partial q_k} = \{f, H\}_{\{q,p\}}, \tag{2.8.58}$$

known as Poisson's bracket. The subscript $\{q, p\}$ denotes the set of variables of the phase space. Inserting this bracket, the temporal change of f becomes

$$\frac{df}{dt} = \frac{\partial f}{\partial t} + \{f, H\}_{\{q,p\}}. \tag{2.8.59}$$

This relation allows us to calculate the temporal changes of any function f depending on the phase space variables. The Poisson bracket itself has some remarkable properties which will be discussed below.

We already encountered conserved quantities which have the property that temporal changes of this quantity vanish. This vanishing can be expressed by Poisson brackets in a very convenient way. Because the conservation of a quantity f assures that

$$\frac{df}{dt} = 0 \qquad\qquad (2.8.60)$$

which is identical with

$$\frac{\partial f}{\partial t} + \{f, H\}_{\{q,p\}} = 0. \qquad\qquad (2.8.61)$$

If the conserved quantity f is independent of time, we find

$$\{f, H\}_{\{q,p\}} = 0, \qquad\qquad (2.8.62)$$

(i.e., the Poisson bracket of the conserved quantity f and the Hamiltonian vanishes).

Let us consider two functions f and g depending on the phase space variables. Using these functions in the Poisson bracket, we can derive some of the general properties of this kind of brackets

$$\{f, g\}_{\{q,p\}} = \sum_{k=1}^{s} \frac{\partial f}{\partial q_k} \frac{\partial g}{\partial p_k} - \frac{\partial f}{\partial p_k} \frac{\partial g}{\partial q_k} = \{f, g\}. \qquad (2.8.63)$$

In the following, we use also the short notation $\{f, g\}$ for the representation of the Poisson bracket $\{f, g\}_{\{q,p\}}$. This notation is used when no confusion on the phase space variables is possible. The Poisson bracket owns the following properties

$$\{f, g\} = -\{g, f\} \qquad \text{antisymmetry.} \qquad\qquad (2.8.64)$$

If one of the functions f or g are constants the Poisson bracket vanishes

$$\{f, c\} = 0 = \{c, g\}. \qquad\qquad (2.8.65)$$

If we have three functions f, g, and k which arepart of the phase space, then, we can check the properties

$$\{f + h, g\} = \{f, g\} + \{h, g\} \qquad \text{linearity} \qquad\qquad (2.8.66)$$
$$\{f\,h, g\} = f\,\{h, g\} + h\,\{f, g\} \qquad \text{Leibniz's rule} \qquad\qquad (2.8.67)$$

$$\frac{\partial}{\partial t} \{f, g\} = \{\frac{\partial f}{\partial t}, g\} + \{f, \frac{\partial g}{\partial t}\} \quad \text{differentiation rule} \quad (2.8.68)$$

If one of the two functions f or g reduces to a phase space variable the Poisson bracket reduces to the partial derivative of the function with respect to the conjungate coordinate. For example if g equals either q_k or p_k the result of the Poisson bracket is

$$\{f, q_k\} = -\frac{\partial f}{\partial p_k} \quad (2.8.69)$$

$$\{f, p_k\} = \frac{\partial f}{\partial q_k}. \quad (2.8.70)$$

If we chose for both f and g coordinates of the phase space then we gain the fundamental Poisson brackets

$$\{q_i, q_j\} = 0 \quad (2.8.71)$$
$$\{p_i, p_j\} = 0 \quad (2.8.72)$$
$$\{q_i, p_i\} = \sum_{k=1}^{s} \frac{\partial q_i}{\partial q_k} \frac{\partial p_j}{\partial p_k} - \frac{\partial q_i}{\partial p_k} \frac{\partial p_j}{\partial q_k}$$
$$= \sum_{k=1}^{s} \delta_{ik} \delta_{jk} = \delta_{ij}. \quad (2.8.73)$$

These relations of the fundamental Poisson brackets are the basis of quantum mechanics. For three functions of the phase space there exists a special relation the so called Jacobi identity

$$\{f, \{g, h\}\} + \{g, \{h, f\}\} + \{h, \{f, g\}\} = 0. \quad (2.8.74)$$

The above properties determine the algebraic properties of the Poisson bracket. Especially, linearity, antisymmetry, and the Jacobi identity define the related Lie algebra of the bracket.

Another important property of the Poisson bracket is the ability to derive, from two conserved quantities J_1 and J_2, another conserved quantity

$$\{J_1, J_2\} = \text{const.} \quad (2.8.75)$$

This behavior is known as Poisson's theorem. A direct proof is feasible if we assume that J_1 and J_2 are independent of time. Let us replace in the Jacobi identity the third function by the Hamiltonian of the system; then, we get

$$\{H, \{J_1, J_2,\}\} + \{J_1, \{J_2, H\}\} + \{J_2 \{H, J_1\}\} = 0. \quad (2.8.76)$$

Since $\{J_2, H\} = 0$ and $\{H, J_1\} = 0$, we find

$$\{H, \{J_1, J_2\}\} = 0. \quad (2.8.77)$$

Thus, the bracket $\{J_1, J_2\}$ is also a conserved quantity. We note that the application of Poisson's theorem will not always provide new conserved quantities because the number of conserved quantities of a standard mechanical system is finite. It is known that the total number of conserved quantities is given by $2n - 1$ such quantities if n is the degree of freedom in the phase space. Thus, Poisson's theorem sometimes delivers trivial constants or the resulting conserved quantity is a function of the original conserved quantities J_1 and J_2. If both cases fail, we obtain a new conserved quantity.

The main application of Poisson brackets is the formulation of equations of motion. The derivation of conserved quantities is a special property of these brackets. To see how equations of motion follow by Poisson's bracket, let us consider that the first argument is one of the phase space variables. As second argument, we use the Hamiltonian. The resulting relations are

$$q'_k = \{q_k, H\} =$$
$$\sum_{i=1}^{s} \frac{\partial q_k}{\partial p_i} - \frac{\partial q_k}{\partial p_i} \frac{\partial H}{\partial q_i} = \sum_{i=1}^{s} \frac{\partial H}{\partial p_i} \delta_{ki} = \frac{\partial H}{\partial p_k}, \qquad (2.8.78)$$

$$p'_k = \{p_k, H\} =$$
$$\sum_{i=1}^{s} \frac{\partial p_k}{\partial q_i} \frac{\partial H}{\partial p_i} - \frac{\partial p_k}{\partial p_i} \frac{\partial H}{\partial q_i} = -\frac{\partial H}{\partial q_k}. \qquad (2.8.79)$$

However, these equations are Hamilton's equation of motion:

$$q'_k = \{q_k, H\} = \frac{\partial H}{\partial P_k}, \qquad (2.8.80)$$

$$p'_k = \{p_k, H\} = -\frac{\partial H}{\partial q_k}. \qquad (2.8.81)$$

Thus the dynamic of a Hamiltonian system follows by means of the Poisson bracket if we know the Hamiltonian

$$q'_k = \{q_k, H\}, \qquad k = 1, 2, \ldots, \qquad (2.8.82)$$
$$p'_k = \{p_k, H\}. \qquad (2.8.83)$$

This system of equations defines the phase space flow.

The following *Mathematica* lines define the Poisson bracket in such a way that some of the above properties are incorporated:

$$\{f, g\}_{\{q,p\}} = \sum_{k=1}^{s} \frac{\partial f}{\partial q_k} \frac{\partial g}{\partial p_k} - \frac{\partial f}{\partial p_k} \frac{\partial g}{\partial q_k}. \qquad (2.8.84)$$

First, we define a notation for the Poisson bracket in such a way that the symbolic use in *Mathematica* is related to the use in the text. The following line defines such a notation:

```
<< Utilities`Notation`
```

```
Notation[
  {f_, g_}(q_,p_) ⟺ PoissonBracket[f_, g_, q_, p_]]
```

The next few cells are representations for the bilinearity of the Poisson bracket. First, we define properties of the bracket for symbols occurring in a product that are independent of the phase space variables.

```
PoissonBracket[a_ f_, g_,
  coordinates_List, momenta_List] :=
a PoissonBracket[ f, g, coordinates, momenta] /;
  (Apply[And, Map[FreeQ[a, #] &, coordinates]] ∧
    Apply[And, Map[FreeQ[a, #] &, momenta]])
```

```
PoissonBracket[ f_, a_ g_,
  coordinates_List, momenta_List] :=
a PoissonBracket[ f, g, coordinates, momenta] /;
  (Apply[And, Map[FreeQ[a, #] &, coordinates]] ∧
    Apply[And, Map[FreeQ[a, #] &, momenta]])
```

The next two cells define the linearity in the first and second argument of the Poisson bracket (PB):

```
PoissonBracket[a_ + f_, g_,
  coordinates_List, momenta_List] :=
 PoissonBracket[ a, g, coordinates, momenta] +
  PoissonBracket[ f, g, coordinates, momenta]
```

```
PoissonBracket[ f_, a_ + g_,
  coordinates_List, momenta_List] :=
 PoissonBracket[ f, a, coordinates, momenta] +
 PoissonBracket[ f, g, coordinates, momenta]
```

The following two cells stand for Leibniz` rule:

```
PoissonBracket[ f_ h_, g_,
  coordinates_List, momenta_List] :=
 f PoissonBracket[ h, g, coordinates, momenta] +
 h PoissonBracket[ f, g, coordinates, momenta]
```

```
PoissonBracket[ g_, f_ h_,
  coordinates_List, momenta_List] :=
 f PoissonBracket[g, h, coordinates, momenta] +
 h PoissonBracket[ g, f, coordinates, momenta]
```

The next cell is related to differentiations:

```
Unprotect[D];
D[PoissonBracket[ f_, g_,
   coordinates_List, momenta_List], indep1_] :=
 PoissonBracket[D[f, indep1], g, coordinates,
   momenta] + h PoissonBracket[ f,
    D[g, indep1], coordinates, momenta]
Protect[
  D];
```

So far, no specific calculation was defined for the PB. The following cell defines how the actual calculations are carried out in the PB:

```
PoissonBracket[f_, g_, coordinates_List, momenta_List,
  indep_ : t] := Block[{}, Fold[Plus, 0, MapThread[
   (∂#1 f ∂#2 g - ∂#2 f ∂#1 g) &, {coordinates, momenta}]]]
```

Now, the application of the function demonstrates the action. Let us check the linearity first. Assume that we have three functions defined on the phase space. Linearity is then demonstrated by

```
{α f[θ[t], pθ[t]] + β g[θ[t], pθ[t]],
   h[θ[t], pθ[t]]}{{θ[t]},{pθ[t]}}
```

$$\alpha \left(h^{(0,1)}(\theta(t), p\theta(t))\, f^{(1,0)}(\theta(t), p\theta(t)) - f^{(0,1)}(\theta(t), p\theta(t))\, h^{(1,0)}(\theta(t), p\theta(t)) \right) +$$
$$\beta \left(h^{(0,1)}(\theta(t), p\theta(t))\, g^{(1,0)}(\theta(t), p\theta(t)) - g^{(0,1)}(\theta(t), p\theta(t))\, h^{(1,0)}(\theta(t), p\theta(t)) \right)$$

An example for Leibniz' rule is given next:

```
{α f[q[t]] h[q[t], p[t]], ζ[q[t], p[t]]
   (β g[p[t]] + γ H[q[t], p[t]])}{{q[t]},{p[t]}}
```

$$\alpha \left(h(q(t), p(t)) \left(\zeta(q(t), p(t)) \left(\beta\, f'(q(t))\, g'(p(t)) + \gamma\, f'(q(t))\, H^{(0,1)}(q(t), p(t)) \right) + \right.\right.$$
$$\left(\beta\, g(p(t)) + \gamma\, H(q(t), p(t)) \right) f'(q(t))\, \zeta^{(0,1)}(q(t), p(t)) \Big) +$$
$$f(q(t)) \left(\zeta(q(t), p(t)) \left(\beta\, g'(p(t))\, h^{(1,0)}(q(t), p(t)) + \gamma\, (H^{(0,1)}(q(t), p(t)) \right.\right.$$
$$\left. h^{(1,0)}(q(t), p(t)) - h^{(0,1)}(q(t), p(t))\, H^{(1,0)}(q(t), p(t)) \right) +$$
$$\left(\beta\, g(p(t)) + \gamma\, H(q(t), p(t)) \right) \left(\zeta^{(0,1)}(q(t), p(t))\, h^{(1,0)}(q(t), p(t)) - \right.$$
$$\left.\left.\left. h^{(0,1)}(q(t), p(t))\, \zeta^{(1,0)}(q(t), p(t)) \right)\right)\right)$$

An example for the derivation rule is given by

```
∂t{f[q[t]], h[q[t], p[t]]}{{q[t]},{p[t]}}
```

$$q'(t)\, f''(q(t))\, h^{(0,1)}(q(t), p(t)) +$$
$$f'(q(t)) \left(p'(t)\, h^{(0,2)}(q(t), p(t)) + q'(t)\, h^{(1,1)}(q(t), p(t)) \right)$$

2.8.7 Manifolds and Classes

So far, we defined a few functions for the Poisson bracket. However, a PB is an object possessing some properties and some methods. The properties are the phase space variables and the methods are the algebraic relations defined in Section 2.8.6. From a theoretical point of view, a PB is part of a dynamic structure incorporating phase space properties and algebraic methods. We already know that a PB is intrinsically connected with the phase space, which is, on its own, a differentiable manifold. The manifold, respectively the phase space, is defined by the phase space variables q_k and p_k. In this phase space, there are functions depending on the phase space variables, such as energy, momentum, angular momentum, and o forth. The PB for the set of variables q_k and p_k generates an algebraic structure on this manifold. Thus, it is natural to separate the total phase space into the algebraic structure and the coordinates defined by the phase space variables. This separation allows us to introduce a concept known as object-oriented representation. Objects in this representation are derived from classes that define a general view of the system. A class consists of properties and methods. In our case, the properties are the phase space variables and the methods are the algebraic structure of the manifold. Thus, we can use an object-oriented representation of the PB which is defined by the class PoissonB:

```
PB = Class["PoissonB", Class["Element"],
  {description = "Poisson Bracket",
   {P = Null, Description → "momentas"},
   {Q = Null, Description → "coordinates"},
   {T = t, Description → "independent variable"},
   {A = {α → α}, Description → "set of
       parameters (given by a list of rules)"}},
  {(* --- constant factor extraction --- *)
   PoissonBracket[a_ f_ , g_] :=
    a PoissonBracket[ f, g] /;
     (Apply[And, Map[FreeQ[a, #] &, Q]] ∧
       Apply[And, Map[FreeQ[a, #] &, P]]),
   (* --- constant factor extraction --- *)
   PoissonBracket[ f_ , a_ g_] :=
    a PoissonBracket[ f, g] /;
     (Apply[And, Map[FreeQ[a, #] &, Q]] ∧
       Apply[And, Map[FreeQ[a, #] &, P]]),
   (* --- Linearity --- *)
   PoissonBracket[a_ + f_ , g_] :=
    PoissonBracket[ a, g] + PoissonBracket[ f, g],
   (* --- Linearity --- *)
   PoissonBracket[ f_ , a_ + g_] :=
    PoissonBracket[ f, a] + PoissonBracket[ f, g],
   (* --- product relation --- *)
   PoissonBracket[ f_ h_ , g_] :=
    f PoissonBracket[ h, g] + h PoissonBracket[ f, g],
   (* --- product relation --- *)
   PoissonBracket[ g_ , f_ h_] :=
    f PoissonBracket[g, h] + h PoissonBracket[ g, f],
   (* --- Calculation of the bracket --- *)
   PoissonBracket[f_ , g_] :=
    Block[{}, Fold[Plus, 0, MapThread[
        (∂_{#1} f ∂_{#2} g - ∂_{#2} f ∂_{#1} g) &, {Q, P}]] /. A],
   PB[Pnew_List, Qnew_List, Tnew_Symbol, Anew_] :=
    Block[{}, P = Pnew; Q = Qnew; A = Anew]}
  ]
```

– Class PoissonB –

The class PB is defined by means of the software package *Elements* allowing one to generate classes for objects. An object here is a specific form of PB designed for a specific phase space. The following examples demonstrate how this software concept can be used to efficiently carry out calculations. Before we give some examples, let us define a simpler notation for a PB.

Since we separated the phase space from its algebraic structure, we are able to replace the phase space coordinates by the phase space object. The following line defines a template for the Poisson bracket combining the poisson manifold as an object and the algebraic properties of the bracket:

```
Notation[
    {f_, g_}_obj_ ⇔ Dot[obj_, PoissonBracket[f_, g_]]]
```

2.8.7.1 A Two-Dimensional Poisson Manifold

Let us first examine phase spaces with two dimensions of freedom. For such a case, we have two phases space variables: the coordinate $q(t)$ and the momentum $p(t)$. Functions in this manifold solely depend on these two coordinates.

The following line defines an object derived from the class PB for these two coordinates. The coordinates p and q are functions of time. The two-dimensional Poisson manifold is represented by the object pm:

```
pm = PB.new[{P → {p[t]}, Q → {q[t]}}]
```

– Object of PoissonB –

The package *Elements* offers a function **GetPropertiesForm[]** to check the properties of a given object. The properties of the defined Poisson manifold are derived by

GetPropertiesForm[pm]	

Property	Value
description	Poisson Bracket
P	$\{p(t)\}$
Q	$\{q(t)\}$
T	t
A	$\{\alpha \to \alpha\}$

This table shows that the momenta are given by the functions $p(t)$ and the coordinates by $q(t)$. In addition, the manifold may depend on parameters which can be collected in the variable A.

Let us assume that we have a physical system characterized by its kinetic energy T and its potential energy V given by

$$h = \frac{p(t)^2}{2m} + V(q(t))$$

$$\frac{p(t)^2}{2m} + V(q(t))$$

This is a Hamiltonian existing on the defined Poisson manifold. Let us apply the Poisson manifold to the two functions $p(t)$ and $q(t)$. The Poisson manifold in the Poisson bracket is given as a subscript to the bracket.

$$\{h, p(t)\}_{\text{pm}}$$

$$V'(q(t))$$

$$\{h, q(t)\}_{\text{pm}}$$

$$-\frac{p(t)}{m}$$

The following is another example for a general Hamiltonian H:

$\{H(p(t), q(t)), q(t)\}_{\text{pm}}$

$-H^{(1,0)}(p(t), q(t))$

A third example deals with a general Hamiltonian H and an arbitrary function f depending on the two coordinates of the manifold. The Poisson bracket of these two functions are

$\{\alpha\, H(q(t), p(t)), f(q(t), p(t))\}_{\text{pm}}$

$\alpha\,(f^{(0,1)}(q(t), p(t))\, H^{(1,0)}(q(t), p(t)) - H^{(0,1)}(q(t), p(t))\, f^{(1,0)}(q(t), p(t)))$

This relation represents Jacobi's identity for three functions H, f, and g:

$\text{Simplify}\big[\{f(q(t), p(t)), \{g(q(t), p(t)), \alpha\, H(q(t), p(t))\}_{\text{pm}}\}_{\text{pm}} +$
$\qquad \{g(q(t), p(t)), \{\alpha\, H(q(t), p(t)), f(q(t), p(t))\}_{\text{pm}}\}_{\text{pm}} +$
$\qquad \{\alpha\, H(q(t), p(t)), \{f(q(t), p(t)), g(q(t), p(t))\}_{\text{pm}}\}_{\text{pm}}\big]$

0

The next example represents linearity in the second argument:

$\{\alpha\, H(q(t), p(t)), f(q(t), p(t)) + g(q(t), p(t))\}_{\text{pm}}$

$\alpha\,(-H^{(0,1)}(q(t), p(t))\, f^{(1,0)}(q(t), p(t)) - H^{(0,1)}(q(t), p(t))\, g^{(1,0)}(q(t), p(t)) +$
$\qquad f^{(0,1)}(q(t), p(t))\, H^{(1,0)}(q(t), p(t)) + g^{(0,1)}(q(t), p(t))\, H^{(1,0)}(q(t), p(t)))$

2.8.7.2 A Four-Dimensional Poisson Manifold

The following line defines a second Poisson manifold for two coordinate pairs q_i and p_i. The manifold is represented by the object

```
pm2 = PB.new[{P → {p1[t], p2[t]}, Q → {q1[t], q2[t]}}]
```

— Object of PoissonB —

The properties of this manifold is gained by

```
GetPropertiesForm[pm2]
```

Property	Value
description	Poisson Bracket
P	$\{p1(t), p2(t)\}$
Q	$\{q1(t), q2(t)\}$
A	$\{\alpha \to \alpha\}$

Let us assume that we know a Hamiltonian in this four-dimensional Poisson manifold given by

$$h4 = \frac{p1(t)^2}{2\,m1} + \frac{p2(t)^2}{2\,m2} + V(q1(t), q2(t))$$

$$\frac{p1(t)^2}{2\,m1} + \frac{p2(t)^2}{2\,m2} + V(q1(t), q2(t))$$

The Hamiltonian consists of two terms: the kinetic energies and a general expression for the potential V. The Poisson brackets for this Hamiltonian and the coordinates in this manifold follow from

$$({\text{h4, }\#1}_{\text{pm2}} \, \&) \, /@ \, \{\text{p1}(t), \text{p2}(t), \text{q1}(t), \text{q2}(t)\}$$

$$\left\{ V^{(1,0)}(\text{q1}(t), \text{q2}(t)), \; V^{(0,1)}(\text{q1}(t), \text{q2}(t)), \; -\frac{\text{p1}(t)}{\text{m1}}, \; -\frac{\text{p2}(t)}{\text{m2}} \right\}$$

Another two-dimensional Hamiltonian with a different potential V gives

$$\text{h41} = \frac{\text{p1}(t)^2}{2\,\text{m1}} + \frac{\text{p2}(t)^2}{2\,\text{m2}} + V(\text{q1}(t))$$

$$\frac{\text{p1}(t)^2}{2\,\text{m1}} + \frac{\text{p2}(t)^2}{2\,\text{m2}} + V(\text{q1}(t))$$

$$({\text{h41, }\#1}_{\text{pm2}} \, \&) \, /@ \, \{\text{p1}(t), \text{p2}(t), \text{q1}(t), \text{q2}(t)\}$$

$$\left\{ V'(\text{q1}(t)), \; 0, \; -\frac{\text{p1}(t)}{\text{m1}}, \; -\frac{\text{p2}(t)}{\text{m2}} \right\}$$

The following is an example incorporating two specific functions of the Poisson manifold:

$$\left\{ \frac{\text{p1}(t)^2}{2} + \frac{\text{p2}(t)^2}{2} + V(\text{q1}(t)), \; \text{q1}(t)^2 - \text{p1}(t)\,\text{q2}(t) \right\}_{\text{pm2}}$$

$$\text{p1}(t)\,\text{p2}(t) - 2\,\text{p1}(t)\,\text{q1}(t) - \text{q2}(t)\,V'(\text{q1}(t))$$

This example demonstrates that the Poisson bracket having two integrals of motion as arguments vanishes:

$$\left\{ \frac{\text{p1}(t)^2}{2} + \frac{\text{p2}(t)^2}{2} + V(\text{q1}(t)), \; \frac{\text{p1}(t)^2}{2} + \frac{\text{p2}(t)^2}{2} + V(\text{q1}(t)) \right\}_{\text{pm2}}$$

$$0$$

2.8.7.3 Hamilton's Equations Derived from the Manifold

Having available an object-based reprsentation, it is convenient to inherit properties of one class to another. This is especially useful in deriving Hamilton's equations based on PBs. In the previous subsection, we introduced class PoissonB collecting all properties and methods of a Poisson manifold. This class can be used by the class HamltonianEquations defined by the phase space variables. Those variables are the basis of the Hamilton manifold. The algebraic structure defined for the PBs is also used by this class.

It is convenient here to define a class for Hamilton's equations which inherits the properties of the Poisson bracket. The properties of the Poisson manifold are equivalent to the properties of the Hamilton manifold.

The following lines define the class HamiltonEquations:

```
HamiltonEquations = Class["HamiltonEquations", PB,
  {description = "Hamilton's equations"},
  {HamEqs[H_, V_ /; FreeQ[V, List]] :=
    ∂_T V == PoissonBracket[H, V],
   HamEqs[H_, V_List] := Map[HamEqs[H, #] &, V],
   HamEqs[H_] := Map[HamEqs[H, #] &, Flatten[{P, Q}]],
   HamiltonEquations[
     Pnew_List, Qnew_List, Tnew_Symbol] :=
     Block[{}, P = Pnew; Q = Qnew; T = Tnew]}
]
```

– Class HamiltonEquations –

To handle the class for Hamiltonian equations and the derived objects in the same way as in a textbooks or in case of Poisson brackets, we introduce the notation

```
Notation[𝓗_𝓔𝓺^obj-[f_] ⟺ Dot[obj_, HamEqs[f_]]]
```

and define the corresponding palette

$$\{\Box, \ \Box\}_\Box$$

$$\mathcal{H}_{\mathcal{E}q}^{\Box}[\Box]$$

Having these tools available, we can apply the classes to specific problems.

2.8.7.4 Hamilton's Equations Derived from the Hamilton–Poisson Manifold

As a first example, let us examine a Hamilton–Poisson (HP) manifold with a single coordinate and a single momentum. The object defining the HP manifold is created by

```
ham1 = HamiltonEquations.new[{P → {p[t]}, Q → {q[t]}}]
```

– Object of HamiltonEquations –

Specifying a single-particle Hamiltonian by kinetic and potential energies, we can derive the set of Hamilton's equations by applying the manifold to the Hamiltonian:

```
GetPropertiesForm[ham1]
```

Property	Value
description	Hamilton's equations
P	$\{p(t)\}$
Q	$\{q(t)\}$
T	t
A	$\{\alpha \to \alpha\}$

$$\mathcal{H}_{\mathcal{E}q}^{haml}\left[\frac{p(t)^2}{2} + V(q(t))\right]$$

$$\{p'(t) == V'(q(t)), q'(t) == -p(t)\}$$

The result is a system of equations defining the dynamic of this particle.

A second example is concerned with a four-dimensional HP manifold. The generalized coordinates and the momenta are primarily given by q_1, q_2, p_1, and p_2.

```
ham2 = HamiltonEquations.
  new[{P → {p1[t], p2[t]}, Q → {q1[t], q2[t]}}]
```

– Object of HamiltonEquations –

As an example, let us consider the double pendulum. The Hamiltonian for this system reads

```
HamDoublePendulum =
                          1
      ───────────────────────────────────────
      2 l₁² l₂² m₂ (m₁ + m₂ Sin[Θ₁[t] - Θ₂[t]]²)
      (l₂² m₂ p₁[t]² + l₁² (m₁ + m₂) p₂[t]² -
        2 m₂ l₁ l₂ p₁[t] p₂[t] Cos[Θ₁[t] - Θ₂[t]]) -
      m₂ g l₂ Cos[Θ₂[t]] - (m₁ + m₂) g l₁ Cos[Θ₁[t]]
```

$$-g\cos(\theta_2(t))\, l_2\, m_2 - g\cos(\theta_1(t))\, l_1\, (m_1 + m_2) +$$
$$\frac{l_2^2\, m_2\, p_1(t)^2 - 2\cos(\theta_1(t) - \theta_2(t))\, l_1\, l_2\, m_2\, p_2(t)\, p_1(t) + l_1^2\, (m_1 + m_2)\, p_2(t)^2}{2\, l_1^2\, l_2^2\, (m_2 \sin^2(\theta_1(t) - \theta_2(t)) + m_1)\, m_2}$$

where p_i ($i = 1, 2$) are the generalized momenta, l_i and m_i are the inertia momenta and the masses of the particles, rspectively, and θ_i are the angles of deviation. The HP manifold *ham2* defined above does not exactly correspond to the variables used in the Hamiltonian. However, we are able to change the coordinate names by setting the properties of the HP manifold using

```
SetProperties[ ham2,
  {P → {p₁[t], p₂[t]}, Q → {θ₁[t], θ₂[t]}}]
```

Now, the HP manifold is defined for the coordinates

```
GetPropertiesForm[ham2]
```

Property	Value
description	Hamilton's equations
P	$\{p_1(t),\ p_2(t)\}$
Q	$\{\theta_1(t),\ \theta_2(t)\}$
T	t
A	$\{\alpha \to \alpha\}$

The four equations of motion can then be obtained using

---

equationsOfMotion = $\mathcal{H}_{\mathcal{E}q}^{ham2}$[HamDoublePendulum]

$$\Big\{ p_1'(t) == g \sin(\theta_1(t))\, l_1\, (m_1 + m_2) +$$

$$\frac{1}{2\, l_1^2\, l_2^2\, m_2} \Big(\frac{2 \sin(\theta_1(t) - \theta_2(t))\, l_1\, l_2\, m_2\, p_1(t)\, p_2(t)}{m_2 \sin^2(\theta_1(t) - \theta_2(t)) + m_1} -$$

$$(2 \cos(\theta_1(t) - \theta_2(t)) \sin(\theta_1(t) - \theta_2(t))\, m_2\, (l_2^2\, m_2\, p_1(t)^2 -$$

$$2 \cos(\theta_1(t) - \theta_2(t))\, l_1\, l_2\, m_2\, p_2(t)\, p_1(t) + l_1^2\, (m_1 + m_2)\, p_2(t)^2)) \Big/$$

$$(m_2 \sin^2(\theta_1(t) - \theta_2(t)) + m_1)^2 \Big), \ p_2'(t) ==$$

$$g \sin(\theta_2(t))\, l_2\, m_2 + \frac{1}{2\, l_1^2\, l_2^2\, m_2}\Big((2 \cos(\theta_1(t) - \theta_2(t)) \sin(\theta_1(t) - \theta_2(t))$$

$$m_2\, (l_2^2\, m_2\, p_1(t)^2 - 2 \cos(\theta_1(t) - \theta_2(t))\, l_1\, l_2\, m_2\, p_2(t)\, p_1(t) +$$

$$l_1^2\, (m_1 + m_2)\, p_2(t)^2)) \big/ (m_2 \sin^2(\theta_1(t) - \theta_2(t)) + m_1)^2 -$$

$$\frac{2 \sin(\theta_1(t) - \theta_2(t))\, l_1\, l_2\, m_2\, p_1(t)\, p_2(t)}{m_2 \sin^2(\theta_1(t) - \theta_2(t)) + m_1}\Big), \ \theta_1'(t) ==$$

$$\frac{2 \cos(\theta_1(t) - \theta_2(t))\, l_1\, l_2\, m_2\, p_2(t) - 2\, l_2^2\, m_2\, p_1(t)}{2\, l_1^2\, l_2^2\, m_2\, (m_2 \sin^2(\theta_1(t) - \theta_2(t)) + m_1)},$$

$$\theta_2'(t) ==$$

$$\frac{2 \cos(\theta_1(t) - \theta_2(t))\, l_1\, l_2\, m_2\, p_1(t) - 2\, l_1^2\, (m_1 + m_2)\, p_2(t)}{2\, l_1^2\, l_2^2\, m_2\, (m_2 \sin^2(\theta_1(t) - \theta_2(t)) + m_1)}\Big\}$$

They represent the dynamics of the double pendulum in the Hamilton–Poisson manifold. This example demonstrates that an object-oriented approach in symbolic computing allows one to mimic the theoretical background as close as possible. It is natural in an object-oriented environment to use the mathematical notions in a one-to-one corespondence. Thus, symbolic computing becomes a basis for theoretical constructs. The ease of use and the close connection to textbook presentations allows for a fast manipulation and reliable calculation of results. In addition to these examples, many other applications of *Elements* to similar subjects are ahead.

2.8.8 Canonical Transformations

The basic idea of canonical transformations is to simplify a Lagrangian system of equations. Canonical transformations convert a Lagrangian by means of a coordinate change to a simpler representation of the Lagrangian. In addition to the simplification of the Lagrangian, it is often observed that the related equations of motion are also simplified. The coordinate change is given by means of a transformation of the following kind:

$$q_i \longrightarrow Q_i = Q(q_i, q_2, ..., q_N). \tag{2.8.85}$$

An example of such a canonical transformation is the introduction of cylindrical coordinates if the problem allows a rotation symmetry around a distinguished axis.

In the Hamiltonian description of mechanics, we not only have coordinates but also generalized momenta to describe the motion of the system. Since the generalized momenta have the same importance in phase space as generalized coordinates, we have to extend the transformation from coordinates to momenta as well. The new coordinates are thus given by

$$q_i \longrightarrow Q_i = Q_i(q_1, q_2 ..., q_N, p_1, ..., p_N), \tag{2.8.86}$$
$$p_i \longrightarrow P_i = P_i(q_1, q_2, ..., q_N, p_1, ..., p_N). \tag{2.8.87}$$

The transformation is executed in such a way that the new coordinates (Q_i, P_i) are functions of the old coordinates (q_i, p_i), $i, j = 1, 2, ..., N$. If the transformations simplify to the form

$$T_k : \begin{cases} q_i \longrightarrow & Q_i = Q_i(q_i) \\ p_i \longrightarrow & P_i = P_i(p_i), \end{cases} \tag{2.8.88}$$

where the coordinates depend only on coordinates and momenta depend only on momenta; we call this kind of transformation a point transformation. The general relation of a transformation for Q_i and P_i incorporating both the coordinates and the momenta are called canonical transformations. A specific feature of canonical transformations is that the Hamiltonian equations of motion are invariant with respect to the transformation; that is

$$p'_i = -\frac{\partial H}{\partial q_i} \qquad\qquad q'_i = -\frac{\partial H}{\partial p_i}$$

$$\xrightarrow{\hspace{0.5cm} T_k \hspace{0.5cm}} \qquad\qquad\qquad (2.8.89)$$

$$P'_i = -\frac{\partial \tilde{H}}{\partial Q_i} \qquad\qquad Q'_i = \frac{\partial \tilde{H}}{\partial P_i}$$

with $\tilde{H} = \tilde{H}(Q_i(q_k, p_k), P_i(q_k, q_k))$ as the new Hamiltonian.

The application of canonical transforms to a Hamiltonian always saves the structure of the Hamiltonian equations but simplifies the resulting representation of the equations of motion. This simplification aims at a reduction of the equation in such a way that a straightforward integration of the equations is possible. An optimum of a canonical transformation is gained in such a case when all new coordinates are cyclic; that is there exist a transformation of the kind

$$H(p_1, \ldots, p_N, q_1, \ldots, q_N) \longrightarrow \tilde{H}(P_1, \ldots, P_N). \qquad (2.8.90)$$

The Hamilton equations of motion are then given by

$$P'_i = -\frac{\partial \tilde{H}}{\partial Q_i} = 0 \quad \text{i.e., } P_i = \text{const.} \quad i = 1, \ldots, N \qquad (2.8.91)$$

$$Q'_i = \frac{\partial \tilde{H}}{\partial P_i} = f_i(P_1, \ldots, P_N), \qquad\qquad\qquad (2.8.92)$$

where f_i are functions depending only on the new momenta and do not show any explicit time dependence. The consequence of this representation is that the solution for the generalized coordinates follows by

$$Q_i = f_i t + \delta_i, \qquad i = 1, \ldots, N, \qquad\qquad (2.8.93)$$

with $\delta_i = Q_i(0)$ the initial condition for the coordinates. The momenta are just conserved quantities in this representation. If we are able to uncover these momenta or transformations, we are able to solve the corresponding equations of motion. The P_i and δ_i are then integrals of motion. The N momenta P_i are the distinguished integrals of motion allowing us to carry out a complete integration. The δ_i allow us to complete the integration and terminate the nontrivial solution process. If we know the solution, we are able to invert the transformation and represent the solution in the original coordinates. For an optimal canonical transformation two facts must exist:

1) Find the new variables

2) Transform the Hamiltonian to the new representation.

2.8.9 Generating Functions

Canonical transformations are determined by generating functions. To demonstrate the meaning of a generating function, let us consider again the Liouville theorem. Simplifying things, we consider a mechanical system with a single degree of freedom. The original canonical variables are (p, q) and the target variables are (P, Q). The theorem by Liouville states the conservation of the phase space volume B

$$\int_B \int \alpha_p \, d_q = \int_B \int dP \, dQ. \tag{2.8.94}$$

From Stokes theorem on volume integrals it is obvious that an integral on the space B is replaced by a contour integral along c in such a way that we have

$$\oint_c p \, dq = \oint_c P \, dQ. \tag{2.8.95}$$

In addition, we assume that the target coordinates P and Q depend on the original coordinates q and p; that is, $P = P(p, q)$ and $Q = Q(p, q)$. The dependence of the target coordinates on the original coordinates may be different from this assumption. It is also possible that we have a relation like $P = P(Q, q)$ and $p = p(Q, q)$ where now Q and q are the independent variables. If we assume such a relation, we find from the line integral the following relation:

$$\oint_c \{P\,(Q,\ q)\,dq - P\,(Q,\ q)\,dQ\} = 0. \tag{2.8.96}$$

This kind of representation suggests that the integrand is given by a total differential of the function $F_1 = F_1\,(Q, q)$; that is,

$$\oint_c (p \, dq - P \, dQ) = \oint_c d\,F_1\,(Q,\ q)$$
$$= \oint_c \frac{\partial F_1}{\partial Q} \, dQ + \frac{\partial F_1}{\partial q} \, dq. \tag{2.8.97}$$

Comparing the coefficients of the total differentials, we find

$$p = \frac{\partial F_1}{\partial q}, \tag{2.8.98}$$

$$P = -\frac{\partial F_1}{\partial Q}. \tag{2.8.99}$$

The first of these equations provide a relation between p and (q, Q) which must be inverted to gain the functional dependence of $Q = Q(p, q)$. The inversion is possible if

$$\frac{\partial^2 F_1}{\partial q \, \partial Q} \neq 0. \tag{2.8.100}$$

Inserting the derived relation $Q = Q(p, q)$ into the second equation, we get an expression for the target momentum: $P = P(q, p)$.

The first example deals with (q, Q) as independent variables. It is also possible to use other combinations of variable pairs such as (P, q), (Q, p) and (P, Q) for independent variables. Let us consider the case when (P, q) are independent variables. Then, the conservation of the phase space volume provides

$$\oint_{\mathscr{C}} (p \, dq - P \, dQ) = \oint_{\mathscr{C}} (p \, dq + Q \, dP) \tag{2.8.101}$$

with $\Phi d(PQ) = \Phi P dQ + \Phi Q dP$. On the other hand, the generating function is now $F_2 = F_2(P, q)$; thus, the line integral is

$$\oint_{\mathscr{C}} \left(\frac{\partial F_2}{\partial P} \, dP + \frac{\partial F_2}{\partial q} \, dq \right) = \oint_{\mathscr{C}} p \, dq + Q \, dP. \tag{2.8.102}$$

From this relations, it follows that

$$p = \frac{\partial F_2}{\partial q}, \tag{2.8.103}$$

$$Q = \frac{\partial F_2}{\partial P}. \tag{2.8.104}$$

An example for this kind of generating function is $F_2 = pq$, which simplifies the two determining transformations to identical transformations:

$$p = \frac{\partial F_2}{\partial q} = P \tag{2.8.105}$$

$$Q = \frac{\partial F_2}{\partial P} = q. \tag{2.8.106}$$

The combination of the independent variables allows two other generating functions given by

$$F_3 = F_3(Q, p) \tag{2.8.107}$$

and

$$F_4 = F_4(P, p). \tag{2.8.108}$$

If the canonical transformation is independent of time, then the representation of the Hamiltonian is gained just by coordinate transformations; that is,

$$\tilde{H} = \tilde{H}(P, Q) = H(p(P, Q), q(P, Q)). \qquad (2.8.109)$$

As an example, let us consider the harmonic oscillator with its Hamiltonian:

```
H =   p²    k
     ───  + ─  q²
     2 m    2
```

$$\frac{p^2}{2m} + \frac{k\,q^2}{2}$$

By substituting $\omega^2 = k/m$, we get the representation

```
Ht = H /. k -> m ω²
```

$$\frac{p^2}{2m} + \frac{1}{2}\,m\,q^2\,\omega^2$$

The Hamiltonian in the present representation suggests that the canonical transformation is designed in such a way that the target variable Q is a cyclic variable. We assume that the canonical transformation is given by the following relation:

```
                                    f[P]
canonTrafo = {p -> f[P] Cos[Q], q ->  ──── Sin[Q]};
                                    m ω

canonTrafo // TableForm
```

$$p \to \cos(Q)\,f(P)$$
$$q \to \frac{f(P)\sin(Q)}{m\,\omega}$$

where $f(P)$ is an arbitrary function of P. Applying this transformation to the Hamiltonian, we get

```
hth = Ht /. canonTrafo // Simplify
```

$$\frac{f(P)^2}{2m}$$

It is obvious that Q is a cyclic variable and, thus, P represents a conserved quantity. The unknown function $f(P)$ is determined by the following procedure. First, represent the canonical transformation for the original momentum p by

```
s1 = p == ( p /. canonTrafo) // Solve[#, p] &
     q      q
```

$$\{\{p \to m\,q\,\omega\cot(Q)\}\}$$

This relation suggests that the generating function is of type $F = F(q, Q) = F_1$ because we have

```
eq1 = ∂_q F[q, Q] == (p /. Flatten[s1])
```

$$F^{(1,0)}(q, Q) == m\,q\,\omega\cot(Q)$$

This relation can be solved to provide

```
s2 = DSolve[eq1, F, {q, Q}] // Flatten
```

$$\left\{F \to \mathrm{Function}\!\left[\{q, Q\}, \frac{1}{2}\,m\,\omega\cot(Q)\,q^2 + c_1[Q]\right]\right\}$$

The simplest solution is generated by setting the arbitrary function $c_1(Q)$ equal to zero which allows us to write

$$F_1 = \frac{m\omega}{2}\,q^2\cot Q. \tag{2.8.110}$$

The second relation defining the target momentum is solved with respect to the old coordinate:

```
solCoordinates = (Solve[P == -∂_Q F[q, Q] /. s2, q] /.
    C[1] -> Function[Q, 0]) // PowerExpand
```

$$\left\{\left\{q \to -\frac{\sqrt{2}\ \sqrt{P}\ \sin(Q)}{\sqrt{m}\ \sqrt{\omega}}\right\}, \left\{q \to \frac{\sqrt{2}\ \sqrt{P}\ \sin(Q)}{\sqrt{m}\ \sqrt{\omega}}\right\}\right\}$$

$$P = -\frac{\partial F_1}{\partial Q} = \frac{m\omega q^2}{2\sin^2 Q} \tag{2.8.111}$$

Using the ansatz for the canonical transformation and the gained results for the old coordinate, we can compare the two results to determine the unknown function $f(P)$ by

```
solF =
  Solve[(q /. canonTrafo) == (q /. solCoordinates[[2]]),
    f[P]] // Flatten
```

$$\{f(P) \to \sqrt{2}\ \sqrt{m}\ \sqrt{P}\ \sqrt{\omega}\}$$

The target Hamiltonian then becomes

```
targetHamiltonian = hth /. solF
```

$$P\omega$$

Since Q is a cyclic variable, we immediately observe that P is a constant of motion. The value of this constant is determined by the total energy E and the frequency ω by

$$P = \frac{E}{\omega} \tag{2.8.112}$$

The equation of motion for the Q coordinate reduces to

```
teqQ = ∂_t Q[t] == ∂_P targetHamiltonian
```

$$Q'(t) == \omega$$

The solution of this equation is derived by

```
solQ = DSolve[teqQ, Q, t] /. C[1] -> α
```

$\{\{Q \rightarrow \text{Function}[\{t\}, \alpha + t\,\omega]\}\}$

where α is the constant of integration. The final solution for the coordinates can be derived by inverting the transformations. Using the introduced representations, we find

```
q = q /. canonTrafo /. Q -> Q[t] /. solQ /. solF /.
    P -> ε/ω // PowerExpand
```

$\left\{ \dfrac{\sqrt{2}\,\sqrt{\mathcal{E}}\,\sin(\alpha + t\,\omega)}{\sqrt{m}\,\omega} \right\}$

However, this solution is the well-known solution of a harmonica oscillator. The above example demonstrates how the generating function can be determined if one is able to guess a basic representation of the canonical transformation.

2.8.10 Action Variables

The method used in the previous subsection demonstrated that the generating function is the basic tool to determine canonical transformations. However, the presented procedure in this section is not a systematic procedure and connected with guesswork. This section is concerned with a systematic approach to derive and determine the generating function in a systematic way. To demonstrate the method let us consider the generating function of the type $F_2 = F_2(q_i, P_i) = F_2(q_i, \alpha_i)$, with $\alpha_i = P_i$. This generating function is denoted by S in the following:

$$S = S(q_i, \ldots, q_N, \alpha_i, \ldots, \alpha_N) = F_2. \qquad (2.8.113)$$

This kind of generating function defines the momenta p_j and coordinates $Q_i = \beta_i$ in the known way by

$$p_i = \frac{\partial S\,(q_i, \alpha_i)}{\partial q_i}, \tag{2.8.114}$$

$$\beta_i = \frac{\partial S\,(q_i, \alpha_i)}{\partial \alpha_i}. \tag{2.8.115}$$

The β_i's are the target coordinates conjungate to the α_i's. The relation between the original and the target Hamiltonian is given by

$$\tilde{H} = \tilde{H}(\alpha_i) = H(q_i, p_i) = H\left(q_i, \frac{\partial S}{\partial q_i}\right). \tag{2.8.116}$$

Since the total energy is a conserved quantity for standard Hamiltonian systems (i.e., $\tilde{H}\,(\alpha_i) = $ const.), the relation

$$\tilde{H}\,(\alpha_i) = H\left(q_i, \frac{\partial S}{\partial q_i}\right) \tag{2.8.117}$$

defines a hypersurface in phase space. On the other hand, this relation defines the generating function. The relation defining the S function is a partial differential equation of first order. The generating function S depends on N independent coordinates q_i ($i = 1, 2, ..., N$). Relation (2.8.117) is known as the time-independent Hamilton–Jacobi equations.

In the case of a time-dependent Hamiltonian, the Hamilton–Jacobi equation also becomes time dependent and generalizes to

$$\frac{\partial S}{\partial t} + H\left(q_i, \frac{\partial S}{\partial q_i}\right) = 0. \tag{2.8.118}$$

In this case, the generating function also depends on the time t. If the system is a conserved system, then the time is separated from the function by

$$S = S\,(q_i, \alpha_i) - E\,t. \tag{2.8.119}$$

In this case, the time-dependent Hamilton–Jacobi equation reduces to

$$\frac{\partial S}{\partial t} + H\left(q_i, \frac{\partial S}{\partial q_i}\right) = -E + H\left(q_i, \frac{\partial S}{\partial q_i}\right) = 0$$
$$\Longleftrightarrow H\left(q_i, \frac{\partial S}{\partial q_i}\right) = E = \tilde{H}\,(\alpha_i). \tag{2.8.120}$$

It is well known that first-order partial differential equations (PDEs) of the above type need N independent integrals of motion for their solution. However, this integrals are given by the target momenta $P_i = \alpha_i$ ($i = 1, 2, ..., N$), which are constants of motion. The problem of finding the generating function now reduces to solving the Hamilton–Jacobi equations, which is equivalent to the solution of the canonical equations of motion. The derivation of an explicit solution for the Hamilton–Jacobi equations in its most general form is a very difficult task. This tasks simplifies if we prescribe the property of separation to the Hamiltonian. The functional dependence of S on the coordinates suggests

$$dS = \sum_{i=1}^{N} \frac{\partial S}{\partial q_i} \, dq_i = \sum_{i=1}^{N} p_i \, dq_i, \tag{2.8.121}$$

which results in the general representation

$$S = \int_{q_0}^{q} p_i \, dq_i, \tag{2.8.122}$$

where $q_0 = (q_1(0), q_2(0), \ldots, q_N(0))$ are the initial conditions of a trajectory in phase space. If the quantities α_i are known, the trajectory $q = (q_1(t), q_2(t), \ldots, q_N(t))$ for times greater than zero are also known. It is obvious that for a determination of S, the trajectories $q_i = q_i(t)$ must be known beforehand. At this point, the question arises of whether the Hamilton–Jacobi theory is a useful theory to derive practical results. This question will be resolved in the following subsections.

2.8.10.1 One-Dimensional Hamilton–Jacobi Equation

In case of a one-dimensional Hamiltonian system, the Hamilton–Jacobi equation is solvable. Let us assume that we are dealing with a Hamiltonian depending on the variables (q, p):

$$H = H(q, p). \tag{2.8.123}$$

The target Hamiltonian $\tilde{H}$ is thus a function in a single variable: the canonical momentum. In addition, the target momentum is a conserved quantity:

$$\tilde{H} = \tilde{H}(\alpha). \tag{2.8.124}$$

For time-independent Hamiltonians, the solution step is the equivalence of this conserved quantity with the Hamiltonian:

$$\tilde{H} = \alpha. \tag{2.8.125}$$

Thus, α is just the total energy of the system. The Hamilton–Jacobi equation then becomes

$$H\left(q, \frac{\partial S}{\partial q}\right) = \alpha. \tag{2.8.126}$$

At the same time, the two relations for the generating functions hold:

$$p = \frac{\partial S(q,\alpha)}{\partial q}, \tag{2.8.127}$$

$$\beta = \frac{\partial S(q,\alpha)}{\partial \alpha}. \tag{2.8.128}$$

Since the transformation are canonical transformations, the equations of motion in the target variables become

$$\alpha' = -\frac{\partial \tilde{H}}{\partial \beta} = 0, \tag{2.8.129}$$

$$\beta' = \frac{\partial \tilde{H}}{\partial \alpha} = 1. \tag{2.8.130}$$

These two equations can be solved by

$$\alpha = \text{const.} = \tilde{H}, \tag{2.8.131}$$

$$\beta = t - t_0. \tag{2.8.132}$$

Knowing the solution in the target variables, we are able to express the solutions in the original variables by

$$t - t_0 = \frac{\partial S(q,\alpha)}{\partial \alpha} = \frac{\partial}{\partial \alpha} \int_{q_0}^{q} p(q, \alpha)\, dq, \tag{2.8.133}$$

$$t - t_0 = \int_{q_0}^{q} \frac{\partial p(q,\alpha)}{\partial \alpha}\, dq. \tag{2.8.134}$$

As an example, let us examine the motion of a particle in the potential $V = V(q)$. The Hamiltonian then becomes

$$H(q, p) = \frac{p^2}{2m} + V(q). \tag{2.8.135}$$

Since the Hamiltonian satisfies the relation

$$H(q, p) = \tilde{H}(\alpha = \alpha), \tag{2.8.136}$$

we find

$$\frac{p^2}{2m} + V(q) = \alpha \tag{2.8.137}$$

or

$$p(q, \alpha) = \pm \sqrt{2m(\alpha - V(q))}. \tag{2.8.138}$$

The final solution of the problem thus results from

$$t - t_0 = \int_{q_0}^{q} \frac{\partial}{\partial \alpha} \left(\pm \sqrt{2\, m(\alpha - V(q))} \right) dq$$

$$= \sqrt{\frac{m}{2}} \int_{q}^{q} \frac{dq}{\sqrt{\alpha - V(q)}}. \tag{2.8.139}$$

Since for a conserved system, α is equal the total energy, the solution reduces to a simple quadrature

$$t - t_0 = \sqrt{\frac{m}{2}} \int_{q_0}^{q} \frac{dq}{q\sqrt{E - V(q)}}. \tag{2.8.140}$$

However, this result is already known from the integration procedures we discussed in Section 2.4. The question of what is the advantage of this procedure compared with the standard quadrature arises. The main advantage is that we are now in a position to introduce variables, action angle variables, allowing us to simplify the problem.

2.8.10.2 Action Angle Variables for one Dimension

The examinations so far demonstrated that the trajectories in phase space are closed curves. The period a particle needed to traverse the complete path is given by $2\pi/\omega$, where ω denotes the cycle frequency of a trajectory. The idea here is to use the periodicity to introduce coordinates which possess this 2π periodicity. We are looking for coordinates which increase their values by 2π if the particle traverses the total path. The targeted variables are denoted by J and θ; J is the conjungate momentum to θ. The set of defining equations for the generating function now reads

$$p = \frac{\partial S(q,J)}{\partial q}, \tag{2.8.141}$$

$$\theta = \frac{\partial S(q,J)}{\partial J}. \tag{2.8.142}$$

The related Hamilton–Jacobi equation is

$$H\left(q, \frac{\partial S}{\partial q}\right) = \alpha = \tilde{H}(J). \tag{2.8.143}$$

For a trajectory with fixed α (i.e., fixed J, $\alpha = \tilde{H}(J)$), we find by differentiating θ with respect to q that

$$\frac{d\theta}{dq} = \frac{\partial}{\partial J}\left(\frac{\partial S}{\partial q}\right). \tag{2.8.144}$$

Our assumption on θ is that it should increase by 2π if the trajectory is completely traversed; that is,

$$2\pi = \oint d\theta = \frac{\partial}{\partial J} \oint \left(\frac{\partial S}{\partial q}\right) dq = \frac{\partial}{\partial J} \oint p\, dq. \qquad (2.8.145)$$

This condition is satisfied if

$$J = \frac{1}{2\pi} \oint p(q, \alpha)\, dq. \qquad (2.8.146)$$

Relation (2.8.146) is also known as the definition of the action variable. The integration is carried out along the trajectory C, which is determined by the total energy $\alpha = \tilde{H}(J) = E$.

The related canonical equations of motion are

$$J' = -\frac{\partial \tilde{H}(J)}{\partial \theta} = 0, \qquad (2.8.147)$$

$$\theta' = \frac{\partial \tilde{H}(J)}{\partial J} = \omega(J). \qquad (2.8.148)$$

The two equations are solved by

$$J = \text{const.}, \qquad (2.8.149)$$
$$\theta = \omega(J)\, t + \delta, \qquad (2.8.150)$$

where $\omega(J)$ is the characteristic frequency of the motion and $\delta = \theta(0)$ is determined by the initial condition.

Example 1: Harmonic Oscillator

As an example, let us examine the harmonic oscillator to demonstrate the derivation of the action angle variables. The Hamiltonian is given by

$$H = \frac{1}{2}(p^2 + \omega^2 q^2).$$

The Hamilton–Jacobi equation reads

$$\frac{1}{2}\left(\frac{\partial S}{\partial q}\right)^2 + \frac{1}{2}\omega^2 q^2 = \alpha, \qquad (2.8.151)$$

where α is an integration constant equal to the total energy $E = H$. The action variable thus follows by

$$J = \frac{1}{2\pi} \oint_C \sqrt{2\left(E - \frac{1}{2}\omega^2 q^2\right)}\, dq, \qquad (2.8.152)$$

with C the closed trajectory in the phase space. This trajectory possesses two turning points at $q = \pm \sqrt{2E}\,/\,\omega$. A direct calculation shows

$$J = \frac{E}{\omega}. \tag{2.8.153}$$

This relation connects the constant of integration $\alpha = E$ with the quantity J; that is,

$$\alpha = E \tilde{H}(J) = J\omega.$$

The generating function S is then given as

$$S(q, J) = \int_{q_0}^{q} \sqrt{2\left(J\omega - \frac{1}{2}\omega^2 q^2\right)}\, dq.$$

Using the original coordinates, we can represent the solution for the generating function by

$$q = \frac{\partial S}{\partial J} = \frac{\partial}{\partial J} \int_{q_0}^{q} \sqrt{2\left(J\omega - \frac{1}{2}\omega^2 q^2\right)}\, dq$$

$$= \int_{q_0}^{q} 2\,\omega\,\frac{1}{2}\,\frac{1}{\sqrt{2\left(J\omega - \frac{1}{2}\omega^2 q^2\right)}}\, dq$$

$$= \omega \int_{q_0}^{q} \frac{1}{\sqrt{2\left(J\omega - \frac{1}{2}\omega^2 q^2\right)}}\, dq \tag{2.8.154}$$

$$= \sqrt{\frac{2J}{\omega}}\,\sin(\theta + \delta),$$

where $\delta = \arcsin\left(q_0\,\omega / \sqrt{2E}\right)$.

The introduction of action angle variables is not only restricted to a two-dimensional phase space. This concept can be generalized to the $2N$-dimensional case. For our example, it was essential to use the total energy as a conserved quantity in the calculations. In the case of a $2N$-dimensional Hamiltonian system, the knowledge of N integrals of motion allows one to separate the Hamiltonian in appropriate coordinates. In any case in which this separation exists, the solution of the problem simplifies dramatically.

2.8.10.3 Separation of Hamiltonians

Based on the Hamilton–Jacobi equation, we discuss here the separation of Hamiltonian systems. The Hamilton–Jacobi equation for a N-dimensional system is

$$H\!\left(q_1, \;...\, q_N, \; \tfrac{\partial S}{\partial q_1}, \; ..., \; \tfrac{\partial S}{\partial q_N}\right) = \tilde{H}\,(\alpha_1, \; ..., \; \alpha_N) \qquad (2.8.155)$$

where q_i are the generalized coordinates and $p_i = \partial S / \partial q_i$ are the generalized momenta generating the phase space. The α_i are the conserved quantities in this space. Thus, the Hamilton–Jacobi equation is the determining equation of S in N independent coordinates q_i.

First-order PDEs allow N independent integrals of motion which determine the solution. We will show that these constants of motion are related to the α_i's. The Hamilton–Jacobi equation in general is not solvable in a closed analytic form until the Hamiltonian is separable.

If the Hamiltonian separates, then the generating function S also separates. On the other hand, this means that S is a direct sum of the separated components depending only on a single coordinate:

$$S\,(q_i, \; \alpha_i) = \textstyle\sum_{k=1}^{N} S_k\,(q_k, \; \alpha_1, \; ..., \; \alpha_N). \qquad (2.8.156)$$

A simple class of Hamiltonians satisfying this condition is those which decay in N subsystems by

$$H\,(p_i, \; q_i) = \textstyle\sum_{k=1}^{N} H_k\,(p_k, \; q_k)$$

(i.e., a system of N decoupled oscillators). In this case, the Hamilton–Jacobi equation reduces to the one-dimensional case discussed earlier:

$$H_k\!\left(q_k, \; \tfrac{\partial S}{\partial q_k}\right) = \alpha_k \qquad\qquad k = 1, \; ..., \; N. \qquad (2.8.157)$$

The integrals α_k of this special case are connected to the Hamiltonian by the sum

$$\alpha = \alpha_1 + \alpha_2 + ... + \alpha_N = \tilde{H}, \qquad (2.8.158)$$

where $\tilde{H}$ is the transformed Hamiltonian. For practical cases, this separation is very seldom usd and thus it is very rare to apply this kind of theory to a problem. However, if we are able to introduce appropriate coordinate transformations, we gain a representation a few steps apart from the solution.

Let us assume that the generating function separates; then, the following relations for generalized momenta hold:

$$P_k = \frac{\partial S(q_k, \alpha_1, \dots \alpha_N)}{\partial q_k}. \tag{2.8.159}$$

The meaning of this relation is that each target momentum P_k only depends on a single coordinate q_k. If we assume, in addition, that the motion in q_k is periodic, we can introduce a set of action variables I_k by

$$I_k = \frac{1}{2\pi} \oint_{C_k} P_k(q_k, \alpha_1, \dots, \alpha_N) \, dq_k, \tag{2.8.160}$$

where C_k is a closed loop in phase space. This relation establishes a relation between the action I_k and the integrals α_k. This relation is used to replace the α_k's by the actions I_k in the generating function S. After the replacement, we can evaluate the two relations for the target coordinates. The angle variables follow from

$$\theta_k = \frac{\partial S}{\partial I_k} = \sum_{m=1}^{N} \frac{\partial S_M(q_m, I_1, \dots, I_N)}{\partial I_k}. \tag{2.8.161}$$

By definition, the angles θ_k are automatically conjungate to the actions I_k. If we know the variables in the transformed phase space, we can derive the canonical equations from the Hamiltonian $\tilde{H} = \tilde{H}(I_1, \dots, I_N)$ by

$$I'_k = - \frac{\partial \tilde{H}(I_1, \dots, I_N)}{\partial \theta_k} = 0, \tag{2.8.162}$$

$$\theta'_k = \frac{\partial \tilde{H}}{\partial I_k} = \omega_k(I_1, \dots, I_N), \tag{2.8.163}$$

where ω_k is the cycle frequency of the kth coordinate. Since all equations are decoupled, the solution is accessible by an integration:

$$I_k = \text{const.,} \tag{2.8.164}$$
$$Q_k = \omega_k(I_1, \dots, I_N) t + \delta_k, \tag{2.8.165}$$

where δ_k is the initial condition of the angles at $t = 0$. Different examinations demonstrate that the knowledge of the action angle variables are a basic tool to solve the Hamilton–Jacobi equations. The main point of this procedure is the uncovering of a sufficient number of integrals of motion.

Let us assume that $I_i(p_k, q_k)$ is an integral of motion; then, we know that along a trajectory, the value of this integral does not change; that is,

$$I_i(p_k, q_k) = \alpha_i. \tag{2.8.166}$$

If we know, in addition, the total energy, then we have

$$\{I_i, H\} = 0, \tag{2.8.167}$$

independent of the coordinates used. If we add H to the set of integrals, we can introduce the term "completely integrable Hamiltonian systems".

Definition: Complete Integrability

A Hamiltonian with N degrees of freedom is said to be completely integrable if N integrals of motion, $I_1, I_2, ..., I_N$, exist. These integral of motion are in involution with each other by

$$\{I_i, I_j\} = 0 \qquad \text{for} \quad i, j = 1, 2, ..., N. \blacksquare \tag{2.8.168}$$

The meaning of this definition becomes obvious if we remember the meaning of an integral of motion. The existence of N integrals of motion I_j restricts the motion to an N-dimensional manifold $\mathcal{M}$. The total motion in the $2N$-dimensional phase space is restricted to an N-dimensional submanifold. An example was the harmonic oscillator which demonstrated this behavior clearly: that is,. the motion of the two-dimensional phase space is restricted to a one-dimensional curve. Knowing the N integrals, we are able to show that the geometric structure of the manifold $\mathcal{M}$ is a N-dimensional torus.

Let us assume that one of the integrals is given by the Hamiltonian $I_i = H$. Then, we know that the equations of motion follow from the Poisson brackets:

$$q'_i = \{q_i, H\}, \tag{2.8.169}$$
$$p'_i = \{p_i, H\}. \tag{2.8.170}$$

The system of equations defines a Hamiltonian flow in phase space. This flow is restricted to the manifold $\mathcal{M}$ because there are I_j integrals of motion known. The velocity field of the flow is defined by

$$\vec{\xi_i} = J \cdot \nabla J_i \qquad\qquad i = 1, 2, ..., N \tag{2.8.171}$$

where $\nabla = (\partial_{q_1}, \partial_{q_2}, ..., \partial_{q_N}, \partial_{p_1}, ..., \partial_{p_N})$ and J is the symplectic matrix

$$J = \begin{pmatrix} 0 & 1 \\ -1 & 0 \end{pmatrix} \tag{2.8.172}$$

with 1 a $N \times N$ identity matrix. This representation of the equations of motion is possible by introducing a set of coordinates with equal standing $\vec{z} = (z_1, z_2, ..., z_N) = (q_1, q_2, ..., q_N, p_1, ..., p_N)$. The Hamiltonian in these coordinates is then

$$H = H(q_i, ..., p_i) = H(z_i). \tag{2.8.173}$$

The Poisson bracket simplifies to

$$\{f, g\}_z = \nabla f . J . \nabla g \tag{2.8.174}$$

and the equations of motion result from

$$\begin{aligned} \vec{z}' &= \{\vec{z}, H\} = \nabla \vec{z} . J . \nabla H, \\ \vec{z}' &= J . \nabla H. \end{aligned} \tag{2.8.175}$$

The symplectic formulation of the equations of motion simplifies the representation but not the physical meaning. The mathematical representation becomes more compact and clear. The velocity field of the Hamiltonian system is then

$$\vec{\xi} = J . \nabla H. \tag{2.8.176}$$

The main property of the velocity field or flow of the Hamiltonian is that the flow is always tangential to the manifold $\mathcal{M}$. For each of the N integrals of motion, the flow is defined by

$$\vec{\xi}_i = J . \nabla_i, \qquad\qquad i = 1, 2, ..., N. \tag{2.8.177}$$

Each of the flow fields are tangential to the manifold $\mathcal{M}$. Because the completely integrable system is characterized by the independent integrals I_j.

Now, we switch to a topology argument contained in the Poincaré–Hopf theorem. Each N-dimensional manifold $\mathcal{M}$ characterized by N integrals of motion with the related flows establishes the topology of a N-dimensional torus.

For two dimensions, we can plot such a torus with the flow fields on top of the surface. In this case, the flows are just the coordinates on the surface (see Figure 2.8.5).

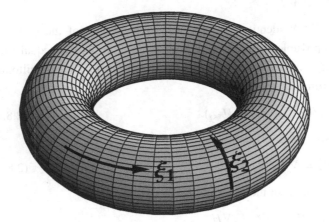

Figure 2.8.5.   Flow fields on a two-dimensional torus. ξ_1 and ξ_2 are the two possible velocity fields.

A practical interpretation of the flow fields on a torus is that the fields can be combed. In each direction of the flow, you can pervade along the flow. In contrast to a torus, a sphere cannot be combed (see Figure 2.8.6). The fields on the poles destroy this property on a sphere.

Figure 2.8.6.   Flow fields on a sphere. Here, the flow field cannot be combed.

On a sphere there is always a velocity field, hair, which prevents a comb from moving on the total surface. Knowing this topological interpretation and the existence of integrals of motion allows us to present a coordinate-free definition of action angle variables.

An N-torus is a natural object which can be generated as a direct product of N independent 2π periodic phase space curves $\mathbf{C}_k$ (see Figure 2.8.7). The phase space curves are designed in such a way that they cannot be transformed to other curves or shrunk to a point.

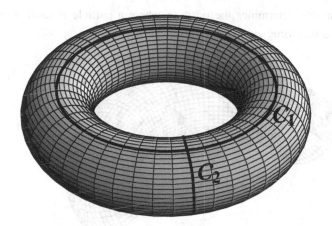

Figure 2.8.7.   A two-dimensional torus as an example for 2π periodic phase space curves.

The set of action variables is thus defined by

$$I_k = \frac{1}{2\pi} \oint_{\mathbf{C}_k} \Sigma_{m=1}^N p_m \, dq_m. \tag{2.8.178}$$

The related generating function

$$S = S(q_1, \ ..., \ q_N, \ I_1, \ ..., \ I_N) \tag{2.8.179}$$

allows the derivation of the angle variables:

$$\theta_k = \frac{\partial S (q_1,..., q_N, I_1,..., I_N)}{\partial I_k}. \tag{2.8.180}$$

Both sets of variables are related to the Hamilton equations of motion:

$$J'_k = - \frac{\partial \tilde{H} (I_1, ..., I_N)}{\partial \theta_k} = 0, \tag{2.8.181}$$

$$\theta'_k = \frac{\partial \tilde{H} (I_1, ..., I_N)}{\partial I_k} = \omega_k (I_1, \ ..., \ I_N). \tag{2.8.182}$$

Knowing this set of equations, the solution for the problem can be derived. Note that the transformation to action angle variables is a global transformation; that is, the total phase space is covered by tori and the trajectories are located on top of the surface.

The initial conditions $(q_1(0), q_2(0), ..., q_N(0), p_1(0), ..., p_N(0))$ determine the specific values of the integrals of motion:

$$I_k(p_i(0), q_i(0)) = \alpha_k, \qquad k = 1, ..., N. \qquad (2.8.183)$$

The I_k's determine on which torus a trajectory is located. The value of the angle variable determines the position where a particle is located on the torus for a two-torus see Figure 2.8.8.

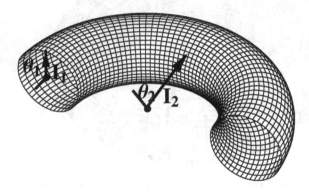

Figure 2.8.8.   Action angle variables on a 2-torus.

A conserved Hamilton system is determined by the dimensions collected in Table 2.8.1.

Phase space dimension	$2N - D_P$
Hyper surface of the energy	$2N - 1 = D_E$
Tori dimension	$N = D_T$

Table 2.8.1.   Definition of different dimensions.

Thus, for N-degrees of freedom, we get Table 2.8.2.

N	1	2	3	4	5
D_P	2	4	6	8	10
D_E	1	3	5	7	9
D_T	1	2	3	4	5

Table 2.8.2.   Collection of different dimensions related to an N-degrees, of freedom, system.

The numbers given allow the following conclusions:

In case of a single degree of freedom the hypersurface of the energy and the torus surface are identical.

For $N = 2$, the two-dimensional tori are embedded in the three-dimensional energy hypersurface. Especially the energy hypersurface divides the phase space in an inner and outer region. If there is a gap between   these regions, a trajectory in this region will stay forever in this gap. Gaps occur for nonintegrable Hamiltonians.

For $N \geqslant 3$, trajectories in gaps can escape into other portions of the energy hypersurface. This phenomenon is known as Arnold diffusion.

The Hamilton equations of motion show that the motion of the angle coordinates is periodic:

$$\theta'_k = \frac{\partial \tilde{H}}{\partial I_k} = \omega_k. \tag{2.8.184}$$

For a multidimensional Hamiltonian system there exist N frequencies of revolution. The ratios of these frequencies determine whether the trajectories in phase space have a closed rational ratio and thus the motion is periodic, or the ratio is irrational and the motion is aperiodic. In the last case, the tori are completely covered by the trajectories and there is no return to the starting point. This case is also known as quasiperiodic. If a trajectory completely covers a torus the system is denoted as ergodic. The discussed properties are obvious for a two-dimensional system. In such a case, we have two frequencies: ω_1 and ω_2. If the ratio

$$\frac{\omega_1}{\omega_2} = \text{irrational}, \tag{2.8.185}$$

then we have an ergodic system. In case of a rational ratio with

$$\frac{\omega_1}{\omega_2} = \frac{n}{m}, \qquad \text{with } n, \; m \in \mathbb{N}, \tag{2.8.186}$$

the trajectories are closed. This behavior is graphically represented by the torus itself or by an angle chart containing the paths (Figure 2.8.9).

 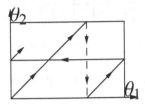

Figure 2.8.9.   Path on a torus and the corresponding angle chart.

Up to now, we discussed completely integrable systems. In such cases, we have N integrals for N degrees of freedom. All of the integrals of motion are in involution (i.e., $\{I_i, I_j\} = 0$, $i, j = 1, ..., N$). For a nonintegrable system, the question arises of what happens if a single integral of motion does not exist. This nonexistence of an integral causes tremendous problems in the process of integration. The questions related to this topic are as old as mechanics itself. Generations of physicists and mathematicians are hunting for the facts of nonintegrable systems. However, the problem was partially solved by Kolmogorov, Arnold, and Moser in 1960 by their famous theorem:

Theorem: KAM Theorem

If the ration ω_1/ω_2 of two frequencies ω_1 and ω_2 is sufficiently irrational (i.e.,

$$\left| \frac{\omega_1}{\omega_2} - \frac{r}{s} \right| > \frac{c}{s^{2+\delta}} \tag{2.8.187}$$

with fixed c and δ), and if the disturbance of the Hamilton system is sufficiently small, then there exists a torus which is the center for spinning trajectories with ω_1 and ω_2. If the disturbance of the Hamiltonian slightly increases $\varepsilon H_1 = 0$, then the torus is twisted and exists up to a critical value $\varepsilon_{max} H_1$.∎

However, the KAM theorem does not provide an upper limit for the critical parameter and thus only delivers a qualitative estimation. Due to Henon (1966), the disturbance of a Hamiltonian system can be of the magnitude $\varepsilon H_1 = 10^{-48}$, where a Moser torus is dislocated.

2.8.11 Exercises

1. An harmonic oscillator is described by the Lagrangian $L = \frac{1}{2} m (x'^2 - \omega^2 x^2)$. Construct the Hamiltonian and write out the equations of motion.

2. A particle moves vertically in a uniform gravitational field g, the Lagrangien being $L = \frac{1}{2} z'^2 - g z$. Construct the Hamiltonian. Hint: Add a total time derivative such as $\frac{1}{2} d(\lambda z^2)/dt = \lambda z z'$ to the Lagrangian.

3. A particle of mass m moves under the influence of gravity along the spiral $z = k \theta$, $r = $ const., where k is a constant and z is vertical. Obtain the Hamiltonian equations of motion.

4. A particle of mss m moves in one diimension under the influence of a force

$$F(x, t) = \frac{k}{x^2} t^{-q},$$

where k and q are positive constants. compute the Lagrangian and Hamiltonian functions. Compare the Hamiltonian and the total energy, and discuss the conservation of energy for the system.

2.8.12 Packages and Programs

Elements Package

The package Elements provides an object-oriented environment. The notations and definitions are described in the help text of the package. In short, Elements allows one to define classes and derive objects from these classes. Each class is divided into two sections containing properties and methods. Simply speaking, properties are parameters of the class and methods are the functions used to calculate some mathematical expressions. Classes are able to inherit properties and methods. For a detailed discussion of the package, see the help text.

```
AppendTo[$Path,
  "C:\\Mma\\Work\\TUMObjects\\Elements05"];
(* -- change the path above to the location
  where the package Elements is located -- *)
<< Elements`
Off[General::spell]; Off[General::spell1];
```

```
GetProperties[o_] :=
  Thread[Map[ToExpression[#] &, Properties[o]] →
    (o.# & /@ Properties[o])]
```

```
GetPropertiesForm[obj_] :=
  DisplayForm[GridBox[Prepend[GetProperties[obj],
    {StyleForm["Property", FontWeight -> Bold],
      StyleForm["Value", FontWeight -> Bold]}],
    RowLines → True, ColumnLines → True,
    GridFrame → True, ColumnAlignments → {Left}]]
```

```
<< Utilities`Notation`
```

Define some notations for Poisson brackets:

```
Notation[
  {f_, g_}_obj_ ⟺ Dot[obj_, PoissonBracket[f_, g_]]]
```

```
Notation[
  {f_, g_}_obj_ ⟺ Dot[obj_, PoissonBracket[f_, g_]],
  WorkingForm → TraditionalForm]
```

Define some notations for Hamilton's operator:

```
Notation[H_{εq}^{obj_}-[f_] ⟺ Dot[obj_, HamEqs[f_]]]
```

```
Notation[𝓗_{ℰq}^{obj}-[f_] ⟺ Dot[obj_, HamEqs[f_]],
 WorkingForm → TraditionalForm]
```

Euler–Lagrange Package

The Euler–Lagrange package allows one to derive the Euler–Lagrange
equations for a given Lagrangian.

```
If[$MachineType == "PC",
  $EulerLagrangePath = $TopDirectory <>
    "/AddOns/Applications/EulerLagrange/";
  AppendTo[$Path, $EulerLagrangePath],
  $EulerLagrangePath =
   StringJoin[$HomeDirectory, "/.Mathematica/3.0/
      AddOns/Applications/EulerLagrange", "/"];
  AppendTo[$Path, $EulerLagrangePath]];
```

```
Needs["EulerLagrange`"]
```

```
LegendreTransform[A_, x_List, momenta_List,
  indep_: {t}] := Block[{momentaRelations},
  momentaRelations =
   MapThread[∂_{#1}A == #2 &, {x, momenta}];
  sol = Flatten[Solve[momentaRelations, x]];
                      Length[x]
  Simplify[Expand[     ∑      x[[i]] ∂_{x[[i]]}A - A] /. sol]]
                      i=1
```

2.9 Chaotic Systems

2.9.1 Introduction

We discussed the structure of the phase space in the last section. The main structuring component was the existence of integrals of motion. Each integral added a certain amount to the tori representing the surfaces where the regular solutions live. The trajectories in phase space exist on these tori and are either periodic or at least quasiperiodic. A fundamental characteristic of a trajectory living on a tori is that it intersects a plane cutting the tori in a characteristic way. The closed or quasiclosed trajectory generates a characteristic pattern on this plane. Figure 2.9.1 demonstrates the global behavior in phase space.

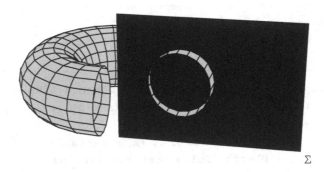

Σ

Figure 2.9.1.   Phase space structure intersected with a plane.

The pattern generated on the intersecting plane will show dots representing the position of the trajectory of the torus. If the trajectory is closed and thus periodic, the pattern will consist of a finite number of points. The number of points is related to the frequency with which a point cycles on the trajectory on the torus. If the trajectory is not closed (the trajectory is quasiperiodic), the points are continuously distributed on the surface of the torus. The pattern then is given as a quasiconnected line on the intersecting plane. Figure 2.9.2 shows a periodic trajectory on a torus.

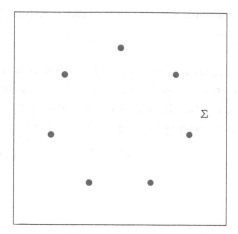

Figure 2.9.2.   Periodic trajectory projected on a phase space intersection.

Let us consider a single torus for a two-dimensional system. The geometric structure of the torus is determined by the two action variables J_1 and J_2. These quantities are completely determined by the total energy fixed by the initial conditions for the system. The flow on the torus (the dynamics) is determined by the two conjugate angle variables θ_1 and θ_2 (see Figure 2.9.3). The evolution in time for these two quantities are given by

$$\theta_1 = \omega_1 t + \delta_1, \tag{2.9.1}$$
$$\theta_2 = \omega_2 t + \delta_2. \tag{2.9.2}$$

The two frequencies ω_1 and ω_2 are determined by the Hamiltonian $\tilde{H} = \tilde{H}(I_1, I_2)$ by

$$\omega_1 = \frac{\partial \tilde{H}}{\partial I_1}, \tag{2.9.3}$$
$$\omega_2 = \frac{\partial \tilde{H}}{\partial I_2}. \tag{2.9.4}$$

The time T_2 to traverse the complete angle range θ_2 given by 2π is determined by the relation

$$T_2 = \frac{2\pi}{\omega_2}. \tag{2.9.5}$$

During this time interval, the angle θ_1 changes by

$$\theta_1 (t + T_2) = \theta_1 (t) + \omega_1 T_2$$
$$= \theta_1 (t) + 2\pi \frac{\omega_1}{\omega_2} \qquad (2.9.6)$$
$$= \theta_1 (t) + 2\pi \alpha(I_1, I_2),$$

where $\alpha = \alpha (I_1)$ denotes the winding number of the trajectory defined by

$$\alpha = \frac{\omega_1}{\omega_2}. \qquad (2.9.7)$$

The winding number is expressed as a function of I_1 because it is always possible to express I_2 by I_1 since the total energy $E = \tilde{H} (I_1, I_2)$ establishes a relation between the two quantities. If we now consider the (I_1, θ_1)-plane as the intersecting plane, the intersecting points are determined by

$$P_i = (\theta_1 (t + i T_2), I_1). \qquad (2.9.8)$$

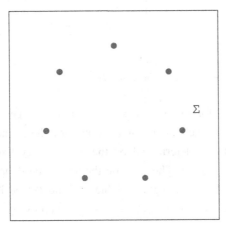

Figure 2.9.3.   Intersection plane of a two-dimensional torus described in action angle variables.

The intersecting plane is also known as the Poincaré plane. The mapping in this plane is represented by the following iterative mapping:

$$\theta_{i+1} = \theta_i + 2\pi \alpha (I_i), \qquad (2.9.9)$$
$$I_{i+1} = I_i. \qquad (2.9.10)$$

The mapping shows that the action variable is not changed during the iteration, whereas the angle continuously increases by a fixed amount given by the winding number. The map given is known as the twist map of the system. A twist mapping performs a mapping of the torus to itself. A fundamental property of the twist mapping is the conservation of the mapping area. This property is closely related to Liouville's theorem, the conservation of space volume. The conservation of the mapping area means that the Jacobi determinant has a fixed value:

$$\frac{\partial (\theta_{i+1}, I_{i+1})}{\partial (\theta_i, I_i)} = 1. \tag{2.9.11}$$

Thus, we expect that the intersections with a torus are regular curves more or less filled with points of the trajectory.

For nonintegrable Hamiltonians, there is a lack of integrals that fix the structure in phase space. For such systems, there is the common assumption that the Hamiltonian is separated into an integrable and into an nonintegrable part. The integrable part is denoted by

$$\tilde{H} = \tilde{H}_0 (I_i). \tag{2.9.12}$$

The total system consists of this integrable part extended by a nonintegrable part $\varepsilon \tilde{H}_1(I_j, \theta_j)$, which is considered as a disturbance. The nonintegrable Hamiltonian thus becomes

$$\tilde{H} = \tilde{H}_0 (I_i) + \varepsilon \tilde{H}_1 (I_i, \theta_i). \tag{2.9.13}$$

The disturbance $\varepsilon \tilde{H}_1$ is the origin of the nonexisting integrals which suppress the integrability and, thus, the torus structure of the phase space. The missing integrals allow a more flexible choice of paths for the trajectories. In the case of the twist mapping, this means that both sets of variables are disturbed. The angle as well as the action variables are thus given by

$$\theta_{i+1} = \theta_i + 2 \pi \alpha (I_i) + \varepsilon f (\theta_i, I_i), \tag{2.9.14}$$
$$I_{i+1} = I_i + \varepsilon g (\theta_i, I_i). \tag{2.9.15}$$

The functions f and g are generated by the Hamiltonian $\varepsilon \tilde{H}_1$. The functions must be chosen in such a way that the conservation of the intersection area is guaranteed.

An example for an area-conserved twist mapping is the Henó map introduced in 1969 by Henó to examine a nonlinear oscillating system. The Henó map is given by

$$\theta_{i+1} = \theta_i \cos (2 \pi \alpha) - (I_i - \theta_i^2) \sin (2 \pi \alpha), \tag{2.9.16}$$
$$I_{i+1} = \theta_i \sin (2 \pi \alpha) + (I_i - \theta_i^2) \cos (2 \pi \alpha). \tag{2.9.17}$$

The parameter α denoting the winding number of the twist map is the critical parameter. We can check the area conservation by defining the Jacobi matrix for the functions by

```
JacobiMatrix[fun_List, vars_List] :=
 Outer[D, fun, vars]
```

The Henó map is realized by

```
Clear[HenonMap]
```

```
HenonMap[{θ_, W_}, α_] := Block[{},
        {θ Cos[2 π α] - (W - θ²) Sin[2 π α],
     θ Sin[2 π α] + (W - θ²) Cos[2 π α]}]
```

The Jacobi determinant is thus defined via the Jacobi matrix:

```
JacobiMatrix[HenonMap[{θ, W}, α], {θ, W}] // MatrixForm
```

$$\begin{pmatrix} \cos(2\pi\alpha) + 2\theta\sin(2\pi\alpha) & -\sin(2\pi\alpha) \\ \sin(2\pi\alpha) - 2\theta\cos(2\pi\alpha) & \cos(2\pi\alpha) \end{pmatrix}$$

The determinant is calculated by

```
JacobiMatrix[HenonMap[{θ, W}, α], {θ, W}] // Det //
 Simplify
```

```
1
```

demonstrating that the Henó map is an area-conserving map. In the following we will use the Henó map to examine the structure of the related phase space. In a first step, we change the total energy of the system by changing the initial angle θ continuously. An increase of the angle gives the following picture:

```
initial = Table[{i, 0.0}, {i, .1, .84, .015}];
```

```
henonPlot = {};
```

The list of initial values are used to calculate the intersecting points in the Poincaré plane. Each initial point is connected with a series of point represented in the Poincaré plane:

```
Do[AppendTo[henonPlot, ListPlot[
    NestList[HenonMap[#, .2114] &, initial[[k]], 255],
    PlotStyle → Hue[k / Length[initial]], Frame → True,
    AspectRatio → 1, AxesLabel → {"θ", "W"},
    PlotRange → {{-1, 1}, {-1, 1}}]],
  {k, 1, Length[initial]}]
```

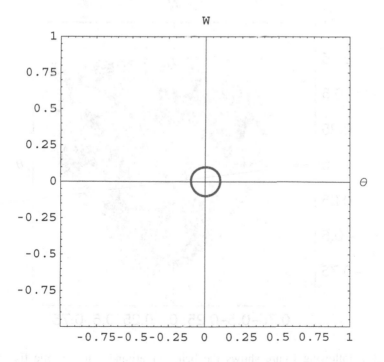

The generated sequence of figures allows one to study the evolution process of the torus by increasing the energy. We observe that the initial circular torus deforms to a more egg-shaped structure. At a very low energy, we observe a granular structure in the Poincaré plane. This discrete structure represents periodic solutions. Increasing the energy, the discrete structure disappears and a quasicontinuous covering of the torus is observed. At this point, we reach the quasiperiodic regime. At a certain threshold of the energy, the torus splits to five eggs. A single torus merges to a fivefold torus. If we further increase the energy, the fivefold torus again becomes a single torus which disintegrates into a broad band of points. This disintegration is the start of the torus destruction. The disintegration of the torus also happens at lower energies, especially in the neighborhood of so-called hyperbolic points. An overview of the different kind of tory is given in the following:

```
Show[henonPlot];
```

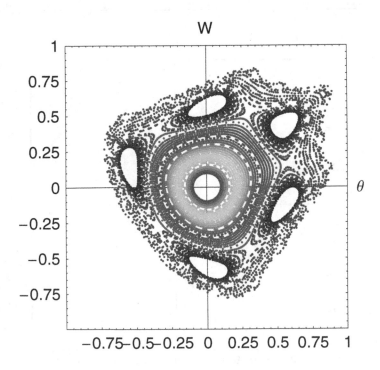

The following figure shows the behavior around a hyperbolic fix point. Here, the disintegration of the tori as well as the occurrence of different tory structures are seen.

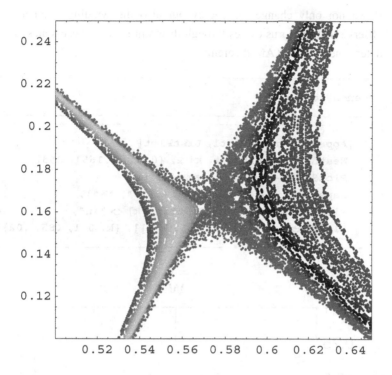

It is clearly shown that the torus around the hyperbolic fix point is demolished. The destruction of the tori becomes more and more diluted. We also realize in the above figure that in the neighborhood of the hyperbolic fix point are several elliptic fixpoints. The existence of elliptic fix points indicates that the tori continue to exist in these neigborhoods. The transition between the regular to the chaotic state seems to be a continuous process. The transition is controlled by the KAM theorem. A similar picture is gained at each hyperbolic point in the Poincaré plane. Hyperbolic fix points occur in between two elliptic fixpoints. This similarity of the pictures led to the term "self-similar structure of the Poincaré plane". Each magnification of the surrounding of a hyperbolic fixpoint looks similar to the above figure. The geometric structure of the Poincaré plane at these points will posses a scaling symmetry representing the self-similarity. In other words, the neighborhood of hyperbolic fixpoints shows the same structure on different scales.

If we not only change the energy but also the winding number α, we observe that the torus cycles through different states. These states are also determined by the KAM theorem:

```
henonPlot1 = {};
```

```
Do[AppendTo[henonPlot1, ListPlot[
    NestList[HenonMap[#, k] &, {0.51, 0.165}, 255],
    PlotStyle → Hue[k], Frame → True,
    AspectRatio → 1, AxesLabel → {"θ", "W"},
    PlotLabel → "k = " <> ToString[k] <> "\n",
    PlotRange → {{-1, 1}, {-1, 1}}]], {k, 0.1, .85, .02}]
```

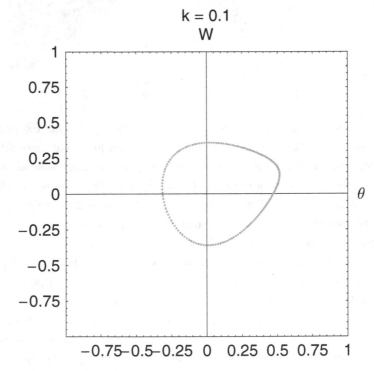

An overview of the different states is given in the following figure:

```
Show[henonPlot1, PlotLabel -> ""];
```

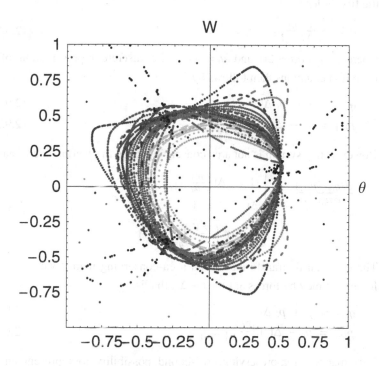

The different colors are related to the different winding numbers.

2.9.2 Discrete Mappings and Hamiltonians

The last subsection introduced the Henó map. Although Henó's map is area conserving, it is not derivable from a Hamiltonian. This subsection is concerned with the question of deriving area-conserving maps from a Hamiltonian. As a first example, let us consider the one-dimensional Hamiltonian:

$$H(p, q) = \tfrac{1}{2} p^2 + V(q). \tag{2.9.18}$$

The related Hamilton equations are

$$q' = p, \tag{2.9.19}$$

$$p' = -\frac{\partial V}{\partial q}. \tag{2.9.20}$$

The left-hand side of the differential equation can be approximated by introducing first-order discrete approximations by a difference scheme of the first order:

$$\dot{q} = \frac{q_{i+1} - q_i}{\Delta t}, \tag{2.9.21}$$

where $q_{i+1} = q(t + \Delta t)$ and $q_i = q(t)$. The discrete representation of the Hamilton equations then follows by

$$q_{i+1} = q_i + p_i \, \Delta t, \tag{2.9.22}$$

$$p_{i+1} = p_i - \Delta t \left(\frac{\partial V}{\partial q} \right) \Big|_{q=q_i}. \tag{2.9.23}$$

However, this system is not area conserving since the Jacobi determinant is

$$\frac{\partial (q_{i+1}, p_{i+1})}{\partial (q_i, p_i)} = \begin{vmatrix} 1 & -\Delta t \left(\frac{\partial^2 V}{\partial q} \right) \big|_{q=q_i} \\ \Delta t & 1 \end{vmatrix}$$

$$= 1 + (\Delta t)^2 \left(\frac{\partial^2 V}{\partial q} \right)_{q=q_i} \neq 1 \tag{2.9.24}$$

The map can be transformed to an area-conserving map if we replace the forces at time t by forces at time $t + \Delta t$; that is,

$$q_{i+1} = q_i + p_i \, \Delta t, \tag{2.9.25}$$

$$p_{i+1} = p_i - \Delta t \left(\frac{\partial V}{\partial q} \right) \Big|_{q=q_{i+1}}. \tag{2.9.26}$$

This map is area-preserving. A second possibility to represent an area preserving map for the above Hamiltonian is

$$q_{i+1} = q_i + \Delta t \, p_{i+1}, \tag{2.9.27}$$

$$p_{i+1} = p_i - \Delta t \left(\frac{\partial V}{\partial q} \right) \Big|_{q=q_i} \tag{2.9.28}$$

This representation is used in the following example.

Example 1: Mathematical Pendulum

Let us consider the example of a mathematical pendulum. The potential of this system is given by

$$V(q) = \frac{k}{(2\pi)^2} (1 - \cos(2\pi q)). \tag{2.9.29}$$

Assuming that the time step $\Delta t = 1$, we get the following map

$$q_{i+1} = q_i + p_{i+1}, \tag{2.9.30}$$

$$p_{i+1} = p_i + \frac{k}{2\pi} \sin(2\pi q_i). \tag{2.9.31}$$

Both equations are examined on a restricted range modulo 1. The mapping is known as the Taylor–Chiricov or standard mapping.

The transition from regular to chaotic behavior discussed earlier for the Henó map can be examined for the standard map on a Poincaré section. The mapping generates a discrete flow of the Hamilton system and can be used to follow the temporal evolution of the system. First, let us define the standard mapping by

```
Clear[Standard]
```

```
Standard[{xi_, yi_}, k_] := Block[{},
    y = Mod[yi - k Sin[2 π xi] / (2 π), 1];
       x = Mod[xi + y, 1];
       {x, y}      ]
```

The mapping is iterated for a certain amount of steps with different initial conditions changing the total energy of the Hamiltonian.

```
Do[h = {0, .54};
    ListPlot[Table[h = Standard[h, k], {i, 1, 1000}],
    PlotRange → {{0, 1}, {0, 1}}, Frame → True,
    PlotStyle → RGBColor[0.996109, 0, 0],
    AspectRatio → 1], {k, .5, 2.8, .1}]
```

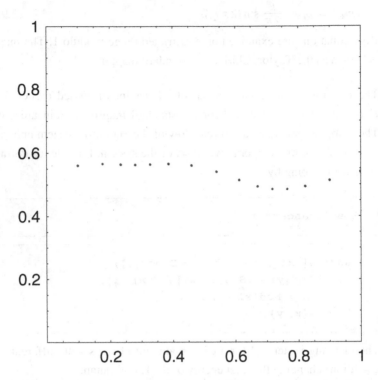

The illustration of the results shows that different dynamical regimes exist. The patterns range from discrete points, to looped curves, to scattered points in the Poincaré section. These different regimes are initiated by different initial energies. It is clearly seen that an increase of the energy changes the dynamical behavior from regular to chaotic behavior. The following subsection discusses the different regimes in connection with a measure to quantify the different states.

2.9.3 Lyapunov Exponent

A basic behavior of the chaotic dynamic is that the infinitesimal change of initial conditions results in an unpredictable state for long times. This deviation of closely related initial trajectories is measured by the so-called Lyapunov exponent. The Lyapunov exponent represents an estimation of the degree of divergence of initially closely related trajectories. The exponential increase of the distance of neighboring trajectories is measured by the Lyapunov exponent. He measured the mean increase of the enlargement of the distance between the trajectories. The Lyapunov exponent is a numerical property of the Hamiltonian system but is not restricted to this kind. This measure can be also applied to non-Hamiltonian systems or maps. To get some insight into the theoretical background, let us consider an n-dimensional autonomous system

$$\frac{dx_i}{dt} = F_i(x_1, \ldots, x_n), \quad i = 1, 2, \ldots, n \tag{2.9.32}$$

Our aim is to estimate the rate of deviation for two initially closely related trajectories. To accomplish this task, we linearize the system in Equation (2.9.32) by considering an infinitesimal neighboring trajectory $\bar{x} = (\bar{x}_1, \ldots, \bar{x}_n)$. The linearization provides the tangent representation of the equations of motion:

$$\frac{d\delta x_i}{dt} = \sum_{i=1}^{n} \delta x_j \left(\frac{\partial F_i}{\partial x_j} \right)_{x=\bar{x}(t)}. \tag{2.9.33}$$

The distance or norm of the distortion δx_i is

$$d(t) = \sqrt{\sum_{i=1}^{n} \delta x_i^2(t)}. \tag{2.9.34}$$

This quantity is the basis for the estimation of the Lyapunov exponent λ. The Lyapunov exponent measures the divergent of two trajectories: a reference trajectory $\bar{x}$ and a neighboring trajectory $\bar{x}(0) + \delta x(0)$. The mean divergence rate is defined by

$$\lambda = \lim_{\substack{t \to \infty \\ d(0) \to 0}} \left(\frac{1}{t} \right) \ln \left(\frac{d(t)}{d(0)} \right), \tag{2.9.35}$$

where $d(0)$ is the norm of the initial state. One characteristic property of the Lyapunov exponent is that λ vanishes for a regular motion because $d(t)$ increases linearly or, at least, algebraically in time.

The relation between the Lyapunov exponent and the trajectory become more obvious if we restrict our examinations to a one-dimensional map:

$$x_{i+1} = f(x_i). \tag{2.9.36}$$

As an example for f let us take the logistic function $f(x) = 4\,\sigma\,x\,(1-x)$. The tangent maps defined in Equation (2.9.33) is given by

$$\delta x_{i+1} = \left(\frac{df(x)}{dx}\right)_{x=x_i} \delta x_i. \tag{2.9.37}$$

Assuming that the distance δx_i is fixed in each iteration, we can simplify the relation to

$$\delta x_{i+1} = \prod_{j=0}^{i} f'(x_i)\,\delta x_0, \tag{2.9.38}$$

where $f'(x_i)$ is the derivative of f at $x = x_i$. The related Lyapunov exponent (2.9.35) then is

$$\lambda = \lim_{N\to\infty} \frac{1}{N} \ln[\prod_{j=1}^{N} f'(x_j)\,\delta x_0]$$

$$\tag{2.9.39}$$

$$= \lim_{N\to\infty} \frac{1}{N} \sum_{j=0}^{N} \ln(f'(x_j)).$$

This relation demonstrate that the Lyapunov exponent is independent of the initial condition x_0. The relation given is implemented as follows:

```
Clear[f, x]
```

```
f[x_, σ_] = 4 σ x (1 - x)
```

$$4\,(1-x)\,x\,\sigma$$

The derivation of the logistic function is

```
g[x_, σ_] = ∂x f[x, σ]
```

$$4\,(1-x)\,\sigma - 4\,x\,\sigma$$

Iterating relation (2.9.33) and calculating the derivative at x_i are the basic calculations for determining the Lyapunov exponent. Since the logistic function depends on a parameter σ, we are also able to study the influence of σ on λ. The following figure shows this dependence:

```
logpl = ListPlot[ Table[
    {σ, Last[FoldList[Plus, 0, Map[Log[Abs[g[#, σ]]] &,
        NestList[f[#, σ] &, .6, 250]]]] / 252},
    {σ, .01, 1, .005}], PlotStyle →
    RGBColor[0.996109, 0, 0],
    PlotJoined → True, AxesLabel → {"σ", "λ"}];
```

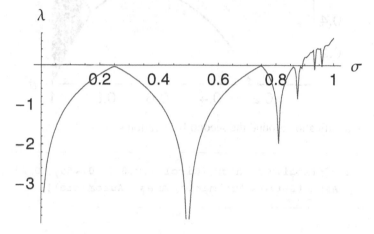

The iteration of the logistic map is as follows:

```
logi = Flatten[
    Table[Map[{σ, #} &, Sort[Take[NestList[f[#, σ] &, .6,
        115], {25, 115}]]], {σ, .01, 1, .005}], 1];
```

```
pllogi = ListPlot[logi, AxesLabel → {"σ", "x"},
   PlotStyle → RGBColor[0, 0, 0.996109]];
```

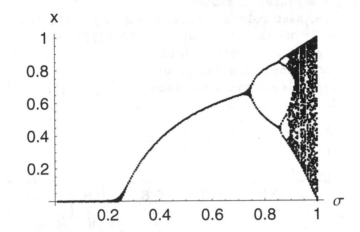

A magnification around the second bifurcation shows

```
Show[pllogi, Graphics[{Circle[{0.852, 0.469}, 0.1]},
   AspectRatio → Automatic, Axes → Automatic]];
```

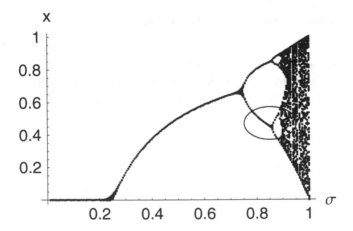

Our interest is the section of this figure marked by a circle. The representation of this selected part in a magnification shows that we get a similar picture:

```
logi1 = Flatten[Table[Map[{σ, #} &, Sort[
      Take[NestList[f[#, σ] &, .6, 215], {75, 215}]]],
   {σ, 0.84, 0.91, .0005}], 1];
```

```
pllogi1 = ListPlot[logi1, AxesLabel → {"σ", "x"},
   PlotStyle → RGBColor[0.996109, 0, 0],
   PlotRange → {{0.84, 0.91}, {.29, .69}}];
```

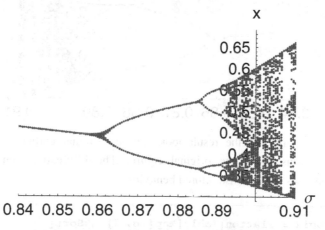

Again, a selection and magnification marked by a circle

```
Show[pllogi1, Graphics[{Circle[{0.886, 0.528}, 0.01]},
    AspectRatio → Automatic, Axes → Automatic]];
```

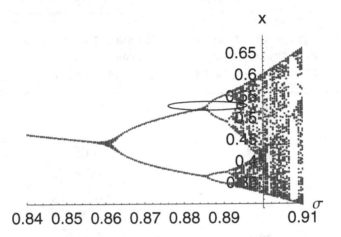

shows again that the result looks similar to that earlier. We observe bifurcations as in the original figure. The bifurcation continues and transverses into an unstructured behavior.

```
logi2 = Flatten[Table[Map[{σ, #} &, Sort[
    Take[NestList[f[#, σ] &, .6, 215], {75, 215}]]],
    {σ, 0.883, 0.896, .00005}], 1];
```

```
pllogi2 = ListPlot[logi2, AxesLabel → {"σ", "x"},
   PlotStyle → RGBColor[0, 0.500008, 0],
   PlotRange → {{0.883, 0.896}, {0.429, 0.603}}];
```

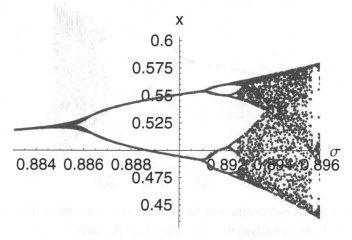

The repeated pattern indicates that the bifurcations occur again and again until a critical value σ_c is reached. At this value, the bifurcating behavior skips to chaos. We define chaos as such a state where the Lyapunov exponent is positive. The different magnification ranges are summarized in the following figure:

```
glist = {pllogi, pllogi1, pllogi2};
```

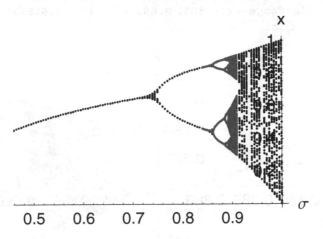

The different colors represent the regions of magnification. The following illustration gives a dynamic view of the magnification.

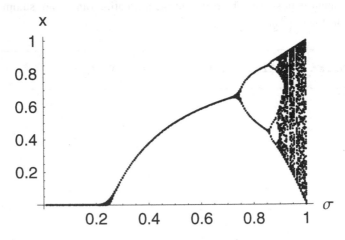

A combination of the bifurcation diagram with the Lyapunov exponent demonstrates that the bifurcation regime is reached at the border of $\sigma_c \sim 0.9$. It is also obvious that the chaotic regime is intermitted by regions where a purely periodic behavior is recover.

```
Show[logpl, pllogi, PlotRange → All];
```

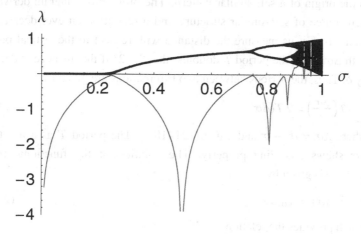

In the periodic regime, the Lyapunov exponent periodically increases and decreases up to zero. At the critical value σ_c, λ transcends the border line at zero. We also observe that the Lyapunov exponent possesses singularities at $-\infty$. These singularities are supercyclic periods of the logistic map. Above the critical value σ_c, the Lyapunov exponent wobbles between the chaotic and the periodic state in ever shorter cycles. However, the supercyclic periods also exist in that regime above σ_c. The transition between the regular and chaotic states is a major characteristic of a nonlinear chaotic system.

Feigenbaum in 1975 extensively studied the transition to chaos. He observed that the period doubling skips to chaos at a critical value of $\sigma_c = 0.892486\ldots$. Below this value, he demonstrated that the ratio of interval lengths has a fixed value determined by the relation

$$\delta = \lim_{n \to \infty} \frac{\sigma_{n+2} - \sigma_{n+1}}{\sigma_{n+1} - \sigma_n} = 4.669\ldots, \tag{2.9.40}$$

which is now called the Feigenbaum constant. The ratio δ exists because the bifurcations occur in decreasing σ intervals. Such a bifurcation pattern is the origin of a self-similar pattern. The bifurcation diagram derived is a rich source of self-similar structure and a repetition on ever decreasing σ intervals. Let us measure the distance with respect to the critical point σ_c with $\Delta\sigma$; then, the period T doubles like $T = 2^n$ if the distance σ decreases by δ. Thus, the periods between two bifurcations are given by

$$T\left(\frac{\Delta\sigma}{\delta}\right) = 2\,T(\Delta\sigma), \tag{2.9.41}$$

where $\Delta\sigma = \sigma_c - \sigma$ and $1/\delta = 0.21418....$ The period T as a function of $\Delta\sigma$ shows a scaling property. The solution of the functional relation (2.9.41) is given by

$$T(\Delta\sigma) = c_0\,\Delta\sigma^\nu, \tag{2.9.42}$$

which provides the relation

```
skal = c0  Δσ^ν   ==  2 c0 Δσ^ν
           ----
            δ^ν
```

$$\delta^{-\nu}\,\Delta\sigma^\nu\,c_0 == 2\,\Delta\sigma^\nu\,c_0$$

The solution is given by

```
sskal = Solve[skal, ν]
```

$$\left\{\left\{\nu \to -\frac{\log(2)}{\log(\delta)}\right\}\right\}$$

The replacement of δ by its numerical value provides the scaling exponent as

```
sskal /. δ → 4.669
```

$$\{\{\nu \to -0.44982\}\}$$

Exactly by this scaling law the period doubles. The relation can be seen as a self-similar scaling behavior before chaos sets in. Since the scaling exponent is a fractional value, some authors call the periodic regime a fractal. Despite the supercyclic periods, the Lyapunov exponent is positive for $\sigma > \sigma_c$. Huberman and Rudnick observed that the envelope $\tilde{\lambda}$ above σ_c also follows a scaling law of the form

$$\tilde{\lambda} \sim (\sigma - \sigma_c)^{-\nu},$$

where, again, $\nu = -\ln(2)/\ln(\delta)$. Because of the change of sign at σ_c and the fact that the increase of $\tilde{\lambda}$ is given by a power law, this transition is called a phase transition of the second kind. The terms arc borrowed from the theory of critical phenomena and statistical physics.

The mathematical relations discussed so far are also presentable in graphical form. The main feature of the logistic function is its self-similarity given by the scaling period doubling. The self-similar behavior of the mapping is also seen in its algebraic structure. The following lines show different state of iteration and the generated polynomial:

```
its = NestList[f[#, σ] &, x, 3]; TableForm[its]
```

x
$4(1-x)x\sigma$
$16(1-x)x\sigma^2(1-4(1-x)x\sigma)$
$64(1-x)x\sigma^3(1-4(1-x)x\sigma)(1-16(1-x)x\sigma^2(1-4(1-x)x\sigma))$

where the first item of this table has no meaning other than to initiate the iteration. The iteration of the logistic map generates at each step a new value for x. This value is the starting point for the next value in x ans so forth. The iteration process can be depicted by means of the function **logistic[]**, which generates a mapping consisting of n iterations for a given σ and x_0:

```
logistic[σ_, x0_, n_] := Block[{pl1, dli1, dlh},
      lh = f[x, σ];
      li1 = NestList[f[#, σ] &, x0, n];
      pl1 = Plot[Evaluate[{x, lh}], {x, 0, 1},
   PlotLabel → "σ =" <> ToString[σ], AspectRatio → 1,
   PlotStyle → {RGBColor[0.996109, 0, 0],
     RGBColor[0.996109, 0, 0]},
   DisplayFunction → Identity];
      Show[pl1, Graphics[Table[{Line[
     {{li1[[i]], li1[[i + 1]]}, {li1[[i + 1]], li1[[i + 1]]}}],
   Line[{{li1[[i + 1]], li1[[i + 1]]}, {li1[[i + 1]],
       li1[[i + 2]]}}]}], {i, 1, Length[li1] - 2}]],
   AspectRatio → Automatic, PlotRange → All,
   DisplayFunction → $DisplayFunction]
      ]
```

To show the changes of fix points $f(x^*) = x^*$, we change the parameter $σ$:

```
Do[logistic[σ, .01, 70], {σ, .7, 1, .025}]
```

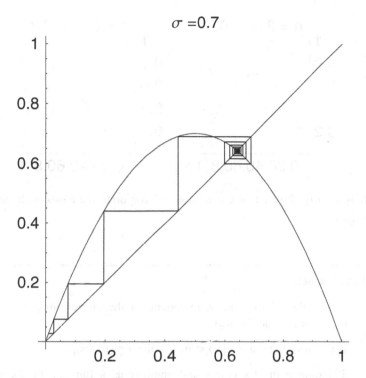

The result of the generated sequence shows how a series of fix points emerge from a single point. The creation of these fixpoints can be observed if we plot the higher iterations $f^{(n)}$ of the logistic mapping. The intersection with the bisector shows how the fixpoints are generated. The following illustration shows the iteration up to order $n = 5$:

```
Do[Show[GraphicsArray[Partition[
    Table[Plot[Evaluate[{its[[1]], its[[i]] } /. σ → σσ],
        {x, 0, 1}, PlotStyle → RGBColor[0.996109, 0, 0],
        PlotRange → All, PlotLabel → "n = " <>
            ToString[i] <> "   " <> "σ = " <> ToString[σσ],
        AspectRatio → 1, DisplayFunction → Identity],
        {i, 2, Length[its]}], 2]],
    DisplayFunction → $DisplayFunction],
    {σσ, .7, 1, .025}];
```

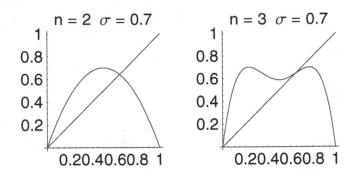

It is clearly shown that the number of fixpoints increases with larger σ values.

2.9.4 Exercises

1. Calculate the Lyapunov exponent for the discrete map $x_{i+1} = 2\,x_i$. Demonstrate that $\lambda = \ln(2)$.

2. Examine the scaling properties of the logistic map.

3. Examine the fix points and stability as a function of the control parameter λ of the cubic map

$$x_{n+1} = \lambda\,x_n(1 - x_n^2).$$

4. Consider a ball bouncing between two walls (neglect gravity) for which one wall has a small periodic motion. Show that the dynamics is not governed by a linear operator.

2.10 Rigid Body

2.10.1 Introduction

All bodies around us consist of atoms or molecules. These basic elements of the matter are either in a regular or irregular order forming the rigid bodies. The rigid bodies are very resistant to mechanical loads. The diameter of atoms and molecules in a solid are small compared with the interatom or intermolecular distances. To a good approximation, solids can be represented as a collection of mass points with fixed distances between atoms. Bodies with the property of fixed interatomic distances are defined as rigid bodies.

To describe the motion of a rigid body, we introduce two kinds of coordinate system:

1. An inertial coordinate system

2. A body-centered coordinate system

The description of the motion is related to six coordinates. These quantities are the coordinates of the mass center and three angles determining the orientation with respect to the inertial system. For the three angles, we choose the Euler angles already introduced in Section 2.2.2. Related to these coordinates are two basic types of motion: a translation and a rotation. These kinds of motion can be motivated by considering infinitesimal small movements of the rigid body. In addition, if we locate the center of mass in the origin of the coordinate system, we are able to separate the energy terms by translation and rotation energy components, meaning the motion is separated by a center of mass movement and a movement around the center of mass.

If, in addition, the potential energy is also separable, the total Lagrangian splits into two parts: the translation and the rotation parts. Each part is independent of the other and determines an independent solution and, as such, an independent state of motion. This behavior was first realized by Euler in 1749.

2.10.2 The Inertia Tensor

Let us examine a rigid body consisting of n particles with masses m_α, $\alpha = 1, 2, ..., n$. Let us assume that this rigid body is rotating with angular velocity $\vec{\omega}$ around a fixed point with respect to the body-centered coordinate system. In addition, let us assume that the total rigid body is moving with a velocity $\vec{V}$ with respect to the inertial coordinate system. Then, the velocity of the αth particle is determined by Equation (2.10.2). For a rigid body, the coordinates are fixed in the rotating frame and thus

$$\vec{v}_r = \left(\frac{d\vec{r}}{dt}\right)_r = 0. \tag{2.10.1}$$

The velocity of the αth particle in the inertial coordinate system is thus given by

$$\vec{v}_\alpha = \vec{V} + \vec{\omega} \times \vec{r}_\alpha. \tag{2.10.2}$$

The velocities are all measured in the inertial system because the velocities in the rotating system are zero because of the rigidity of the body. The kinetic energy of the αth particle is thus determined by

$$T_\alpha = \frac{m_\alpha}{2} \vec{v}_\alpha^2 \tag{2.10.3}$$

which results in the total kinetic energy, including translations and rotations of the rigid body, being

$$T = \frac{1}{2} \sum_{\alpha=1}^{n} m_\alpha \left(\vec{V} + \vec{\omega} \times \vec{r}_\alpha\right)^2. \tag{2.10.4}$$

Expansion of the quadratic term results in

$$T = \frac{1}{2} \sum_{\alpha=1}^{m} m_\alpha \left(\vec{V}^2 + 2\vec{V} \cdot (\vec{\omega} \times \vec{r}_\alpha) + (\vec{\omega} \times \vec{r}_\alpha)^2\right)$$

$$= \frac{1}{2} \sum_{\alpha=1}^{m} m_\alpha \vec{V}^2 + \frac{1}{2} \sum_{\alpha=1}^{m} \vec{V} \cdot (\vec{\omega} \times \vec{r}_\alpha) m_\alpha + \tag{2.10.5}$$

$$\frac{1}{2} \sum_{\alpha=1}^{n} m_\alpha (\vec{\omega} \times \vec{r}_\alpha)^2.$$

This expression represents the general representation of the total kinetic energy. This expression is valid for any choice of origin from which the location of the $\alpha$the particle $\vec{r}_\alpha$ is measured.

Locating the origin of the coordinate system into the center of mass, this expression is simplified to a much shorter expression. The second term is rewritten as

$$\sum_{\alpha=1}^{n} \vec{V} \cdot (\vec{\omega} \times \vec{r}_{\alpha}) \, m_{\alpha} = \vec{V} \cdot \vec{\omega} \times \sum_{\alpha=1}^{n} m_{\alpha} \vec{r}_{\alpha}. \tag{2.10.6}$$

With

$$\sum_{\alpha=1}^{n} m_{\alpha} \vec{r}_{\alpha} = M \vec{R}, \tag{2.10.7}$$

it follows that

$$\sum_{\alpha=1}^{n} \vec{V} \cdot (\vec{\omega} \times \vec{r}_{\alpha}) \, m_{\alpha} = \vec{V} \cdot \vec{\omega} \times M \vec{R}. \tag{2.10.8}$$

Since the center of mass is located in the origin, we must set $\vec{R} = 0$, which reduced the total kinetic energy to

$$T = \frac{1}{2} \sum_{\alpha=1}^{n} m_{\alpha} \, \vec{V}^2 + \frac{1}{2} \sum_{\alpha=1}^{n} m_{\alpha} \, (\vec{\omega} \times \vec{r}_{\alpha})^2$$
$$= T_{\text{trans}} + T_{\text{rot}}, \tag{2.10.9}$$

with

$$T_{\text{trans}} = \frac{1}{2} \vec{V}^2 M \tag{2.10.10}$$

and

$$T_{\text{rot}} = \frac{1}{2} \sum_{\alpha=1}^{n} m_{\alpha} \, (\vec{\omega} \times \vec{r}_{\alpha})^2. \tag{2.10.11}$$

T_{trans} and T_{rot} are expressions for the translation and rotation part of the kinetic energy, respectively.

In the following, we will specifically look at the rotation part of the motion:

$$T_{\text{rot}} = \frac{1}{2} \sum_{\alpha=1}^{n} m_{\alpha} \, (\vec{\omega} \times \vec{r}_{\alpha})^2. \tag{2.10.12}$$

Applying the vector identity $(\vec{A} \times \vec{B})^2 = (\vec{A} \times \vec{B}) \cdot (\vec{A} \times \vec{B}) = A^2 B^2 - (\vec{A} \cdot \vec{B})$ to the rotation energy, we are able to write

$$T_{\text{rot}} = \frac{1}{2} \sum_{\alpha=1}^{n} m_{\alpha} \left\{ \omega^2 \, r_{\alpha}^2 - (\vec{\omega} \times \vec{r}_{\alpha})^2 \right\}. \tag{2.10.13}$$

Replacing the vectors $\vec{\omega} = (\omega_1, \omega_2, \omega_3)$ and $\vec{r}_{\alpha} = (x_{\alpha 1}, x_{\alpha 2}, x_{\alpha 3})$ by their components, we get

$$T_{\text{rot}} = \frac{1}{2} \sum_{\alpha=1}^{n} m_{\alpha} \, (\textstyle\sum_{i=1}^{3} \omega_i^2)(\sum_{k=1}^{3} x_{\alpha k}^2) -$$
$$(\textstyle\sum_{i=1}^{3} \omega_i \, x_{\alpha i})(\sum_{j=1}^{3} \omega_j \, x_{\alpha j}). \tag{2.10.14}$$

The frequencies ω_i are represented by introducing Kronecker's symbol

$$\omega_i = \sum_{j=1}^{3} \delta_{ij} \, \omega_j. \tag{2.10.15}$$

The insertion of the frequencies in this form allows us to combine the sums over i and j as a common sum and extract them from the expression

$$T_{\text{rot}} = \frac{1}{2} \sum_{\alpha=1}^{n} m_\alpha \sum_{i,j=1}^{3} \{\omega_i \, \omega_j \, \delta_{ij} \sum_{k=1}^{3} x_{\alpha k}^2$$

$$- \omega_i \, \omega_j \, x_{\alpha i} \, x_{\alpha j}\} \tag{2.10.16}$$

$$= \sum_{i,j=1}^{3} \omega_i \, \omega_j \left\{ \sum_{\alpha=1}^{n} m_\alpha \left[\delta_{ij} \sum_{k=1}^{3} x_{\alpha k}^2 - x_{\alpha i} \, x_{\alpha j} \right] \right\}.$$

If we introduce the definition

$$\Theta_{ij} = \sum_{\alpha=1}^{n} m_\alpha \left\{ \delta_{ij} \sum_{k=1}^{3} x_{\alpha k}^2 - x_{\alpha i} \, x_{\alpha j} \right\}, \tag{2.10.17}$$

then the rotation energy is the simple form

$$T_{\text{rot}} = \frac{1}{2} \sum_{i,j=1}^{3} \omega_i \, \Theta_{i,j} \, \omega_j, \tag{2.10.18}$$

where $\Theta_{i,j}$ is known as the inertia tensor. The components of this tensor are

$$\Theta =$$

$$\begin{pmatrix} \sum_{\alpha=1}^{n} m_\alpha (x_{\alpha 2}^2 + x_{\alpha 3}^2) & -\sum_{\alpha=1}^{n} m_\alpha \, x_{\alpha 1} \, x_{\alpha 2} & \sum_{\alpha=1}^{n} m_\alpha \, x_{\alpha 2} \, x_{\alpha 3} \\ -\sum_{\alpha=1}^{n} m_\alpha \, x_{\alpha 2} \, x_{\alpha 1} & \sum_{\alpha=1}^{n} m_\alpha (x_{\alpha 1}^2 + x_{\alpha 3}^2) & -\sum_{\alpha=1}^{n} m_\alpha \, x_{\alpha 2} \, x_{\alpha 3} \\ -\sum_{\alpha=1}^{n} m_\alpha \, x_{\alpha 3} \, x_{\alpha 1} & -\sum_{\alpha=1}^{n} m_\alpha \, x_{\alpha 3} \, x_{\alpha 2} & \sum_{\alpha=1}^{n} m_\alpha (x_{\alpha 1}^2 + x_{\alpha 2}^2) \end{pmatrix} \tag{2.10.19}$$

The elements on the diagonal Θ_{ii} are known as main inertia moments, whereas the Θ_{ij} in the off-diagonal elements are known as deviation moments. From the structure of the elements in Θ_{ij}, it is obvious that this tensor is a symmetrical tensor; that is,

$$\Theta_{ij} = \Theta_{ji}. \tag{2.10.20}$$

Taking this property into account, it is clear that only six components of the tensor are independent of each other. Another essential property of the inertia tensor is that the sum over particles is extractable from the tensor structure. In other words, we can replace the masses by a continuous mass distribution $\rho(\vec{r}) = \rho(x_1, x_2, x_3)$ and replace the sum by an integral over the spatial coordinates. This replacement results in the continuous representation of the inertia tensor:

$$\Theta_{ij} = \int_V \rho(\vec{r}) \, \{ \delta_{ij} \sum_k x_k^2 - x_i, \, x_j \} \, dx_1 \, dx_2 \, dx_3, \tag{2.10.21}$$

where V is the total volume of the body under consideration.

2.10.3 The Angular Momentum

The angular momentum of a rigid body with respect to a fixed point O in the body-centered coordinate system is given by

$$\vec{L} = \sum_{\alpha=1}^{n} \vec{r}_\alpha \times \vec{p}_\alpha. \tag{2.10.22}$$

Appropriate choices for such a point are as follws

1. A fixed point in the body and inertial system around which the body circles (top)

2. The center of mass

In the body-centered coordinate system, the momentum $\vec{p}_\alpha$ is

$$\vec{p}_\alpha = m_\alpha \vec{v}_\alpha = m_\alpha (\vec{\omega} \times \vec{r}_\alpha). \tag{2.10.23}$$

Thus, the angular momentum becomes

$$\vec{L} = \sum_{\alpha=1}^{n} m_\alpha \vec{r}_\alpha \times (\vec{\omega} \times \vec{r}_\alpha). \tag{2.10.24}$$

The vector identity $\vec{A} \times (\vec{B} \times \vec{A}) = A^2 \vec{B} - \vec{A} \times (\vec{B}.\vec{A})$ allows us to simplify $\vec{L}$ to

$$\vec{L} = \sum_{\alpha=1}^{n} m_\alpha \{ \vec{r}_\alpha^2 \vec{\omega} - \vec{r}_\alpha (\vec{r}_\alpha . \vec{\omega}) \}. \tag{2.10.25}$$

The replacement of vectors by their components provides the ith component of the angular momentum:

$$
\begin{aligned}
L_i &= \sum_{\alpha=1}^{n} m_\alpha \{ \omega_i \sum_{k=1}^{3} x_{\alpha k}^2 - x_{\alpha i} \sum_{j=1}^{3} x_{\alpha j} \omega_j \} \\
&= \sum_{\alpha=1}^{n} m_\alpha \sum_{j=1}^{3} \{ \omega_j \delta_{ij} \sum_{k=1}^{3} x_{\alpha k}^2 - x_{\alpha i} x_{\alpha j} \omega_j \} \\
&= \sum_{j=1}^{3} \omega_j \sum_{\alpha=1}^{n} m_\alpha \{ \delta_{ij} \sum_{k=1}^{3} x_{\alpha k}^2 - x_{\alpha i} x_{\alpha j} \} \\
&= \sum_{j=1}^{3} \omega_j \Theta_{ij}.
\end{aligned}
\tag{2.10.26}
$$

In tensor notation, we write

$$\vec{L} = \vec{\Theta}.\vec{\omega}. \tag{2.10.27}$$

Multiplying the *i*th component of the angular momentum by $\frac{1}{2}\omega_i$ and summing up the components, we get

$$\sum_{i=1}^{3} \frac{1}{2} \omega_i L_i = \frac{1}{2} \sum_{j,i=1}^{3} \omega_i \omega_j \Theta_{ij} = T_{\text{rot}} = \frac{1}{2} \vec{\omega} \cdot \vec{L}. \quad (2.10.28)$$

2.10.4 Principal Axes of Inertia

If we consider the angular momentum and the kinetic energy as a function of the inertia tensor, we observe that these expressions simplify if the inertia tensor takes on a special form such as

$$\Theta_{ij} = \Theta_i \, \delta_{ij} \qquad\qquad\qquad\qquad\qquad\qquad (2.10.29)$$

or

$$\overline{\Theta} = \begin{pmatrix} \Theta_1 & 0 & 0 \\ 0 & \Theta_2 & 0 \\ 0 & 0 & \Theta_3 \end{pmatrix}. \qquad\qquad\qquad\qquad (2.10.30)$$

This simplification is known as the principal axes representation of the inertial tensor. If we are able to write down the Θ tensor in such a way, it follows for the angular momentum that

$$L_i = \sum_j \Theta_i \, \delta_{ij} \omega_j = \Theta_i \, \omega_i, \qquad\qquad\qquad (2.10.31)$$

and for the rotation energy,

$$T_{\text{rot}} = \frac{1}{2} \sum_{i,j} \Theta_i \, \delta_{ij} \omega_i \omega_j = \frac{1}{2} \sum_i \Theta_i \omega_i^2. \qquad\qquad (2.10.32)$$

This simplification only occurs if we are able to find a body-centered coordinate system in which the deviation moments Θ_{ij} vanish. In this case, the inertia tensor consists of three independent components: the principal inertia moments.

Uncovering this special coordinate system is related to the idea that the rotation around a principal axes is characterized by the alignment of the angular momentum $\vec{L}$ and the angular frequency $\vec{\omega}$; that is,

$$\vec{L} = \overline{\Theta} \cdot \vec{\omega} \qquad\qquad\qquad\qquad\qquad\qquad (2.10.33)$$

The representation of the angular momentum in the principal axes system and in the general system must be identical. This invariance of the physical quantity $\vec{L}$ provides the following set of equations:

$$L_1 = \Theta \omega_1 = \Theta_{11} \omega_1 + \Theta_{12} \omega_2 + \Theta_{13} \omega_3,$$
$$L_2 = \Theta \omega_2 = \Theta_{21} \omega_1 + \Theta_{22} \omega_2 + \Theta_{23} \omega_3, \qquad (2.10.34)$$
$$L_3 = \Theta \omega_3 = \Theta_{31} \omega_1 + \Theta_{32} \omega_2 + \Theta_{33} \omega_3,$$

$$\Longleftrightarrow \qquad (\Theta_{11} - \Theta) \omega_1 + \Theta_{12} \omega_2 + \Theta_{13} \omega_3 = 0,$$
$$\Theta_{21} \omega_1 + (\Theta_{22} - \Theta) \omega_2 + \Theta_{23} \omega_3 = 0, \qquad (2.10.35)$$
$$\Theta_{31} \omega_1 + \Theta_{32} \omega_2 + (\Theta_{33} - \Theta) \omega_3 = 0,$$

$$\Longleftrightarrow \qquad \sum_{j=1}^{3} (\Theta_{ij} - \Theta \delta_{ij}) \omega_j = 0, \quad i = 1, 2, 3. \qquad (2.10.36)$$

A condition to find non-trivial solutions of this system of equations is

$$\det (\Theta_{ij} - \Theta \delta_{ij}) = 0, \qquad (2.10.37)$$

which represents a cubic algebraic relation for Θ. The three different solutions for Θ are related to the principal inertia moments Θ_1, Θ_2, and Θ_3. Knowing these three quantities, it is possible to classify the behavior of the rigid body or top.

With all three components different,

$$\Theta_1 \neq \Theta_2 \neq \Theta_3, \qquad (2.10.38)$$

we call the top unsymmetrical. With two components equal to each other,

$$\Theta_1 = \Theta_2 \neq \Theta_3, \qquad (2.10.39)$$

we call the body a symmetric top. With all three components equal to each other,

$$\Theta_1 = \Theta_2 = \Theta_3, \qquad (2.10.40)$$

we have a spherical top.

The steps discussed above are implemented by a few lines. The inertia tensor with principal diagonal elements is

```
th = IdentityMatrix[3] {θ, θ, θ}
```

$$\begin{pmatrix} \theta & 0 & 0 \\ 0 & \theta & 0 \\ 0 & 0 & \theta \end{pmatrix}$$

The general inertial tensor is represented by a two-dimensional matrix:

```
theta = Table[θ[i, j], {j, 1, 3}, {i, 1, 3}];
theta // MatrixForm
```

$$\begin{pmatrix} \theta(1,\,1) & \theta(2,\,1) & \theta(3,\,1) \\ \theta(1,\,2) & \theta(2,\,2) & \theta(3,\,2) \\ \theta(1,\,3) & \theta(2,\,3) & \theta(3,\,3) \end{pmatrix}$$

The angular velocity $\vec{\omega}$ is given by the vector

```
ω = {ω1, ω2, ω3}
```

$\{\omega 1,\, \omega 2,\, \omega 3\}$

The invariance condition for the angular momentum reads

```
Thread[th.ω == theta.ω, List] // TableForm
```

$\theta\,\omega 1 == \omega 1\,\theta(1,\,1) + \omega 2\,\theta(2,\,1) + \omega 3\,\theta(3,\,1)$

$\theta\,\omega 2 == \omega 1\,\theta(1,\,2) + \omega 2\,\theta(2,\,2) + \omega 3\,\theta(3,\,2)$

$\theta\,\omega 3 == \omega 1\,\theta(1,\,3) + \omega 2\,\theta(2,\,3) + \omega 3\,\theta(3,\,3)$

For a nontrivial solution of this set of equations, the following relation must hold. The determinant defines the third-order polynomial in Θ and allows three solutions depending on the components of the general inertia tensor.

```
Solve[Det[theta - th] == 0, θ] // Simplify
```

$$\Big\{\Big\{\theta \to \frac{1}{6}\,\big(2\,(\theta(1,\,1) + \theta(2,\,2) + \theta(3,\,3)) -$$
$$2^{2/3}\,\big(-2\,\theta(1,\,1)^3 + 3\,(\theta(2,\,2) + \theta(3,\,3))\,\theta(1,\,1)^2 +$$
$$3\,(\theta(2,\,2)^2 - 4\,\theta(3,\,3)\,\theta(2,\,2) + \theta(3,\,3)^2 - 3\,\theta(1,\,2)\,\theta(2,\,1) -$$
$$3\,\theta(1,\,3)\,\theta(3,\,1) + 6\,\theta(2,\,3)\,\theta(3,\,2))\,\theta(1,\,1) -$$
$$2\,\theta(2,\,2)^3 - 2\,\theta(3,\,3)^3 + 3\,\theta(2,\,2)\,\theta(3,\,3)^2 +$$

$$18\,\theta(1,3)\,\theta(2,2)\,\theta(3,1)\,-$$
$$27\,\theta(1,3)\,\theta(2,1)\,\theta(3,2) - 9\,\theta(2,2)\,\theta(2,3)\,\theta(3,2)\,-$$
$$9\,\theta(1,2)\,(3\,\theta(2,3)\,\theta(3,1) + \theta(2,1)\,(\theta(2,2) - 2\,\theta(3,3)))\,+$$
$$3\,\theta(2,2)^2\,\theta(3,3) - 9\,\theta(1,3)\,\theta(3,1)\,\theta(3,3)\,-$$
$$9\,\theta(2,3)\,\theta(3,2)\,\theta(3,3)\,+$$
$$\sqrt{(}(2\,\theta(1,1)^3 - 3\,(\theta(2,2) + \theta(3,3))\,\theta(1,1)^2 + 3\,(-\theta(2,2)^2\,+$$
$$4\,\theta(3,3)\,\theta(2,2) - \theta(3,3)^2 + 3\,\theta(1,2)\,\theta(2,$$
$$1) + 3\,\theta(1,3)\,\theta(3,1) - 6\,\theta(2,3)\,\theta(3,2))$$
$$\theta(1,1) + 2\,\theta(2,2)^3 + 2\,\theta(3,3)^3\,-$$
$$3\,\theta(2,2)\,\theta(3,3)^2 - 18\,\theta(1,3)\,\theta(2,2)\,\theta(3,1)\,+$$
$$27\,\theta(1,3)\,\theta(2,1)\,\theta(3,2) + 9\,\theta(2,2)\,\theta(2,3)$$
$$\theta(3,2) + 9\,\theta(1,2)\,(3\,\theta(2,3)\,\theta(3,1)\,+$$
$$\theta(2,1)\,(\theta(2,2) - 2\,\theta(3,3)))\,-$$
$$3\,\theta(2,2)^2\,\theta(3,3) + 9\,\theta(1,3)\,\theta(3,1)\,\theta(3,3)\,+$$
$$9\,\theta(2,3)\,\theta(3,2)\,\theta(3,3))^2\,-$$
$$4\,(\theta(1,1)^2 - (\theta(2,2) + \theta(3,3))\,\theta(1,1) + \theta(2,2)^2\,+$$
$$\theta(3,3)^2 + 3\,\theta(1,2)\,\theta(2,1) + 3\,\theta(1,3)\,\theta(3,1)\,+$$
$$3\,\theta(2,3)\,\theta(3,2) - \theta(2,2)\,\theta(3,3))^3)\,)\,\hat{}\,(1/3)\,-$$
$$(2\,\sqrt[3]{2}\,(\theta(1,1)^2 - (\theta(2,2) + \theta(3,3))\,\theta(1,1) + \theta(2,2)^2\,+$$
$$\theta(3,3)^2 + 3\,\theta(1,2)\,\theta(2,1) + 3\,\theta(1,3)\,\theta(3,1)\,+$$
$$3\,\theta(2,3)\,\theta(3,2) - \theta(2,2)\,\theta(3,3)))\,\big/$$
$$((-2\,\theta(1,1)^3 + 3\,(\theta(2,2) + \theta(3,3))\,\theta(1,1)^2\,+$$
$$3\,(\theta(2,2)^2 - 4\,\theta(3,3)\,\theta(2,2) + \theta(3,3)^2 - 3\,\theta(1,2)\,\theta(2,1)\,-$$
$$3\,\theta(1,3)\,\theta(3,1) + 6\,\theta(2,3)\,\theta(3,2))\,\theta(1,1)\,-$$
$$2\,\theta(2,2)^3 - 2\,\theta(3,3)^3 + 3\,\theta(2,2)\,\theta(3,3)^2\,+$$
$$18\,\theta(1,3)\,\theta(2,2)\,\theta(3,1)\,-$$
$$27\,\theta(1,3)\,\theta(2,1)\,\theta(3,2) - 9\,\theta(2,2)\,\theta(2,3)\,\theta(3,2)\,-$$
$$9\,\theta(1,2)\,(3\,\theta(2,3)\,\theta(3,1) + \theta(2,1)\,(\theta(2,2) - 2\,\theta(3,3)))\,+$$
$$3\,\theta(2,2)^2\,\theta(3,3) - 9\,\theta(1,3)\,\theta(3,1)\,\theta(3,3)\,-$$
$$9\,\theta(2,3)\,\theta(3,2)\,\theta(3,3)\,+$$
$$\sqrt{(}(2\,\theta(1,1)^3 - 3\,(\theta(2,2) + \theta(3,3))\,\theta(1,1)^2\,+$$
$$3\,(-\theta(2,2)^2 + 4\,\theta(3,3)\,\theta(2,2) - \theta(3,3)^2\,+$$
$$3\,\theta(1,2)\,\theta(2,1) + 3\,\theta(1,3)\,\theta(3,1)\,-$$
$$6\,\theta(2,3)\,\theta(3,2))\,\theta(1,1) + 2\,\theta(2,2)^3\,+$$
$$2\,\theta(3,3)^3 - 3\,\theta(2,2)\,\theta(3,3)^2 - 18\,\theta(1,3)$$
$$\theta(2,2)\,\theta(3,1) + 27\,\theta(1,3)\,\theta(2,1)\,\theta(3,2)\,+$$
$$9\,\theta(2,2)\,\theta(2,3)\,\theta(3,2) + 9\,\theta(1,2)\,(3\,\theta(2,3)$$
$$\theta(3,1) + \theta(2,1)\,(\theta(2,2) - 2\,\theta(3,3)))\,-$$
$$3\,\theta(2,2)^2\,\theta(3,3) + 9\,\theta(1,3)\,\theta(3,1)$$
$$\theta(3,3) + 9\,\theta(2,3)\,\theta(3,2)\,\theta(3,3))^2\,-$$
$$4\,(\theta(1,1)^2 - (\theta(2,2) + \theta(3,3))\,\theta(1,1)\,+$$

$$\theta(2, 2)^2 + \theta(3, 3)^2 + 3\,\theta(1, 2)\,\theta(2, 1) +$$
$$3\,\theta(1, 3)\,\theta(3, 1) + 3\,\theta(2, 3)\,\theta(3, 2) -$$
$$\theta(2, 2)\,\theta(3, 3))^3))\,\hat{}\,(1/3)))\Big\},$$

$$\Big\{\theta \to \frac{1}{12}\,\big(4\,(\theta(1, 1) + \theta(2, 2) + \theta(3, 3)) +$$
$$2^{2/3}$$
$$\big(1 -$$
$$i\,\sqrt{3}\,\big)$$
$$\big(-2\,\theta(1, 1)^3 + 3\,(\theta(2, 2) + \theta(3, 3))\,\theta(1, 1)^2 +$$
$$\quad 3\,(\theta(2, 2)^2 - 4\,\theta(3, 3)\,\theta(2, 2) + \theta(3, 3)^2 - 3\,\theta(1, 2)\,\theta(2, 1) -$$
$$\quad\quad 3\,\theta(1, 3)\,\theta(3, 1) + 6\,\theta(2, 3)\,\theta(3, 2))\,\theta(1, 1) -$$
$$\quad 2\,\theta(2, 2)^3 - 2\,\theta(3, 3)^3 + 3\,\theta(2, 2)\,\theta(3, 3)^2 +$$
$$\quad 18\,\theta(1, 3)\,\theta(2, 2)\,\theta(3, 1) -$$
$$\quad 27\,\theta(1, 3)\,\theta(2, 1)\,\theta(3, 2) -$$
$$\quad 9\,\theta(2, 2)\,\theta(2, 3)\,\theta(3, 2) -$$
$$\quad 9\,\theta(1, 2)\,(3\,\theta(2, 3)\,\theta(3, 1) + \theta(2, 1)\,(\theta(2, 2) - 2\,\theta(3, 3))) +$$
$$\quad 3\,\theta(2, 2)^2\,\theta(3, 3) - 9\,\theta(1, 3)\,\theta(3, 1)\,\theta(3, 3) -$$
$$\quad 9\,\theta(2, 3)\,\theta(3, 2)\,\theta(3, 3) +$$
$$\quad \sqrt{\big((2\,\theta(1, 1)^3 - 3\,(\theta(2, 2) + \theta(3, 3))\,\theta(1, 1)^2 +}$$
$$\quad\quad 3\,(-\theta(2, 2)^2 + 4\,\theta(3, 3)\,\theta(2, 2) - \theta(3, 3)^2 +$$
$$\quad\quad\quad 3\,\theta(1, 2)\,\theta(2, 1) + 3\,\theta(1, 3)\,\theta(3, 1) -$$
$$\quad\quad\quad 6\,\theta(2, 3)\,\theta(3, 2))\,\theta(1, 1) + 2\,\theta(2, 2)^3 +$$
$$\quad\quad 2\,\theta(3, 3)^3 - 3\,\theta(2, 2)\,\theta(3, 3)^2 - 18\,\theta(1, 3)$$
$$\quad\quad \theta(2, 2)\,\theta(3, 1) + 27\,\theta(1, 3)\,\theta(2, 1)\,\theta(3, 2) +$$
$$\quad\quad 9\,\theta(2, 2)\,\theta(2, 3)\,\theta(3, 2) + 9\,\theta(1, 2)\,(3\,\theta(2, 3)$$
$$\quad\quad\quad \theta(3, 1) + \theta(2, 1)\,(\theta(2, 2) - 2\,\theta(3, 3))) -$$
$$\quad\quad 3\,\theta(2, 2)^2\,\theta(3, 3) + 9\,\theta(1, 3)\,\theta(3, 1)$$
$$\quad\quad\quad \theta(3, 3) + 9\,\theta(2, 3)\,\theta(3, 2)\,\theta(3, 3))^2 -$$
$$\quad\quad 4\,(\theta(1, 1)^2 - (\theta(2, 2) + \theta(3, 3))\,\theta(1, 1) + \theta(2, 2)^2 +$$
$$\quad\quad\quad \theta(3, 3)^2 + 3\,\theta(1, 2)\,\theta(2, 1) + 3\,\theta(1, 3)\,\theta(3, 1) +$$
$$\quad\quad\quad 3\,\theta(2, 3)\,\theta(3, 2) - \theta(2, 2)\,\theta(3, 3))^3))\,\hat{}\,(1/3) +$$
$$\quad \big(2\,\sqrt[3]{2}\,\big(1 + i\,\sqrt{3}\,\big)(\theta(1, 1)^2 - (\theta(2, 2) + \theta(3, 3))\,\theta(1, 1) +$$
$$\quad\quad \theta(2, 2)^2 + \theta(3, 3)^2 + 3\,\theta(1, 2)\,\theta(2, 1) +$$
$$\quad\quad 3\,\theta(1, 3)\,\theta(3, 1) + 3\,\theta(2, 3)\,\theta(3, 2) - \theta(2, 2)\,\theta(3, 3))\big)\big/$$
$$\quad \big((-2\,\theta(1, 1)^3 + 3\,(\theta(2, 2) + \theta(3, 3))\,\theta(1, 1)^2 +$$
$$\quad\quad 3\,(\theta(2, 2)^2 - 4\,\theta(3, 3)\,\theta(2, 2) + \theta(3, 3)^2 - 3\,\theta(1, 2)\,\theta(2, 1) -$$
$$\quad\quad\quad 3\,\theta(1, 3)\,\theta(3, 1) + 6\,\theta(2, 3)\,\theta(3, 2))\,\theta(1, 1) -$$
$$\quad\quad 2\,\theta(2, 2)^3 - 2\,\theta(3, 3)^3 + 3\,\theta(2, 2)\,\theta(3, 3)^2 +$$
$$\quad\quad 18\,\theta(1, 3)\,\theta(2, 2)\,\theta(3, 1) -$$
$$\quad\quad 27\,\theta(1, 3)\,\theta(2, 1)\,\theta(3, 2) - 9\,\theta(2, 2)\,\theta(2, 3)\,\theta(3, 2) -$$
$$\quad\quad 9\,\theta(1, 2)\,(3\,\theta(2, 3)\,\theta(3, 1) + \theta(2, 1)\,(\theta(2, 2) - 2\,\theta(3, 3))) +$$

$$3\,\theta(2,\,2)^2\,\theta(3,\,3) - 9\,\theta(1,\,3)\,\theta(3,\,1)\,\theta(3,\,3) -$$
$$9\,\theta(2,\,3)\,\theta(3,\,2)\,\theta(3,\,3) +$$
$$\sqrt{\big((2\,\theta(1,\,1)^3 - 3\,(\theta(2,\,2) + \theta(3,\,3))\,\theta(1,\,1)^2 +}$$
$$3\,(-\theta(2,\,2)^2 + 4\,\theta(3,\,3)\,\theta(2,\,2) - \theta(3,\,3)^2 +$$
$$3\,\theta(1,\,2)\,\theta(2,\,1) + 3\,\theta(1,\,3)\,\theta(3,\,1) -$$
$$6\,\theta(2,\,3)\,\theta(3,\,2))\,\theta(1,\,1) + 2\,\theta(2,\,2)^3 +$$
$$2\,\theta(3,\,3)^3 - 3\,\theta(2,\,2)\,\theta(3,\,3)^2 - 18\,\theta(1,\,3)$$
$$\theta(2,\,2)\,\theta(3,\,1) + 27\,\theta(1,\,3)\,\theta(2,\,1)\,\theta(3,\,2) +$$
$$9\,\theta(2,\,2)\,\theta(2,\,3)\,\theta(3,\,2) + 9\,\theta(1,\,2)\,(3\,\theta(2,\,3)$$
$$\theta(3,\,1) + \theta(2,\,1)\,(\theta(2,\,2) - 2\,\theta(3,\,3))) -$$
$$3\,\theta(2,\,2)^2\,\theta(3,\,3) + 9\,\theta(1,\,3)\,\theta(3,\,1)$$
$$\theta(3,\,3) + 9\,\theta(2,\,3)\,\theta(3,\,2)\,\theta(3,\,3))^2 -$$
$$4\,(\theta(1,\,1)^2 - (\theta(2,\,2) + \theta(3,\,3))\,\theta(1,\,1) +$$
$$\theta(2,\,2)^2 + \theta(3,\,3)^2 + 3\,\theta(1,\,2)\,\theta(2,\,1) +$$
$$3\,\theta(1,\,3)\,\theta(3,\,1) + 3\,\theta(2,\,3)\,\theta(3,\,2) -$$
$$\theta(2,\,2)\,\theta(3,\,3))^3\big)\big)^\wedge(1/3)\big)\Big\},$$

$$\Big\{\theta \to \frac{1}{12}\,\big(4\,(\theta(1,\,1) + \theta(2,\,2) + \theta(3,\,3)) +$$
$$2^{2/3}$$
$$\big(1 +$$
$$i\,\sqrt{3}\big)$$
$$\big(-2\,\theta(1,\,1)^3 + 3\,(\theta(2,\,2) + \theta(3,\,3))\,\theta(1,\,1)^2 +$$
$$3\,(\theta(2,\,2)^2 - 4\,\theta(3,\,3)\,\theta(2,\,2) + \theta(3,\,3)^2 - 3\,\theta(1,\,2)\,\theta(2,\,1) -$$
$$3\,\theta(1,\,3)\,\theta(3,\,1) + 6\,\theta(2,\,3)\,\theta(3,\,2))\,\theta(1,\,1) -$$
$$2\,\theta(2,\,2)^3 - 2\,\theta(3,\,3)^3 + 3\,\theta(2,\,2)\,\theta(3,\,3)^2 +$$
$$18\,\theta(1,\,3)\,\theta(2,\,2)\,\theta(3,\,1) -$$
$$27\,\theta(1,\,3)\,\theta(2,\,1)\,\theta(3,\,2) -$$
$$9\,\theta(2,\,2)\,\theta(2,\,3)\,\theta(3,\,2) -$$
$$9\,\theta(1,\,2)\,(3\,\theta(2,\,3)\,\theta(3,\,1) + \theta(2,\,1)\,(\theta(2,\,2) - 2\,\theta(3,\,3))) +$$
$$3\,\theta(2,\,2)^2\,\theta(3,\,3) -$$
$$9\,\theta(1,\,3)\,\theta(3,\,1)\,\theta(3,\,3) -$$
$$9\,\theta(2,\,3)\,\theta(3,\,2)\,\theta(3,\,3) +$$
$$\sqrt{\big((2\,\theta(1,\,1)^3 - 3\,(\theta(2,\,2) + \theta(3,\,3))\,\theta(1,\,1)^2 +}$$
$$3\,(-\theta(2,\,2)^2 + 4\,\theta(3,\,3)\,\theta(2,\,2) - \theta(3,\,3)^2 +$$
$$3\,\theta(1,\,2)\,\theta(2,\,1) + 3\,\theta(1,\,3)\,\theta(3,\,1) -$$
$$6\,\theta(2,\,3)\,\theta(3,\,2))\,\theta(1,\,1) + 2\,\theta(2,\,2)^3 +$$
$$2\,\theta(3,\,3)^3 - 3\,\theta(2,\,2)\,\theta(3,\,3)^2 - 18\,\theta(1,\,3)$$
$$\theta(2,\,2)\,\theta(3,\,1) + 27\,\theta(1,\,3)\,\theta(2,\,1)\,\theta(3,\,2) +$$
$$9\,\theta(2,\,2)\,\theta(2,\,3)\,\theta(3,\,2) + 9\,\theta(1,\,2)\,(3\,\theta(2,\,3)$$
$$\theta(3,\,1) + \theta(2,\,1)\,(\theta(2,\,2) - 2\,\theta(3,\,3))) -$$
$$3\,\theta(2,\,2)^2\,\theta(3,\,3) + 9\,\theta(1,\,3)\,\theta(3,\,1)$$

$$\theta(3, 3) + 9\,\theta(2, 3)\,\theta(3, 2)\,\theta(3, 3))^2 -$$
$$4\,(\theta(1, 1)^2 - (\theta(2, 2) + \theta(3, 3))\,\theta(1, 1) + \theta(2, 2)^2 +$$
$$\theta(3, 3)^2 + 3\,\theta(1, 2)\,\theta(2, 1) + 3\,\theta(1, 3)\,\theta(3, 1) +$$
$$3\,\theta(2, 3)\,\theta(3, 2) - \theta(2, 2)\,\theta(3, 3))^3))\wedge(1/3) +$$
$$\left(2\sqrt[3]{2}\left(1 - i\sqrt{3}\right)(\theta(1, 1)^2 - (\theta(2, 2) + \theta(3, 3))\,\theta(1, 1) +\right.$$
$$\theta(2, 2)^2 + \theta(3, 3)^2 +$$
$$3\,\theta(1, 2)\,\theta(2, 1) + 3\,\theta(1, 3)\,\theta(3, 1) +$$
$$3\,\theta(2, 3)\,\theta(3, 2) - \theta(2, 2)\,\theta(3, 3)))\Big/$$
$$\big((-2\,\theta(1, 1)^3 + 3\,(\theta(2, 2) + \theta(3, 3))\,\theta(1, 1)^2 +$$
$$3\,(\theta(2, 2)^2 - 4\,\theta(3, 3)\,\theta(2, 2) + \theta(3, 3)^2 - 3\,\theta(1, 2)\,\theta(2, 1) -$$
$$3\,\theta(1, 3)\,\theta(3, 1) + 6\,\theta(2, 3)\,\theta(3, 2))\,\theta(1, 1) -$$
$$2\,\theta(2, 2)^3 - 2\,\theta(3, 3)^3 + 3\,\theta(2, 2)\,\theta(3, 3)^2 +$$
$$18\,\theta(1, 3)\,\theta(2, 2)\,\theta(3, 1) -$$
$$27\,\theta(1, 3)\,\theta(2, 1)\,\theta(3, 2) - 9\,\theta(2, 2)\,\theta(2, 3)\,\theta(3, 2) -$$
$$9\,\theta(1, 2)\,(3\,\theta(2, 3)\,\theta(3, 1) + \theta(2, 1)\,(\theta(2, 2) - 2\,\theta(3, 3))) +$$
$$3\,\theta(2, 2)^2\,\theta(3, 3) - 9\,\theta(1, 3)\,\theta(3, 1)\,\theta(3, 3) -$$
$$9\,\theta(2, 3)\,\theta(3, 2)\,\theta(3, 3) +$$
$$\sqrt{\big((2\,\theta(1, 1)^3 - 3\,(\theta(2, 2) + \theta(3, 3))\,\theta(1, 1)^2 +}$$
$$3\,(-\theta(2, 2)^2 + 4\,\theta(3, 3)\,\theta(2, 2) - \theta(3, 3)^2 +$$
$$3\,\theta(1, 2)\,\theta(2, 1) + 3\,\theta(1, 3)\,\theta(3, 1) -$$
$$6\,\theta(2, 3)\,\theta(3, 2))\,\theta(1, 1) + 2\,\theta(2, 2)^3 +$$
$$2\,\theta(3, 3)^3 - 3\,\theta(2, 2)\,\theta(3, 3)^2 - 18\,\theta(1, 3)$$
$$\theta(2, 2)\,\theta(3, 1) + 27\,\theta(1, 3)\,\theta(2, 1)\,\theta(3, 2) +$$
$$9\,\theta(2, 2)\,\theta(2, 3)\,\theta(3, 2) + 9\,\theta(1, 2)\,(3\,\theta(2, 3)$$
$$\theta(3, 1) + \theta(2, 1)\,(\theta(2, 2) - 2\,\theta(3, 3))) -$$
$$3\,\theta(2, 2)^2\,\theta(3, 3) + 9\,\theta(1, 3)\,\theta(3, 1)$$
$$\theta(3, 3) + 9\,\theta(2, 3)\,\theta(3, 2)\,\theta(3, 3))^2 -$$
$$4\,(\theta(1, 1)^2 - (\theta(2, 2) + \theta(3, 3))\,\theta(1, 1) +$$
$$\theta(2, 2)^2 + \theta(3, 3)^2 + 3\,\theta(1, 2)\,\theta(2, 1) +$$
$$3\,\theta(1, 3)\,\theta(3, 1) + 3\,\theta(2, 3)\,\theta(3, 2) -$$
$$\theta(2, 2)\,\theta(3, 3))^3))\wedge(1/3)))\Big\}\Big\}$$

2.10.5 Steiner's Theorem

In practical calculations, it is more convenient to determine the inertia tensor with respect to a symmetry point or a symmetry line. The point or line of symmetry is usually defined by the rigid body itself. The geometric shape distinguishes such points or lines. Let us assume that the symmetry point Q is in the direction $\vec{a}$ apart from the center of mass. Then, the inertia tensor with respect to the point Q is given by

$$\tilde{\Theta}_{ij} = \sum_{\alpha=1}^{n} m_\alpha \{ \delta_{ij} \sum_{k=1}^{3} \tilde{x}_{\alpha k}^2 - \tilde{x}_{\alpha i} \tilde{x}_{\alpha j} \}, \tag{2.10.41}$$

where

$$\tilde{x}_{\alpha i} = x_{\alpha i} + a_i. \tag{2.10.42}$$

Inserting the new coordinates into $\tilde{\Theta}_{ij}$, we get

$$\tilde{\Theta}_{ij} = \sum_{\alpha=1}^{n} m_\alpha \{ \delta_{ij} \sum_{k=1}^{3} (x_{\alpha k} + a_k)^2 -$$
$$(x_{\alpha i} + a_i)(x_{\alpha j} + a_j) \}$$
$$= \sum_{\alpha=1}^{n} m_\alpha \{ \delta_{ij} \sum_{k=1}^{3} x_{\alpha k}^2 - x_{\alpha i} x_{\alpha j} \} + \tag{2.10.43}$$
$$\sum_{\alpha=1}^{n} m_\alpha \{ \delta_{ij} \sum_{k=1}^{3} 2 x_{\alpha k} +$$
$$a_k + a_k^2 - (a_i x_{\alpha j} + a_j x_{\alpha i} + a_i a_j) \}$$

$$\tilde{\Theta}_{ij} = \Theta_{ij} + \sum_{\alpha=1}^{n} m_\alpha \{ \delta_{ij} \sum_{k=1}^{3} a_k^2 - a_i a_j \} +$$
$$\sum_{\alpha=1}^{n} m_\alpha \{ 2 \delta_{ij} \sum_{k=1}^{3} x_{\alpha k} a_k - a_i x_{\alpha j} - a_j x_{\alpha i} \} \tag{2.10.44}$$

Terms containing sums of the type

$$\sum_{\alpha=1}^{n} m_\alpha x_{k\sigma} = 0 \tag{2.10.45}$$

vanish because we are in the center of mass system. Thus, the inertia tensor reduces to

$$\tilde{\Theta}_{ij} = \Theta_{ij} + \sum_{\alpha=1}^{n} m_\alpha \{ \delta_{ij} \sum_{k=1}^{3} a_k^2 - a_i a_j \}$$
$$= \Theta_{ij} + M (\delta_{ij} a^2 - a_i a_j) \tag{2.10.46}$$

with $M = \sum_{\alpha=1}^{n} m_\alpha$ and $a^2 = \sum_{k=1}^{3} a_k^2$.

The inertia tensor with respect to the center of mass is determined with respect to the symmetry point by

$$\Theta_{ij} = \tilde{\Theta}_{ij} - M (\delta_{ij} a^2 - a_i a_j) \tag{2.10.47}$$

This relation is known as Steiner's theorem.

2.10.6 Euler's Equations of Motion

Let us first examine the motion of a force-free rigid body. As already discussed, the movement of the center of mass does not affect the spinning portion of the top. Thus, it is sufficient to consider the Lagrangian as a function of the rotational components which are given by

$$\mathcal{L} = T_{\text{rot}} = \frac{1}{2} \sum_i \Theta_i \, \omega_i^2. \tag{2.10.48}$$

The representation of the rotation is efficiently described by the three Euler angles φ, θ and ψ. The angular velocity $\vec{\omega}$ can be represented by these three angles as

$$\vec{\omega} = \begin{pmatrix} \varphi' \sin\theta \sin\psi \;+\; \theta' \cos\psi \\ \varphi' \sin\theta \cos\psi \;-\; \theta' \sin\psi \\ \varphi' \cos\theta \;+\; \psi' \end{pmatrix}.$$

$$= \vec{\omega}\,(\varphi,\, \theta,\, \psi,\, \varphi',\, \theta',\, \psi') \tag{2.10.49}$$

The Lagrangian thus depends on the three generalized coordinates φ, θ, and ψ. The Euler–Lagrange equations for the rotating rigid body are thus given by

$$\frac{\partial\mathcal{L}}{\partial\varphi} - \frac{d}{dt}\left(\frac{\partial\mathcal{L}}{\partial\varphi'}\right) = 0, \tag{2.10.50}$$

$$\frac{\partial\mathcal{L}}{\partial\theta} - \frac{d}{dt}\left(\frac{\partial\mathcal{L}}{\partial\theta'}\right) = 0, \tag{2.10.51}$$

$$\frac{\partial\mathcal{L}}{\partial\psi} - \frac{d}{dt}\left(\frac{\partial\mathcal{L}}{\partial\psi'}\right) = 0. \tag{2.10.52}$$

Each of these equations determines the rotation of the top. We note that the Lagrangian and the Lagrange equations are set up in different coordinates. Since both coordinates are related by Equation (2.10.49), it is obvious that the derivatives in the Euler–Lagrange equations are calculated by the following rules. For example, the last equation provides

$$\frac{\partial\mathcal{L}}{\partial\psi} = \sum_{i=1}^{3} \frac{\partial\mathcal{L}}{\partial\omega_i} \frac{\partial\omega_i}{\partial\psi}, \tag{2.10.53}$$

and for the velocities, we have

$$\frac{\partial\mathcal{L}}{\partial\psi'} = \sum_{i=1}^{3} \frac{\partial\mathcal{L}}{\partial\omega_i} \frac{\partial\omega_i}{\partial\psi'}. \tag{2.10.54}$$

The total Euler–Lagrange equations then become

$$\sum_{i=1}^{3} \left\{ \frac{\partial\mathcal{L}}{\partial\omega_i} \frac{\partial\omega_i}{\partial\psi} - \frac{d}{dt}\left(\frac{\partial\mathcal{L}}{\partial\omega_i} \frac{\partial\omega_i}{\partial\psi'} \right) \right\} = 0. \tag{2.10.55}$$

For example, let us demonstrate the calculations for the ψ-coordinate. A differentiation of ω_i with respect to ψ and ψ' delivers the following relations:

$$\frac{\partial \omega_1}{\partial \psi} = \varphi' \sin\theta \cos\psi - \theta' \sin\psi = \omega_2, \tag{2.10.56}$$

$$\frac{\partial \omega_2}{\partial \psi} = -\varphi' \sin\theta \sin\psi - \theta' \cos\psi = -\omega_1, \tag{2.10.57}$$

$$\frac{\partial \omega_3}{\partial \psi} = 0, \tag{2.10.58}$$

and

$$\frac{\partial \omega_1}{\partial \dot\psi} = 0, \tag{2.10.59}$$

$$\frac{\partial \omega_2}{\partial \dot\psi} = 0, \tag{2.10.60}$$

$$\frac{\partial \omega_3}{\partial \dot\psi} = 1. \tag{2.10.61}$$

On the other hand, the Lagrange function provides

$$\frac{\partial \mathcal{L}}{\partial \omega_i} = \frac{\partial T_{\text{rot}}}{\partial \omega_i} = \Theta_i \, \omega_i. \tag{2.10.62}$$

From the Euler–Lagrange equation, we obtain

$$\Theta_i \, \omega_1 \, \omega_2 + \Theta_2 \, \omega_2 \, (-\omega_1) - \frac{d}{dt} \, \Theta_3 \, \omega_3 = 0$$
$$\Longleftrightarrow \quad (\Theta_1 - \Theta_2) \, \omega_1 \, \omega_2 - \Theta_3 \, \omega'_3 = 0. \tag{2.10.63}$$

The other two equations are derived by similar calculations. However, we can short-cut the calculation by permuting the indices of ω and Θ because the x_3-axis was chosen arbitrarily as the rotation axis:

$$(\Theta_3 - \Theta_1) \, \omega_1 \, \omega_3 - \Theta_2 \, \omega'_2 = 0,$$
$$(\Theta_2 - \Theta_3) \, \omega_3 \, \omega_2 - \Theta_1 \, \omega'_1 = 0. \tag{2.10.64}$$

The three equations of motion can be amalgamated in a single relation by using the Levi–Civita tensor ε_{ijk} and a index notation:

$$(\Theta_i - \Theta_j) \, \omega_i \, \omega_j - \sum_k \Theta_k \, \omega'_k \, \varepsilon_{ijk} = 0. \tag{2.10.65}$$

The derived three equations are also known as Euler equations. Leonard Euler derived these equations in 1758 for a force-free top.

To show how the Euler equations look for the Euler angles, let us carry out the same calculations as above in *Mathematica*. First, let us define the angular velocity $\vec{\omega}$ as a vector using Equation (2.10.49):

```
ω = {∂_t φ[t] Sin[θ[t]] Sin[ψ[t]] + ∂_t θ[t] Cos[ψ[t]],
     ∂_t φ[t] Sin[θ[t]] Cos[ψ[t]] - ∂_t θ[t] Sin[ψ[t]],
     ∂_t ψ[t] Cos[θ[t]] + ∂_t φ[t]}; ω // MatrixForm
```

$$\begin{pmatrix} \cos(\psi(t))\,\theta'(t) + \sin(\theta(t))\sin(\psi(t))\,\phi'(t) \\ \cos(\psi(t))\sin(\theta(t))\,\phi'(t) - \sin(\psi(t))\,\theta'(t) \\ \phi'(t) + \cos(\theta(t))\,\psi'(t) \end{pmatrix}$$

The inertia tensor with three different principal values are given by

```
th1 = IdentityMatrix[3] {Θ1, Θ2, Θ3}; th1 // MatrixForm
```

$$\begin{pmatrix} \Theta 1 & 0 & 0 \\ 0 & \Theta 2 & 0 \\ 0 & 0 & \Theta 3 \end{pmatrix}$$

The Lagrangian for the force-free top follows by the relation

```
L = 1/2 ω.th1.ω
```

$$\frac{1}{2}\left(\Theta 2\,(\cos(\psi(t))\sin(\theta(t))\,\phi'(t) - \sin(\psi(t))\,\theta'(t))^2 + \right.$$
$$\left. \Theta 1\,(\cos(\psi(t))\,\theta'(t) + \sin(\theta(t))\sin(\psi(t))\,\phi'(t))^2 + \Theta 3\,(\phi'(t) + \cos(\theta(t))\,\psi'(t))^2\right)$$

Applying the *Mathematica* function EulerLagrange[] introduced in Section 2.7 on Lagrange dynamics, we find three coupled nonlinear ordinary differential equations of second order.

```
EulerLagrange[L, {φ, ψ, θ}, t]
```

$$\Big\{ -\frac{\partial(\Theta1\,\cos(\psi(t))\,\sin(\theta(t))\,\sin(\psi(t))\,\theta'(t))}{\partial t} -$$

$$\frac{\partial(-\Theta2\,\cos(\psi(t))\,\sin(\theta(t))\,\sin(\psi(t))\,\theta'(t))}{\partial t} -$$

$$\frac{\partial(\Theta3\,\phi'(t))}{\partial t} - \frac{\partial(\Theta2\,\cos^2(\psi(t))\,\sin^2(\theta(t))\,\phi'(t))}{\partial t} -$$

$$\frac{\partial(\Theta1\,\sin^2(\theta(t))\,\sin^2(\psi(t))\,\phi'(t))}{\partial t} - \frac{\partial(\Theta3\,\cos(\theta(t))\,\psi'(t))}{\partial t} == 0,$$

$$\Theta1\,\sin(\theta(t))\,\theta'(t)\,\phi'(t)\,\cos^2(\psi(t)) - \Theta2\,\sin(\theta(t))\,\theta'(t)\,\phi'(t)\,\cos^2(\psi(t)) -$$

$$\Theta1\,\sin(\psi(t))\,\theta'(t)^2\,\cos(\psi(t)) + \Theta2\,\sin(\psi(t))\,\theta'(t)^2\,\cos(\psi(t)) +$$

$$\Theta1\,\sin^2(\theta(t))\,\sin(\psi(t))\,\phi'(t)^2\,\cos(\psi(t)) -$$

$$\Theta2\,\sin^2(\theta(t))\,\sin(\psi(t))\,\phi'(t)^2\,\cos(\psi(t)) - \frac{\partial(\Theta3\,\cos(\theta(t))\,\phi'(t))}{\partial t} -$$

$$\frac{\partial(\Theta3\,\cos^2(\theta(t))\,\psi'(t))}{\partial t} - \Theta1\,\sin(\theta(t))\,\sin^2(\psi(t))\,\theta'(t)\,\phi'(t) +$$

$$\Theta2\,\sin(\theta(t))\,\sin^2(\psi(t))\,\theta'(t)\,\phi'(t) == 0,$$

$$\Theta1\,\cos(\theta(t))\,\sin(\theta(t))\,\sin^2(\psi(t))\,\phi'(t)^2 + \Theta2\,\cos(\theta(t))\,\cos^2(\psi(t))\,\sin(\theta(t))\,\phi'(t)^2 +$$

$$\Theta1\,\cos(\theta(t))\,\cos(\psi(t))\,\sin(\psi(t))\,\theta'(t)\,\phi'(t) -$$

$$\Theta2\,\cos(\theta(t))\,\cos(\psi(t))\,\sin(\psi(t))\,\theta'(t)\,\phi'(t) - \Theta3\,\sin(\theta(t))\,\psi'(t)\,\phi'(t) -$$

$$\Theta3\,\cos(\theta(t))\,\sin(\theta(t))\,\psi'(t)^2 - \frac{\partial(\Theta1\,\cos^2(\psi(t))\,\theta'(t))}{\partial t} -$$

$$\frac{\partial(\Theta2\,\sin^2(\psi(t))\,\theta'(t))}{\partial t} - \frac{\partial(\Theta1\,\cos(\psi(t))\,\sin(\theta(t))\,\sin(\psi(t))\,\phi'(t))}{\partial t} -$$

$$\frac{\partial(-\Theta2\,\cos(\psi(t))\,\sin(\theta(t))\,\sin(\psi(t))\,\phi'(t))}{\partial t} == 0 \Big\}$$

The three equations contain the principal inertial moments Θ_1, Θ_2, and Θ_3 as parameters. A much simpler representation of these equations follows if we consider the top to be a spherical top for which all three moments of inertia are equal. For this case, the equations of motion read

```
EulerLagrange[
  L /. {θ1 -> θ, θ2 -> θ, θ3 -> θ}, {φ, ψ, θ}, t]
```

$$\left\{-\frac{\partial(\Theta\,\phi'(t))}{\partial t}-\frac{\partial(\Theta\cos^2(\psi(t))\sin^2(\theta(t))\,\phi'(t))}{\partial t}-\right.$$

$$\frac{\partial(\Theta\sin^2(\theta(t))\sin^2(\psi(t))\,\phi'(t))}{\partial t}-\frac{\partial(\Theta\cos(\theta(t))\,\psi'(t))}{\partial t}==0,$$

$$-\frac{\partial(\Theta\cos(\theta(t))\,\phi'(t))}{\partial t}-\frac{\partial(\Theta\cos^2(\theta(t))\,\psi'(t))}{\partial t}==0,$$

$$\Theta\cos(\theta(t))\sin(\theta(t))\sin^2(\psi(t))\,\phi'(t)^2+\Theta\cos(\theta(t))\cos^2(\psi(t))\sin(\theta(t))\,\phi'(t)^2-$$

$$\Theta\sin(\theta(t))\,\psi'(t)\,\phi'(t)-\Theta\cos(\theta(t))\sin(\theta(t))\,\psi'(t)^2-$$

$$\left.\frac{\partial(\Theta\cos^2(\psi(t))\,\theta'(t))}{\partial t}-\frac{\partial(\Theta\sin^2(\psi(t))\,\theta'(t))}{\partial t}==0\right\}$$

So far, we examined the equations of motion for a force-free top. In such cases when the top is moving in a force field, the equations of motion change. To derive the equations of motion for a top spinning in a force field, let us start with the temporal change of the angular moment which equals the force moment acting on the top:

$$\left(\frac{d\vec{L}}{dt}\right)_{\text{fix}}=\vec{M},\tag{2.10.66}$$

where $\vec{M}$ is the moment generated by the force. However, the change of the angular moment in the inertial system is determined by the expressions in the body system:

$$\left(\frac{d\vec{L}}{dt}\right)_{\text{fix}}=\left(\frac{d\vec{L}}{dt}\right)_{\text{body}}+\vec{\omega}\times\vec{L}\tag{2.10.67}$$

or

$$\frac{d\vec{L}}{dt}+\vec{\omega}\times\vec{L}=\vec{M}.\tag{2.10.68}$$

The components along the x_3-axis are given by

$$L'_3+\omega_1 L_2-\omega_2 L_1=M_3.\tag{2.10.69}$$

Since we selected the coordinate system in such a way that the coordinate axis is identical with the principal axis of the inertia tensor, we can express the angular moments by

$$L_i=\Theta_i\,\omega_i;\tag{2.10.70}$$

then, it follows that

$$\Theta_3 \, \omega'_3 - (\Theta_1 - \Theta_2) \, \omega_1 \, \omega_2 = M_3. \qquad (2.10.71)$$

The general case is expressed by the relation

$$(\Theta_i - \Theta_j) \, \omega_i \, \omega_j - \sum_k (\Theta_k \, \omega'_k - M_k) \, \varepsilon_{ijk} = 0. \qquad (2.10.72)$$

Equations (2.10.72) are the equations of motion for a top moving in a force field.

We note here that the motion of a top is mainly determined by its inertial moments. Consequently, two tops with equal inertia moments but different shapes carry out the same motion. This behavior was first realized by Cauchy in 1827. As a consequence of this observation, Cauchy introduced the equivalent ellipsoid.

2.10.7 Force-Free Motion of a Symmetrical Top

Let us examine the motion of a force-free symmetrical top. For a symmetrical top, we have $\Theta = \Theta_1 = \Theta_2 \neq \Theta_3$. The three Euler equations thus read

```
EulerEquations = {(Θ - Θ3) ω2[t] ω3[t] - Θ ∂ₜω1[t] == 0,
   (Θ3 - Θ) ω3[t] ω1[t] - Θ ∂ₜω2[t] == 0, -Θ3 ∂ₜω3[t] == 0};
EulerEquations // TableForm
```

$(\theta - \theta 3) \, \omega 2(t) \, \omega 3(t) - \theta \, \omega 1'(t) == 0$
$(\theta 3 - \theta) \, \omega 1(t) \, \omega 3(t) - \theta \, \omega 2'(t) == 0$
$-\theta 3 \, \omega 3'(t) == 0$

Let us, in addition, assume that the center of mass is at rest. This simplifies the motion to a pure rotation. In addition, we assume that the angular velocity $\vec{\omega}$ is not pointed in the direction of the principal inertia directions. The solution of the third equation of motion shows that ω_3 is a constant equal to κ_3:

```
sol3 = DSolve[EulerEquations[[3]], ω3, t] /.
   C[1] -> κ3 // Flatten
```

$\{\omega 3 \rightarrow \text{Function}[\{t\}, \kappa 3]\}$

Thus, the first two Euler equations simplify to

```
EulerEquations12 = Take[EulerEquations /. sol3, {1, 2}];
EulerEquations12 // TableForm
```

$(\theta - \theta 3)\, \kappa 3\, \omega 2(t) - \theta\, \omega 1'(t) == 0$

$(\theta 3 - \theta)\, \kappa 3\, \omega 1(t) - \theta\, \omega 2'(t) == 0$

The two first-order coupled equations can be solved by

```
sol12 =
   DSolve[{EulerEquations12}, {ω1, ω2}, t] // Flatten
```

$$\left\{\omega 1 \rightarrow \text{Function}\!\left[\{t\}, c_1 \cos\!\left(t\,\kappa 3 - \frac{t\,\theta 3\,\kappa 3}{\theta}\right) + c_2 \sin\!\left(t\,\kappa 3 - \frac{t\,\theta 3\,\kappa 3}{\theta}\right)\right],\right.$$
$$\left.\omega 2 \rightarrow \text{Function}\!\left[\{t\}, c_2 \cos\!\left(t\,\kappa 3 - \frac{t\,\theta 3\,\kappa 3}{\theta}\right) - c_1 \sin\!\left(t\,\kappa 3 - \frac{t\,\theta 3\,\kappa 3}{\theta}\right)\right]\right\}$$

where c_1 and c_2 are constants of integration. The angular velocity then becomes

```
ω = {ω1[t], ω2[t], ω3[t]} /. sol3 /. sol12
```

$$\left\{c_1 \cos\!\left(t\,\kappa 3 - \frac{t\,\theta 3\,\kappa 3}{\theta}\right) + c_2 \sin\!\left(t\,\kappa 3 - \frac{t\,\theta 3\,\kappa 3}{\theta}\right),\right.$$
$$\left.c_2 \cos\!\left(t\,\kappa 3 - \frac{t\,\theta 3\,\kappa 3}{\theta}\right) - c_1 \sin\!\left(t\,\kappa 3 - \frac{t\,\theta 3\,\kappa 3}{\theta}\right), \kappa 3\right\}$$

It is obvious that the length of the angular velocity vector remains constant by checking

```
√ω.ω // Simplify
```

$$\sqrt{\kappa 3^2 + c_1^2 + c_2^2}$$

The three constants of integration c_1, c_2, and κ_3 fix the value of the angular velocity. As defined in section 2.10.1, we are talking about a symmetrical top with an inertia tensor:

```
Θ = IdentityMatrix[3] {θ, θ, θ3}; Θ // MatrixForm
```

$$\begin{pmatrix} \theta & 0 & 0 \\ 0 & \theta & 0 \\ 0 & 0 & \theta 3 \end{pmatrix}$$

The corresponding angular momentum is given by the vector

```
L = Θ.ω // Simplify
```

$$\left\{ \theta \left(c_1 \cos\left(\frac{t(\theta - \theta 3)\kappa 3}{\theta} \right) + c_2 \sin\left(\frac{t(\theta - \theta 3)\kappa 3}{\theta} \right) \right), \right.$$
$$\left. \theta \left(c_2 \cos\left(\frac{t(\theta - \theta 3)\kappa 3}{\theta} \right) - c_1 \sin\left(\frac{t(\theta - \theta 3)\kappa 3}{\theta} \right) \right), \theta 3 \, \kappa 3 \right\}$$

Examining the angular momentum in the inertial system, we have to check the relation

```
Simplify[ω × L + ∂t L]
```

$$\{0, 0, 0\}$$

The result demonstrates that the angular momentum is a conserved quantity in the inertial system. Another conservation law is given by

```
        1
  T = ───  ω.Θ.ω // Simplify
        2
```

$$\frac{1}{2}\left(\theta 3\,\kappa 3^2 + \theta\left(c_1^2 + c_2^2\right)\right)$$

The kinetic energy of the force-free top is also a quantity which is purely determined by the constants of integration and the values of the inertial tensor.

In conclusion, our observations are that the value of the angular velocity, the angular momentum in the inertial system, and the kinetic energy are conserved quantities. The conservation of angular velocity and angular momentum cause the projection of the angular momentum to the angular velocity to also be a conserved quantity. Thus, the motion of the angular velocity is executed in such a way that ϖ precesses with a constant angle between the angular momentum around the x_3-axis. This behavior is shown in the following illustration:

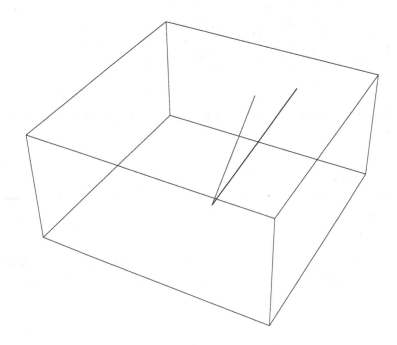

The red line is related to the angular momentum, whereas the blue line represents the angular velocity.

2.10.8 Motion of a Symmetrical Top in a Force Field

Let us examine the motion of a forced symmetrical top. The inertia tensor of a symmetrical top is characterized by $\Theta = \Theta_1 = \Theta_2 \neq \Theta_3$. We define this tensor by

```
Θ = IdentityMatrix[3] {θ, θ, θ3}; θ // MatrixForm
```

$$\begin{pmatrix} \theta & 0 & 0 \\ 0 & \theta & 0 \\ 0 & 0 & \theta 3 \end{pmatrix}$$

The three Euler equations thus read

```
EulerEquations = {(θ - θ3) ω2[t] ω3[t] - θ ∂ₜω1[t] == M1,
    (θ3 - θ) ω3[t] ω1[t] - θ ∂ₜω2[t] == M2,
    -θ3 ∂ₜω3[t] == M3}; EulerEquations // TableForm
```

$(\theta - \theta 3)\, \omega 2(t)\, \omega 3(t) - \theta\, \omega 1'(t) == M1$
$(\theta 3 - \theta)\, \omega 1(t)\, \omega 3(t) - \theta\, \omega 2'(t) == M2$
$-\theta 3\, \omega 3'(t) == M3$

where M_1, M_2, and M_3 are the acting moments of the force. Let us assume that the center of mass is at rest. This simplifies the motion to a pure rotation. In addition, we assume that the angular velocity $\vec{\omega}$ is not pointed in the direction of the principal inertia directions. The solution of this coupled system of equations is given by

```
sol = DSolve[EulerEquations, {ω1, ω2, ω3}, t] // Flatten
```

$$\left\{\omega3 \to \text{Function}\left[\{t\},\, c_1 - \frac{\text{M3}\,t}{\theta3}\right],\right.$$

$$\omega1 \to \text{Function}\left[\{t\},\, c_2 \cos\left(\frac{t\,\sqrt{\theta^2 - 2\,\theta3\,\theta + \theta3^2}\,(2\,\theta3\,c_1 - \text{M3}\,t)}{2\,\theta\,\theta3}\right) + \right.$$

$$\frac{1}{\theta\,\sqrt{(\theta - \theta3)^2}}\left(\cos\left(\frac{t\,\sqrt{(\theta - \theta3)^2}\,(2\,\theta3\,c_1 - \text{M3}\,t)}{2\,\theta\,\theta3}\right)\int_{\text{K\$306}}^{t}\right.$$

$$\left(\text{M2}\,(\theta - \theta3)\sin\left(\frac{\text{K\$305}\,\sqrt{(\theta - \theta3)^2}\,(2\,\theta3\,c_1 - \text{K\$305}\,\text{M3})}{2\,\theta\,\theta3}\right) - \right.$$

$$\text{M1}\,\sqrt{(\theta - \theta3)^2}$$

$$\left.\cos\left(\frac{\text{K\$305}\,\sqrt{(\theta - \theta3)^2}\,(2\,\theta3\,c_1 - \text{K\$305}\,\text{M3})}{2\,\theta\,\theta3}\right)\right)$$

$$d\,\text{K\$305} + \left(\int_{\text{K\$1694}}^{t}\left(-\text{M2}\,(\theta - \theta3)\right.\right.$$

$$\cos\left(\frac{\text{K\$1693}\,\sqrt{(\theta - \theta3)^2}\,(2\,\theta3\,c_1 - \text{K\$1693}\,\text{M3})}{2\,\theta\,\theta3}\right) - $$

$$\text{M1}\,\sqrt{(\theta - \theta3)^2}$$

$$\left.\sin\left(\frac{\text{K\$1693}\,\sqrt{(\theta - \theta3)^2}\,(2\,\theta3\,c_1 - \text{K\$1693}\,\text{M3})}{2\,\theta\,\theta3}\right)\right)$$

$$\left.d\,\text{K\$1693}\right)\sin\left(\frac{t\,\sqrt{(\theta - \theta3)^2}\,(2\,\theta3\,c_1 - \text{M3}\,t)}{2\,\theta\,\theta3}\right)\right) + $$

$$\left.c_3 \sin\left(\frac{t\,\sqrt{\theta^2 - 2\,\theta3\,\theta + \theta3^2}\,(2\,\theta3\,c_1 - \text{M3}\,t)}{2\,\theta\,\theta3}\right)\right],$$

$$\omega2 \to \text{Function}\left[\right.$$

$$\{t\},$$

$$\left(\theta\,\theta3\left(\frac{\sqrt{\theta^2 - 2\,\theta3\,\theta + \theta3^2}\,(2\,\theta3\,c_1 - \text{M3}\,t)}{2\,\theta\,\theta3} - \frac{\text{M3}\,t\,\sqrt{\theta^2 - 2\,\theta3\,\theta + \theta3^2}}{2\,\theta\,\theta3}\right)\right.$$

$$c_3 \cos\left(\frac{t\sqrt{\theta^2 - 2\,\theta3\,\theta + \theta3^2}\;(2\,\theta3\,c_1 - M3\,t)}{2\,\theta\,\theta3}\right)\Bigg)\Bigg/$$

$$(-c_1\,\theta3^2 + M3\,t\,\theta3 + \theta\,c_1\,\theta3 - M3\,t\,\theta) +$$

$$\frac{1}{\theta\,(\theta - \theta3)}\left(\cos\left(\frac{t\sqrt{(\theta - \theta3)^2}\;(2\,\theta3\,c_1 - M3\,t)}{2\,\theta\,\theta3}\right)\int_{K\$1694}^{t}\left(-M2\,(\theta - \theta3)\right.\right.$$

$$\cos\left(\frac{K\$1693\,\sqrt{(\theta - \theta3)^2}\;(2\,\theta3\,c_1 - K\$1693\,M3)}{2\,\theta\,\theta3}\right) -$$

$$M1\,\sqrt{(\theta - \theta3)^2}$$

$$\sin\left(\frac{K\$1693\,\sqrt{(\theta - \theta3)^2}\;(2\,\theta3\,c_1 - K\$1693\,M3)}{2\,\theta\,\theta3}\right)\Bigg)$$

$$d\,K\$1693 - \left(\int_{K\$306}^{t}\left(M2\,(\theta - \theta3)\right.\right.$$

$$\sin\left(\frac{K\$305\,\sqrt{(\theta - \theta3)^2}\;(2\,\theta3\,c_1 - K\$305\,M3)}{2\,\theta\,\theta3}\right) -$$

$$M1\,\sqrt{(\theta - \theta3)^2}$$

$$\cos\left(\frac{K\$305\,\sqrt{(\theta - \theta3)^2}\;(2\,\theta3\,c_1 - K\$305\,M3)}{2\,\theta\,\theta3}\right)\Bigg)$$

$$d\,K\$305\Bigg)\sin\left(\frac{t\sqrt{(\theta - \theta3)^2}\;(2\,\theta3\,c_1 - M3\,t)}{2\,\theta\,\theta3}\right)\Bigg) +$$

$$\left(\theta\,\theta3\left(\frac{M3\,t\,\sqrt{\theta^2 - 2\,\theta3\,\theta + \theta3^2}}{2\,\theta\,\theta3} - \right.\right.$$

$$\frac{\sqrt{\theta^2 - 2\,\theta3\,\theta + \theta3^2}\;(2\,\theta3\,c_1 - M3\,t)}{2\,\theta\,\theta3}\right)$$

$$c_2 \sin\left(\frac{t\sqrt{\theta^2 - 2\,\theta3\,\theta + \theta3^2}\;(2\,\theta3\,c_1 - M3\,t)}{2\,\theta\,\theta3}\right)\Bigg)\Bigg/$$

$$(-c_1\,\theta3^2 + M3\,t\,\theta3 + \theta\,c_1\,\theta3 - M3\,t\,\theta)\Bigg]\Bigg\}$$

where c_1, c_2, and c_3 are constants of integration. We realize that the solution is determined up to an integration. The angular velocity for this kind of motion is thus given by

```
ω = {ω1[t], ω2[t], ω3[t]} /. sol
```

$$\left\{ c_2 \cos\left(\frac{t\sqrt{\theta^2 - 2\theta3\,\theta + \theta3^2}\;(2\theta3\,c_1 - M3\,t)}{2\theta\theta3} \right) + \right.$$

$$\frac{1}{\theta\sqrt{(\theta-\theta3)^2}} \left(\cos\left(\frac{t\sqrt{(\theta-\theta3)^2}\;(2\theta3\,c_1 - M3\,t)}{2\theta\theta3} \right) \int_{K\$306}^{t} \right.$$

$$\left(M2\,(\theta-\theta3)\sin\left(\frac{K\$305\sqrt{(\theta-\theta3)^2}\;(2\theta3\,c_1 - K\$305\,M3)}{2\theta\theta3} \right) - \right.$$

$$M1\sqrt{(\theta-\theta3)^2}$$

$$\left. \cos\left(\frac{K\$305\sqrt{(\theta-\theta3)^2}\;(2\theta3\,c_1 - K\$305\,M3)}{2\theta\theta3} \right) \right)$$

$$d\,K\$305 + \left(\int_{K\$1694}^{t} \left(-M2\,(\theta-\theta3) \right. \right.$$

$$\cos\left(\frac{K\$1693\sqrt{(\theta-\theta3)^2}\;(2\theta3\,c_1 - K\$1693\,M3)}{2\theta\theta3} \right) - $$

$$M1\sqrt{(\theta-\theta3)^2}$$

$$\left. \sin\left(\frac{K\$1693\sqrt{(\theta-\theta3)^2}\;(2\theta3\,c_1 - K\$1693\,M3)}{2\theta\theta3} \right) \right)$$

$$\left. \left. d\,K\$1693 \right) \sin\left(\frac{t\sqrt{(\theta-\theta3)^2}\;(2\theta3\,c_1 - M3\,t)}{2\theta\theta3} \right) \right) + $$

$$\left. c_3 \sin\left(\frac{t\sqrt{\theta^2 - 2\theta3\,\theta + \theta3^2}\;(2\theta3\,c_1 - M3\,t)}{2\theta\theta3} \right) \right\},$$

$$\left(\theta\theta3 \left(\frac{\sqrt{\theta^2 - 2\theta3\,\theta + \theta3^2}\;(2\theta3\,c_1 - M3\,t)}{2\theta\theta3} - \frac{M3\,t\sqrt{\theta^2 - 2\theta3\,\theta + \theta3^2}}{2\theta\theta3} \right) \right.$$

$$c_3$$

$$\cos\left(\frac{t\sqrt{\theta^2 - 2\,\theta 3\,\theta + \theta 3^2}\;(2\,\theta 3\,c_1 - M3\,t)}{2\,\theta\,\theta 3}\right)\Bigg)\Bigg/$$

$$(-c_1\,\theta 3^2 + M3\,t\,\theta 3 + \theta\,c_1\,\theta 3 - M3\,t\,\theta) +$$

$$\frac{1}{\theta\,(\theta - \theta 3)}$$

$$\left(\cos\left(\frac{t\sqrt{(\theta - \theta 3)^2}\;(2\,\theta 3\,c_1 - M3\,t)}{2\,\theta\,\theta 3}\right)\int_{K\$1694}^{t}\left(-M2\,(\theta - \theta 3)\right.\right.$$

$$\cos\left(\frac{K\$1693\,\sqrt{(\theta - \theta 3)^2}\;(2\,\theta 3\,c_1 - K\$1693\,M3)}{2\,\theta\,\theta 3}\right) -$$

$$M1\,\sqrt{(\theta - \theta 3)^2}$$

$$\sin\left(\frac{K\$1693\,\sqrt{(\theta - \theta 3)^2}\;(2\,\theta 3\,c_1 - K\$1693\,M3)}{2\,\theta\,\theta 3}\right)\Bigg)$$

$$d\,K\$1693 - \left(\int_{K\$306}^{t}\right.$$

$$\left(M2\,(\theta - \theta 3)\sin\left(\frac{K\$305\,\sqrt{(\theta - \theta 3)^2}\;(2\,\theta 3\,c_1 - K\$305\,M3)}{2\,\theta\,\theta 3}\right) -\right.$$

$$M1\,\sqrt{(\theta - \theta 3)^2}\,\cos\Bigg($$

$$\left.\frac{K\$305\,\sqrt{(\theta - \theta 3)^2}\;(2\,\theta 3\,c_1 - K\$305\,M3)}{2\,\theta\,\theta 3}\right)\Bigg)\,d\,K\$305\Bigg)$$

$$\sin\left(\frac{t\sqrt{(\theta - \theta 3)^2}\;(2\,\theta 3\,c_1 - M3\,t)}{2\,\theta\,\theta 3}\right)\Bigg) +$$

$$\left(\theta\,\theta 3\left(\frac{M3\,t\,\sqrt{\theta^2 - 2\,\theta 3\,\theta + \theta 3^2}}{2\,\theta\,\theta 3} - \frac{\sqrt{\theta^2 - 2\,\theta 3\,\theta + \theta 3^2}\;(2\,\theta 3\,c_1 - M3\,t)}{2\,\theta\,\theta 3}\right)\right.$$

$$c_2$$

$$\sin\left(\frac{t\sqrt{\theta^2 - 2\,\theta3\,\theta + \theta3^2}\,(2\,\theta3\,c_1 - M3\,t)}{2\,\theta\,\theta3}\right)\bigg)\bigg/$$

$$(-c_1\,\theta3^2 + M3\,t\,\theta3 + \theta\,c_1\,\theta3 - M3\,t\,\theta),\ c_1 - \frac{M3\,t}{\theta3}\bigg\}$$

Contrary to the force-free case, it is clear that the length of the angular velocity is not a constant. However, the value of the length is now determined by the principal values of the inertia tensor, the integration constants, and the acting moments:

$\sqrt{\omega.\omega}$ // Simplify

$$\sqrt{\Bigg(\frac{1}{\theta^2\,(\theta - \theta3)^2\,\theta3^2}\Bigg(\theta^2\,(M3^2\,t^2 - 2\,M3\,\theta3\,c_1\,t + \theta3^2\,(c_1^2 + c_2^2 + c_3^2))\,(\theta - \theta3)^2 +}$$

$$\theta3^2\Bigg(\int_{K\$1694}^{t}\Bigg(-M2\,(\theta - \theta3)\Bigg.$$

$$\cos\left(\frac{K\$1693\,\sqrt{(\theta - \theta3)^2}\,(2\,\theta3\,c_1 - K\$1693\,M3)}{2\,\theta\,\theta3}\right) -$$

$$M1\,\sqrt{(\theta - \theta3)^2}$$

$$\sin\left(\frac{K\$1693\,\sqrt{(\theta - \theta3)^2}\,(2\,\theta3\,c_1 - K\$1693\,M3)}{2\,\theta\,\theta3}\right)\Bigg)$$

$$d\,K\$1693\Bigg)^2 + \theta3^2\Bigg(\int_{K\$306}^{t}\Bigg(M2\,(\theta - \theta3)\Bigg.$$

$$\sin\left(\frac{K\$305\,\sqrt{(\theta - \theta3)^2}\,(2\,\theta3\,c_1 - K\$305\,M3)}{2\,\theta\,\theta3}\right) -$$

$$M1\,\sqrt{(\theta - \theta3)^2}$$

$$\cos\left(\frac{K\$305\,\sqrt{(\theta - \theta3)^2}\,(2\,\theta3\,c_1 - K\$305\,M3)}{2\,\theta\,\theta3}\right)\Bigg)\Bigg)$$

$$d\,\text{K\$305}\Bigg)^2 + 2\,\theta\,\sqrt{(\theta - \theta 3)^2}$$

$$\theta 3^2\, c_3 \int_{\text{K\$1694}}^t \Bigg(-\text{M2}\,(\theta - \theta 3)\cos\Bigg($$

$$\frac{\text{K\$1693}\,\sqrt{(\theta - \theta 3)^2}\,(2\,\theta 3\,c_1 - \text{K\$1693 M3})}{2\,\theta\,\theta 3}\Bigg) -$$

$$\text{M1}\,\sqrt{(\theta - \theta 3)^2}$$

$$\sin\Bigg(\frac{\text{K\$1693}\,\sqrt{(\theta - \theta 3)^2}\,(2\,\theta 3\,c_1 - \text{K\$1693 M3})}{2\,\theta\,\theta 3}\Bigg)\Bigg)$$

$$d\,\text{K\$1693} + 2\,\theta\,\sqrt{(\theta - \theta 3)^2}\,\,\theta 3^2\, c_2 \int_{\text{K\$306}}^t$$

$$\Bigg(\text{M2}\,(\theta - \theta 3)\sin\Bigg(\frac{\text{K\$305}\,\sqrt{(\theta - \theta 3)^2}\,(2\,\theta 3\,c_1 - \text{K\$305 M3})}{2\,\theta\,\theta 3}\Bigg) -$$

$$\text{M1}\,\sqrt{(\theta - \theta 3)^2}\,\cos\Bigg($$

$$\frac{\text{K\$305}\,\sqrt{(\theta - \theta 3)^2}\,(2\,\theta 3\,c_1 - \text{K\$305 M3})}{2\,\theta\,\theta 3}\Bigg)\Bigg)\,d\,\text{K\$305}\Bigg)\Bigg)$$

The corresponding angular momentum is given by the vector

```
L = 0.ω // Simplify
```

$$\Bigg\{\theta\Bigg(c_2 \cos\Bigg(\frac{t\,\sqrt{(\theta - \theta 3)^2}\,(2\,\theta 3\,c_1 - \text{M3}\,t)}{2\,\theta\,\theta 3}\Bigg) +$$

$$c_3 \sin\Bigg(\frac{t\,\sqrt{(\theta - \theta 3)^2}\,(2\,\theta 3\,c_1 - \text{M3}\,t)}{2\,\theta\,\theta 3}\Bigg) +$$

$$\frac{1}{\theta \sqrt{(\theta - \theta 3)^2}} \left(\cos\left(\frac{t \sqrt{(\theta - \theta 3)^2} \ (2 \theta 3 \ c_1 - M3 \ t)}{2 \theta \theta 3} \right) \int_{K\$306}^{t} \right.$$

$$\left(M2 \ (\theta - \theta 3) \sin\left(\frac{K\$305 \sqrt{(\theta - \theta 3)^2} \ (2 \theta 3 \ c_1 - K\$305 \ M3)}{2 \theta \theta 3} \right) - \right.$$

$$M1 \sqrt{(\theta - \theta 3)^2}$$

$$\left. \cos\left(\frac{K\$305 \sqrt{(\theta - \theta 3)^2} \ (2 \theta 3 \ c_1 - K\$305 \ M3)}{2 \theta \theta 3} \right) \right)$$

$$d\,K\$305 + \left(\int_{K\$1694}^{t} \left(-M2 \ (\theta - \theta 3) \right. \right.$$

$$\cos\left(\frac{K\$1693 \sqrt{(\theta - \theta 3)^2} \ (2 \theta 3 \ c_1 - K\$1693 \ M3)}{2 \theta \theta 3} \right) -$$

$$M1 \sqrt{(\theta - \theta 3)^2}$$

$$\sin\left(\frac{K\$1693 \sqrt{(\theta - \theta 3)^2} \ (2 \theta 3 \ c_1 - K\$1693 \ M3)}{2 \theta \theta 3} \right) \right)$$

$$d\,K\$1693 \left. \sin\left(\frac{t \sqrt{(\theta - \theta 3)^2} \ (2 \theta 3 \ c_1 - M3 \ t)}{2 \theta \theta 3} \right) \right) \right),$$

$$\frac{1}{\theta - \theta 3} \left(\cos\left(\frac{t \sqrt{(\theta - \theta 3)^2} \ (2 \theta 3 \ c_1 - M3 \ t)}{2 \theta \theta 3} \right) \right.$$

$$\int_{K\$1694}^{t}$$

$$\left(-M2 \ (\theta - \theta 3) \cos\left(\frac{K\$1693 \sqrt{(\theta - \theta 3)^2} \ (2 \theta 3 \ c_1 - K\$1693 \ M3)}{2 \theta \theta 3} \right) - \right.$$

$$M1 \sqrt{(\theta - \theta 3)^2}$$

$$\sin\left(\frac{K\$1693 \sqrt{(\theta - \theta 3)^2} \ (2 \theta 3 \ c_1 - K\$1693 \ M3)}{2 \theta \theta 3} \right) \right)$$

$$d\,\text{K\$1693} - \left(\left(\int_{\text{K\$306}}^{t}\right.\right.$$

$$\left(\text{M2}\,(\theta - \theta 3)\sin\left(\frac{\text{K\$305}\sqrt{(\theta - \theta 3)^2}\,(2\,\theta 3\,c_1 - \text{K\$305 M3})}{2\,\theta\,\theta 3}\right) - \right.$$

$$\text{M1}\sqrt{(\theta - \theta 3)^2}\,\cos\left(\right.$$

$$\left.\left.\frac{\text{K\$305}\sqrt{(\theta - \theta 3)^2}\,(2\,\theta 3\,c_1 - \text{K\$305 M3})}{2\,\theta\,\theta 3}\right)\right)d\,\text{K\$305}\right)$$

$$\sin\left(\frac{t\sqrt{(\theta - \theta 3)^2}\,(2\,\theta 3\,c_1 - \text{M3}\,t)}{2\,\theta\,\theta 3}\right) + \theta$$

$$\sqrt{(\theta - \theta 3)^2}$$

$$\left(c_3\cos\left(\frac{t\sqrt{(\theta - \theta 3)^2}\,(2\,\theta 3\,c_1 - \text{M3}\,t)}{2\,\theta\,\theta 3}\right) - \right.$$

$$\left.\left.\left.c_2\sin\left(\frac{t\sqrt{(\theta - \theta 3)^2}\,(2\,\theta 3\,c_1 - \text{M3}\,t)}{2\,\theta\,\theta 3}\right)\right)\right), \theta 3\,c_1 - \text{M3}\,t\right\}$$

Examining the angular momentum in the inertial system, we observe that the angular momentum equals the acting moments:

```
Simplify[ω × L + ∂_t L]
```

$$\{-\text{M1}, -\text{M2}, -\text{M3}\}$$

At this time, the angular momentum is not a conserved quantity in the inertial system. Also the kinetic energy no more is conserved.

```
T = 1/2 ω.θ.ω // Simplify
```

$$\frac{1}{2\,\theta\,(\theta-\theta3)^2\,\theta3}\left(\theta\,(M3^2\,t^2 - 2\,M3\,\theta3\,c_1\,t + \theta3\,(\theta3\,c_1^2 + \theta\,(c_2^2 + c_3^2)))\,(\theta-\theta3)^2 + \right.$$

$$\theta3\left(\left(\int_{K\$1694}^{t}\left(-M2\,(\theta-\theta3)\right.\right.\right.$$

$$\cos\left(\frac{K\$1693\,\sqrt{(\theta-\theta3)^2}\,(2\,\theta3\,c_1 - K\$1693\,M3)}{2\,\theta\,\theta3}\right) -$$

$$M1\,\sqrt{(\theta-\theta3)^2}$$

$$\sin\left(\frac{K\$1693\,\sqrt{(\theta-\theta3)^2}\,(2\,\theta3\,c_1 - K\$1693\,M3)}{2\,\theta\,\theta3}\right)\right)$$

$$d\,K\$1693\Bigg)^2 + \theta3$$

$$\left(\left(\int_{K\$306}^{t}\left(M2\,(\theta-\theta3)\sin\left(\frac{K\$305\,\sqrt{(\theta-\theta3)^2}\,(2\,\theta3\,c_1 - K\$305\,M3)}{2\,\theta\,\theta3}\right)\right.\right.\right. -$$

$$M1\,\sqrt{(\theta-\theta3)^2}$$

$$\cos\left(\frac{K\$305\,\sqrt{(\theta-\theta3)^2}\,(2\,\theta3\,c_1 - K\$305\,M3)}{2\,\theta\,\theta3}\right)\right)$$

$$d\,K\$305\Bigg)^2 + 2\,\theta\,\sqrt{(\theta-\theta3)^2}\,\theta3\,c_3\int_{K\$1694}^{t}$$

$$\left(-M2\,(\theta-\theta3)\cos\left(\frac{K\$1693\,\sqrt{(\theta-\theta3)^2}\,(2\,\theta3\,c_1 - K\$1693\,M3)}{2\,\theta\,\theta3}\right) - \right.$$

$$M1\,\sqrt{(\theta-\theta3)^2}$$

$$\sin\left(\frac{K\$1693\,\sqrt{(\theta-\theta3)^2}\,(2\,\theta3\,c_1 - K\$1693\,M3)}{2\,\theta\,\theta3}\right)\right)$$

$$d\,K\$1693 + 2\,\theta\,\sqrt{(\theta-\theta3)^2}\,\theta3\,c_2\int_{K\$306}^{t}$$

$$\left(M2\,(\theta - \theta 3)\sin\left(\frac{K\$305\,\sqrt{(\theta - \theta 3)^2}\,(2\,\theta 3\,c_1 - K\$305\,M3)}{2\,\theta\,\theta 3} \right) - \right.$$

$$M1\,\sqrt{(\theta - \theta 3)^2}$$

$$\left. \cos\left(\frac{K\$305\,\sqrt{(\theta - \theta 3)^2}\,(2\,\theta 3\,c_1 - K\$305\,M3)}{2\,\theta\,\theta 3} \right) \right) d K\$305 \right)$$

In conclusion, our observations are that neither the values of the angular velocity, the angular momentum in the inertial system, nor the kinetic energy are conserved quantities.

2.10.9 Exercises

1. Investigate the motion of a symmetrical top in a gravitational field, one point on th axis of the top being held fixed. Show that the total energy E and the angular momenta p_ϕ and p_ψ about the vertical axis and about the symmetry axis of the top are constants of the motion.

2. Show that none of the principal moments of inertia can exceed the sum of the other two.

3. Calculate the moments of inertia I_1, I_2, and I_3 for a homogenous sphere of radius R and mass m.

4. A door is constructed of a thin homogenous slab of material; it has a width of 1 m. If the door is opened through $90°$ it is found that upon release, it closes itself in 2 s. Assume that the hinges are frictionless and show that the line of hinges must make an angle of approximately $3°$ with the vertical.

2.10.10 Packages and Programs

Euler–Lagrange Package

The Euler–Lagrange package serves to derive the Euler–Lagrange equations from a given Lagrangian:

```
If[$MachineType == "PC",
  $EulerLagrangePath = $TopDirectory <>
    "/AddOns/Applications/EulerLagrange/";
  AppendTo[$Path, $EulerLagrangePath],
  $EulerLagrangePath =
   StringJoin[$HomeDirectory, "/.Mathematica/3.0/
      AddOns/Applications/EulerLagrange", "/"];
  AppendTo[$Path, $EulerLagrangePath]];
```

The next line loads the package.

```
<< EulerLagrange.m
```

==

EulerLagrange™ 1.0 (Dos/Windows®)

© 1992–2003 Dr. Gerd Baumann

Runs with Mathematica® Version 3.0 or later

Licensed to one machine only, copying prohibited

==

```
Options[EulerLagrange]
```

{eXpand → False}

```
SetOptions[EulerLagrange, eXpand → True]
```

{eXpand → True}

Define some notations

```
<< Utilities`Notation`
```

Define the notation of a variational derivative connected with the function EulerLagrange:

```
Notation[ δ/δu_ ∫t₁^t₂ f_ d t_ ⟺ EulerLagrange[f_, u_, t_]]
```

To access the variational derivative, we define an alias variable *var* allowing one to access the symbolic definition by the escape sequence [ESC] *var* [ESC].

```
AddInputAlias[ δ/δ□ ∫t₁^t₂ □ d □, "var"]
```

The following is an example for an arbitrary Lagrangian:

```
δ/δu ∫t₁^t₂ L[u[t], ∂t u[t]] d t
```

```
-∂{t,1} L^(0,1) [u[t], u'[t]] + L^(1,0) [u[t], u'[t]] == 0
```

We also define an EulerLagrange operator allowing us to access the EulerLagrange functon as a symbol

```
Notation[ℰ_u_^x_ [den_] ⟺ EulerLagrange[den_, u_, x_]]
```

The following is the alias notation for the EulerLagrange operator:

```
AddInputAlias[ℰ_□^□ [□], "ELop"]
```

3
Nonlinear Dynamics

3.1 Introduction

In recent years, nonlinear dynamics became an actual topic of research. Nonlinear models are generic of all sciences. The exception in nature are linear models. However, linear models are useful for examining phenomena with a direct response. A principal theme of the preceding chapter has been nonlinear systems of just a few degrees of freedom showing complex behavior. A natural question to ask is, "What happens to this dynamical systems in the limit of infinite degree of freedom?" In this limit, the model become continuous and the discrete variables are replaced by fields. Thus, the description of a system in terms of a finite number of ordinary differential equations (ODEs), with time as the only independent variable, goes over to a partial differential equation (PDE) with both spatial and temporal variables as the independent variables. If only a few nonlinear ODEs can display complex behavior, it might be thought that a continuum of them could only display more complicated behavior. In many cases, this is indeed so, and nonlinear PDEs will display chaos in both time and space. However, there is also an important class of nonlinear

PDEs whose behavior is remarkably regular. This regular dynamic is the subject of this chapter.

Here, we examine a nonlinear field model by means of purely analytic solution procedures. The symbolic approach is supported by numerical calculations which demonstrate the findings of the symbolic calculations. The model discussed is a standard models in nonlinear dynamics. However, the solution procedures are applicable for different model equations belonging to the same class of regular models. The nonlinear field equation we are going to examine is the Korteweg–de Vries (KdV) equation. This equation is a model with many physical and engineering applications. For example, shallow water waves are the original physical system described by Korteweg and deVries in 1895. The derivation of the KdV equation resolved a long dispute on observations made by Russel in 1844 when he follows a solitary wave on horseback along the Union Canal outside Edinburgh. After Korteweg and deVries's work, the problem disappeared and it was not until the early 1960s that the KdV equation reappeared in certain plasma physics problems. A motivation for studying the KdV equation was provided by the work of Fermi, Ulam, and Pasta (FPU) in 1955 (Figure 3.1.1 and 3.1.2).

Figure 3.1.1. Enrico Fermi born September 29, 1901; died November 29, 1954.

Figure 3.1.2.   Stanislaw Marcin Ulam born April 03, 1909; died May 13,1984.

The question of FPU was the energy distribution in a nonlinear coupled chain of oscillators. FPU initially assumed that a certain amount of energy will be continuously distributed in a chain after a certain time. However, numerical experiments on the Los Alamos MANIAC computer demonstrated that this assumption was wrong. The energy periodically cycled through the initially populated modes and there was little energy sharing. A decade later in 1965, Kruskal and Zabusky picked up the FPU contradiction and examined the discrete FPU model in the continuous limit. One result was that the FPU model can be reduced to the KdV equation if an asymptotic solution approach is used. They studied the KdV equation by numerical integration and observed that for certain initial conditions, stable cycling solutions in the chain exists, which they called solitary waves. The numerical results were derived by the development of a remarkable new solution technique by Kruskal and co-workers [3.3], which led to the development of a whole new area of mathematical physics.

To begin, we first investigate some of the more elementary properties of the KdV equation. The chapter is organized as follows. In Section 3.2, we present a procedure to derive nonlinear field models starting from a dispersion relation. Section 3.3 introduces a general procedure to analytically access a nonlinear equation of motion by means of the inverse scattering method. The method is based on the asymptotic behavior of the solution and uses the Marchenko equation to derive the solutions. Section

3.4 is concerned with the conservation laws for the KdV equation. This section presents general procedures applicable also to other nonlinear field equations. Section 3.5 discusses a numerical procedure to solve the KdV equation. The numerical procedure presented is used to simulate the collision of solitons. We demonstrate that the solution procedure has to satisfy certain restrictions to gain reliable numerical results.

3.2 The Korteweg–de Vries Equation

Weak nonlinear waves can be described by an integro-differential equation of the form

$$u_t - u\,u_x + \int_{-\infty}^{\infty} K(x-\xi)\,u_\xi(\xi,\,t)\,d\xi = 0. \tag{3.2.1}$$

The dispersive behavior of the waves is contained in a kernel K. The dispersion relation K is obtained by a Fourier transform of the related phase velocity $c(k) = \omega(k)/k$ by

$$K(x) = \frac{1}{2\pi} \int_{-\infty}^{\infty} c(k)\,e^{-ikx}\,dk, \tag{3.2.2}$$

where $\omega(k)$ is the dispersion relation of the wave. The Korteweg–deVries (KdV) equation was first derived at the end of the 19th century to describe water waves in shallow channels. Experimental data of the dispersion relation in such channels show that the square of the phase velocity is expressed by a hyperbolic relation:

$$c^2(k) = \frac{g}{k}\,\tanh kh, \tag{3.2.3}$$

where h is the mean depth of the channel measured from the undisturbed surface of the water and g is the acceleration of gravity of the Earth. For waves with large wavelengths, we observe that the argument of tanh is small. Thus, we can use a Taylor expansion to approximate the phase velocity by

$$c(k) = \sqrt{\frac{g}{k}\,\tanh k\,h} \approx \sqrt{g\,h}\left(1 - \frac{h^2 k^2}{6} + O(k^4)\right). \tag{3.2.4}$$

As a consequence, the kernel K given in Equation (3.2.2) is represented by an expansion in the form

$$K(x) = \frac{1}{2\pi} \int_{-\infty}^{\infty} \sqrt{g\,h}\,\left(1 - \frac{h^2\,k^2}{6}\right) e^{ikx}\,dk$$

$$= \sqrt{g\,h}\,\left(\delta(x) + \frac{h^2}{6}\,\delta''(x)\right),$$

(3.2.5)

where $\delta(x)$ is the Dirac's delta function and the primes denote derivatives with respect to the argument. If we consider these relations in our original equation of motion (3.2.1), we get

$$u_t - u\,u_x + \sqrt{g\,h}\,\int_{-\infty}^{\infty} \left(\delta(x-\xi) + \frac{h^2}{6}\,\delta''(x-\xi)\right) u_\xi(\xi,\,t)\,d\,\xi$$

$$= u_t - u\,u_x + \sqrt{g\,h}\,\left(u_x + \frac{h^2}{6}\,u_{xxx}\right) = 0$$

(3.2.6)

Transforming Equation (3.2.6) to a moving coordinate system by $X = x + v\,t$ for $v = -\sqrt{g\,h}$, scaling the time t and the wave amplitude u by $\tau = h^2\,v\,t/6$ and $\tilde{u} = u/(h^2\,v)$, respectively, results in a standard representation of the KdV equation:

$$u_t - 6\,u\,u_x + u_{xxx} = 0.$$

(3.2.7)

In Equation (3.2.7), we use the original variables to denote the transformed quantities.

The derivation of the KdV equation can be supported by *Mathematica* by defining the related functions used in the above calculations. First, we introduce a definition of the dispersion relation using Equation (3.2.4):

```
c[k_] := Block[{g, h}, Sqrt[g Tanh[k h] / k]]
```

which reproduces the square root of the tanh:

```
c[k]
```

$$\sqrt{\frac{g\,\tanh(h\,k)}{k}}$$

The linearized dispersion relation necessary for the kernel definition follows by a Taylor expansion with

```
disperse[k_ , n_] :=
 Block[{}, Normal[Series[c[k], {k, 0, n}]]]
```

providing in fourth-order approximation:

```
disperse[k, 4]
```

$$\frac{19}{360} h^4 \sqrt{g\,h}\, k^4 - \frac{1}{6} h^2 \sqrt{g\,h}\, k^2 + \sqrt{g\,h}$$

The dispersion kernel (3.2.5) is defined by the inverse Fourier transform as

```
𝒦[xi_ , n_] :=
 Block[{k, itrafo, dis, t}, dis = disperse[k, n];
  itrafo = Simplify[
    1/√2 Pi InverseFourierTransform[dis, k, t]];
  itrafo = itrafo /. t → x - xi]
```

providing for a second-order approximation of the dispersion relation:

```
𝒦[ξ, 2] // Expand
```

$$\frac{1}{6} \sqrt{g\,h}\, \delta''(x - \xi)\, h^2 + \sqrt{g\,h}\, \delta(x - \xi)$$

The incorporation of the integral in Equation (3.2.6) defines the resulting equation:

```
Equation[n_] :=
 Block[{gl}, gl = Integrate[𝒦[xi, n] D[u[xi, t], xi],
    {xi, -Infinity, Infinity}];
  gl = Simplify[gl];
  gl = D[u[x, t], t] - u[x, t] D[u[x, t], x] + gl]
```

which, on application, gives

KdV = Equation[3]

$u^{(0,1)}(x,\ t) - u(x,\ t)\ u^{(1,0)}(x,\ t) + \dfrac{1}{6}\ \sqrt{g\,h}\ (u^{(3,0)}(x,\ t)\ h^2 + 6\,u^{(1,0)}(x,\ t))$

We can use this function to derive higher-order dispersive equations by increasing the approximation order. The following is an example for $n = 5$:

Equation[5]

$u^{(0,1)}(x,\ t) - u(x,\ t)\ u^{(1,0)}(x,\ t) +$

$\dfrac{1}{360}\ \sqrt{g\,h}\ (19\,u^{(5,0)}(x,\ t)\ h^4 + 60\,u^{(3,0)}(x,\ t)\ h^2 + 360\,u^{(1,0)}(x,\ t))$

Since only the dispersion effects are used in the calculation, we cannot change the nonlinear character of the equation. The nonlinearity in the present form is crucial for the application of the following solution procedure. The standard version of the KdV equation follows by the following transformation:

```
kd = (Simplify[KdV /. u -> Function[{x, t},

        h² √gh u[x - √gh t, h² (√gh/6) t]]] /.

     {x - √gh t -> x, h² (√gh/6) t -> t}) / (gh⁵/6)
```

$u^{(0,1)}(x,\ t) - 6\,u(x,\ t)\ u^{(1,0)}(x,\ t) + u^{(3,0)}(x,\ t)$

Here, we used a transformation with the general form $u = \alpha\,U(x + v\,t,\ \eta\,t)$, where α, v, and η are constants to be determined in such a way that the equation simplifies.

3.3 Solution of the Korteweg–de Vries Equation

In this section, we derive the analytical solutions of the KdV equations using certain initial and boundary conditions. The KdV equation is given by

$$u_t - 6\,u\,u_x + u_{xxx} = 0 \quad \text{with } t > 0 \text{ and } -\infty < x < \infty \tag{3.3.1}$$

and the initial condition $u(x, t = 0) = u_0(x)$. We assume natural boundary conditions; that is, the solution of the KdV equation (3.3.1) is assumed to vanish sufficiently fast at $|x| \to \infty$. To arrive at our solution, we us the inverse scattering theory (IST). This procedure is closely related to its linear counterpart, the Fourier transform (FT). In Section 5.2, we use the Fourier transform technique to construct solutions of the Schrödinger equation. In addition to its methodical connection with IST and FT, both IST and FT are also logically related to the Sturm–Liouville problem. The main difference between IST and FT is that the Fourier transform is only capable of solving linear problems, whereas the IST can also be applied to nonlinear differential equations.

3.3.1 The Inverse Scattering Transform

The solution steps for the inverse scattering transform are summarized as follows (see Figure 3.3.1):

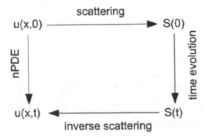

Figure 3.3.1.   Solution procedure of the inverse scattering. Start with a nonlinear PDE. Determine the scattering data from the initial conditions. Carry out a time evolution of the scattering data. Invert the scattering data to the original coordinates.

1. The starting point is a set of nonlinear partial differential equations (nPDEs) for a certain initial condition $u(x, 0)$.

2. By a scattering process, we get the scattering data $S(0)$ at the initial time $t = 0$ from the initial data.

3. Since the characteristic data of the scattering process is related to a linear problem, we can determine the time evolution of the scattering data for the asymptotic behavior $|x| \to \infty$.

4. The inverse scattering process gives us the solution $u(x, t)$. The inverse scattering process is closely related to a linear integro-differential equation, the Marchenko equation, well known in the theory of scattering.

Using these four steps in the solution process, we get a large number of solutions. The most prominent solutions contained in this set are for solitons and multisolitons. We note that the solution process discussed so far is not only applicable to the KdV equation but also delivers solutions for more complicated equations. A collection of equations solvable by IST is given by Calogero and Degasparis [3.1]. Note that the IST procedure is not applicable to all nonlinear initial value problems. There exists, however, a set of equations for which the IST procedure works very well. One of these equations is the KdV equation, which is a completely integrable equation. Other types of nonlinear equation can be solved by Lie's symmetry analysis discussed in the author's book on symmetry analysis of differential equations [7.21].

As mentioned earlier, the starting point of the IST is the initial condition $u(x, 0) = u_0(x)$. In close analogy to the example discussed in the chapter on quantum mechanics (Section 5.5), we examine here a scattering problem with the scattering potential $u(x, 0) = u_0(x)$. To calculate the scattering data $S(0)$, we consider the related Sturm–Liouville problem in the form

$$\psi_{xx} + (\lambda - u_0(x))\,\psi = 0, \qquad -\infty < x < \infty, \qquad (3.3.2)$$

where λ represents the eigenvalue. The time-independent scattering data is derived from the asymptotic behavior of the wave function ψ. Our treatment of Equation (3.3.2) is analogous to our calculations in quantum mechanics. The asymptotic behavior of the wave function is given by

$$\psi(x; k) \sim \begin{cases} e^{-ikx} + b(k)\,e^{ikx} & \text{for} \quad x \to \infty \\ a(k)\,e^{-ikx} & \text{for} \quad x \to -\infty, \end{cases} \qquad (3.3.3)$$

where $\lambda > 0$ and $k = \sqrt{\lambda}$ refer to the case of a continuous spectrum and where

$$\psi_n(x) \sim c_n \, e^{-\kappa_n x} \quad \text{for} \quad x \to \infty \qquad n = 1, 2, \ldots, N \tag{3.3.4}$$

for $\lambda < 0$ and $\kappa_n = \sqrt{-\lambda}$ refers to the case of discrete eigenvalues. The characteristic data of the scattering process is the set of reflection and transmission indices $b(k)$ and $a(k)$ and the normalization constant c_n. This set of data is called the scattering data $S(0)$ and is collected in a list $S(0) = \{a(k), b(k), c_n\}$. The listed data support the theory. The measurable quantities in a scattering process are the reflection and transmission coefficients $b(k)$ and $a(k)$. The question from the experimental point of view is how the measurable quantities can be used to derive the interaction potential. Theoretically, the answer is given by Marchenko [3.2]. He demonstrated that knowledge of the scattering data and eigenvalues of the Sturm–Liouville problem are sufficient to reconstruct the potential of the scattering process by a linear integral equation of the form

$$K(x, z) + M(x + z) + \int_z^\infty K(x, y)\, M(y + z)\, dy = 0, \tag{3.3.5}$$

where M is defined by the scattering data as

$$M(x) = \sum_{n=1}^N c_n^2 \, e^{-\kappa_n x} + \frac{1}{2\pi} \int_{-\infty}^\infty b(k)\, e^{ikx}\, dk. \tag{3.3.6}$$

The solution $K(x, z)$ of the integral equation (3.3.5) delivers the representation of the potential $u_0(x)$:

$$-2\, \frac{d}{dx}\, K(x, x) = u_0(x). \tag{3.3.7}$$

Knowing the scattering data, we are able to reconstruct the potential $u_0(x)$ by means of the Marchenko equation (3.3.5).

Another aspect of solving the KdV equation is how time influences the scattering. Up to now, we have only considered the stationary characteristics of the scattering process. We now consider not only the initial condition $u = u(x, t = 0)$ in the scattering process but also the full time-dependent behavior of the solution $u(x, t)$. We assume that the time-dependent potential $u(x, t)$ in the Sturm–Liouville problem satisfies the natural boundary conditions requiring that for $|x| \to \infty$, the solution vanishes sufficiently fast. In all of the expressions, the time variable t is considered as a parameter. Because of the parametric dependency of the Sturm–Liouville problem on t, we expect that all spectral data also depend on t. We assume the eigenvalues $\lambda = \lambda(t)$ to include a time dependence in the Sturm–Liouville problem which, in this case, reads

$$\psi_{xxx} + (\lambda(t) - u(x; t)) \psi = 0, \tag{3.3.8}$$

where $u(x, t)$ satisfies the KdV equation (3.3.8). Differentiation of Equation (3.3.8) with respect to x as well as with respect to t gives us

$$\psi_{xxx} - u_x \psi + (\lambda - u) \psi_x = 0, \tag{3.3.9}$$
$$\psi_{xxt} + (\lambda_t - u_t) \psi + (\lambda - u) \psi_t = 0. \tag{3.3.10}$$

By introducing the expression

$$R(x, t) = \psi_t + u_x \psi - 2 (u - 2 \lambda) \psi_x, \tag{3.3.11}$$

we find that the current $\psi_x R - \psi R_x$ satisfies the relation

$$\frac{\partial}{\partial x} (\psi_x R - \psi R_x) = \lambda_t \psi^2, \tag{3.3.12}$$

which connects the time derivative of the eigenvalues λ to the gradient of the current. To derive this relation, we have used Equations (3.3.9) and (3.3.10) as well as the KdV equation (3.3.1) itself.

If the eigenvalues λ of the Sturm–Liouville problem are discrete $\left(\kappa_n = \sqrt{-\lambda}\right)$, an integration of Equation (3.3.12) with respect to x yields

$$0 = \psi_x R - \psi R_x \Big|_{-\infty}^{\infty} = \lambda_t \int_{-\infty}^{\infty} \psi^2 \, dx. \tag{3.3.13}$$

Since the wave function ψ and its derivatives vanish for $|x| \to \infty$, the left-hand side of Equation (3.3.13) is gone. Normalizing ψ by $\int_{-\infty}^{\infty} \psi^2 \, dx = 1$ results in

$$\frac{d \kappa_n^2}{dt} = 0 \quad \text{or} \quad \kappa_n = \text{const.} \tag{3.3.14}$$

We therefore have an isospectral problem. We now can use Equation (3.3.11) to determine directly the normalization constants c_n. On the other hand, u and ψ vanish for $x \to \infty$. Using the asymptotic representation of the eigenfunctions ψ, we find, with the help of

$$\psi_n(x; t) \sim c_n(t) e^{-\kappa_n x} \tag{3.3.15}$$

and the asymptotic form (3.3.11)

$$\frac{d c_n}{dt} - 4 \kappa_n^3 c_n = 0. \tag{3.3.16}$$

Integrating this expression gives

$$c_n(t) = c_n(0)\, e^{4\kappa_n^3 t}, \qquad n = 1, 2, \ldots, N, \tag{3.3.17}$$

where $c_n(0)$ are the normalization constants of the time-independent Sturm–Liouville problem. Following these steps, we see how the discrete part of the spectral data follows from the time-independent eigenvalue problem.

The continuous part of the spectral data is derived by an analogous procedure. The integration of relation (3.3.12) with respect to x produces the continuous part of the eigenvalues:

$$\psi_x R - \psi R_x = g(t; k). \tag{3.3.18}$$

The asymptotic representation of the eigenfunctions is now

$$\begin{aligned}
\psi(x; t, k) &\sim a(k; t)\, e^{-ikx} && \text{for} \quad x \to -\infty \\
\psi(x; t, k) &\sim e^{-ikx} + b(k; t)\, e^{ikx} && \text{for} \quad x \to \infty.
\end{aligned} \tag{3.3.19}$$

In the limiting case of $x \to \infty$, we find by using Equation (3.3.11)

$$R(x; t, k) \sim \left(\frac{da}{dt} + 4\, i\, k^3\, a \right) e^{-ikx} \tag{3.3.20}$$

and thus we obtain

$$\psi_x R - \psi R_x \to 0 \qquad \text{for} \qquad x \to -\infty. \tag{3.3.21}$$

This relation allows a further integration, which results in

$$R = h(t; k)\, \psi. \tag{3.3.22}$$

Using Equation (3.3.22) we get the expression

$$\frac{da}{dt} + 4\, i\, k^3\, a = h\, a. \tag{3.3.23}$$

The corresponding relations for $x \to \infty$ are expressed by

$$\frac{db}{dt}\, e^{ikx} + 4\, i\, k^3 (e^{-ikx} - b\, e^{ikx}) = h\, (e^{-ikx} + b\, e^{ikx}). \tag{3.3.24}$$

Since the trigonometric functions are linearly independent functions, we can write

$$\frac{db}{dt} - 4\, i\, k^3\, b = h\, b, \tag{3.3.25}$$

$$h = 4\, i\, k^3. \tag{3.3.26}$$

Equation (3.3.23) is thus reducible to

$$\frac{da}{dt} = 0. \tag{3.3.27}$$

A simultaneous integration of Equations (3.3.27) and (3.3.25) gives

$$a(k; t) = a(k; 0), \qquad\qquad\qquad\qquad (3.3.28)$$
$$b(k; t) = b(k; 0)\, e^{8\,i\,k^3\,t}. \qquad\qquad\qquad (3.3.29)$$

For times $t > 0$, we obtain a time-dependent reflection index $b(k; t)$ and a constant transmission rate $a(k; t)$.

The complete set of scattering data (discrete plus continuous data) for the time-dependent scattering problem of the KdV equation is summarized as follows:

$$S(t) = \{c_n(t) = c_n(0)\, e^{4\,\kappa_n^3\,t},\ a(k; 0),\ b(k; t) = b(k; 0)\, e^{8\,i\,k^3\,t}\}. \qquad (3.3.30)$$

The assumption of a time-dependent potential is reflected in the scattering data through both the time dependent normalization constants c_n in the discrete spectrum and the time-dependent reflection coefficients b in the continuous spectrum.

To complete the solution process of the inverse scattering transform, we need to take into account the time-dependence of the scattering data in Marchenko's integral equation. Since time appears only as a parameter in the relations of the scattering data, we can use the expression from the stationary part of the scattering process and extend it to obtain the equations of the time-dependent scattering. The time-dependent potential and the solution of the KdV equation follow from the time-dependent Marchenko equation. The spectral characteristics are contained in the M term. If we generalize relation (3.3.6) for the time-dependent case of spectral data, we get

$$M(x; t) = \sum_{n=1}^{N} c_n(0)^2\, e^{8\,\kappa_n^3\,t} + \frac{1}{2\pi} \int_{-\infty}^{\infty} b(k; 0)\, e^{i\,(8\,k^3\,t - k\,x)}\, d\,k. \quad (3.3.31)$$

The original Marchenko equation then transforms to

$$K(x, z; t) + M(x + z; t) + \int_{x}^{\infty} K(x, y; t)\, M(y + z; t)\, dy = 0. \qquad (3.3.32)$$

The solution of the KdV equation follows from

$$u(x, t) = -2\, \frac{\partial}{\partial x}\, K(x, x; t). \qquad\qquad\qquad (3.3.33)$$

In principle, Equation (3.3.33) gives the solution for the KdV equation provided the spectral data are known. However, deriving the spectral data is not simple, even for the KdV equation. Calculating the general solution of the Marchenko equation is a second problem in the solution process. This situation is similar to the Fourier technique, for which the inverse transformation is, at times, unrecoverable. Given a spectral density $A(k)$, it is sometimes impossible to analytically invert the representation from Fourier space into real space. However, since our main problem is the application of the IST, we show in the following subsection that the IST can be successfully applied to the solution of the KdV equation.

3.3.2 Soliton Solutions of the Korteweg–de Vries Equation

In the previous subsection, we saw how nonlinear initial value problems can be solved using the inverse scattering method. In this subsection, we construct the solution for a specific problem. As an initial condition, we choose the potential in the Sturm–Liouville problem to be $u_0(x) = -V_0 \operatorname{sech}^2 x$. This famous potential was used by Pöschel and Teller for an anharmonic oscillator. We will discuss this type of potential in Section 5.5 when examining the quantum mechanical Pöschel–Teller problem. We observe there that the reflection index $b(k)$ vanishes if the amplitude of the potential is given by $V_0 = N(N+1)$, with N an integer. In our discussion of solutions for the KdV equation, we restrict our considerations to this case.

We assume that $N = 1$. The initial condition is thus reduced to $u_0(x) = -2 \operatorname{sech}^2 x$. The related Sturm–Liouville problem (3.3.2) for this specific case reads

$$\psi_{xx} + (\lambda - 2 \operatorname{sech}^2 x)\,\psi = 0. \tag{3.3.34}$$

Equation (3.3.34) is identical to Equation (5.5.57) of Chapter 5 with $V_0 = 2$. We will demonstrate in the quantum mechanical treatment of the problem that in this case, the corresponding eigenfunctions are given by the associated Legendre polynomials $P_1^1(x) = \operatorname{sech}(x)/\sqrt{2}$. The corresponding eigenvalue is $\kappa_1 = 1$. The normalization constant follows from the normalization condition $\int_{-\infty}^{\infty} \psi^2\, dx = 1$. According to our considerations in the previous subsection, we can immediately write down the time evolution of the normalization constant c_1 as

$$c_1(t) = \sqrt{2}\, e^{4t}. \tag{3.3.35}$$

Since we are dealing with a reflectionless potential ($b(k) = 0$), we can write the M term of the Marchenko equation as

$$M(x; t) = 2\, e^{8\,t-x}.$$ (3.3.36)

The Marchenko equation itself reads

$$K(x, z; t) + 2\, e^{8\,t-(x+z)} + 2 \int_x^\infty K(x, y; t)\, e^{8\,t-(y+z)}\, dy = 0.$$ (3.3.37)

Solutions of Equation (3.3.37) are derivable by a separation ansatz for the function K in the form $K(x, z; t) = K(x; t)\, e^{-z}$. Substituting this expression into Equation (3.3.37) gives us the relation

$$K(x; t) + 2\, e^{8\,t-x} + 2\, K(x; t) \int_x^\infty e^{8\,t-2\,y}\, dy = 0.$$ (3.3.38)

We have thus reduced an integral equation to an algebraic relation for K. The solution of Equation (3.3.38) is given by

$$K(x; t) = -\frac{2\, e^{8\,t-x}}{1+e^{8\,t-2\,x}}.$$ (3.3.39)

The unknown $K(x, z; t)$ is thus represented by

$$K(x, z; t) = -\frac{2\, e^{8\,t-x}}{1+e^{8\,t-2\,x}}\, e^{-z}.$$ (3.3.40)

In fact, the solution of the KdV can be obtained using Equation (3.3.32) to derive the time-dependent potential $u(x, t)$ from K:

$$u(x, t) = 2\, \frac{\partial}{\partial x}\left(\frac{2\, e^{8\,t-2\,x}}{1+e^{8\,t-2\,x}}\right) = -2\, \mathrm{sech}^2(x - 4\,t).$$ (3.3.41)

This type of solution is known as the soliton solution of the KdV. It was first derived at the end of the 19th century by Korteweg and de Vries. The solution itself describes a wave with constant shape and constant propagation velocity $v = 4$ moving to the right. By choosing the amplitude, we derive one solution out of an infinite set of solutions for the KdV equation. In the following, we discuss more complicated cases where two and more eigenvalues have to be taken into account for the calculation.

To demonstrate how IST can be applied to more complicated situations, consider the case with an initial condition $u_0(x) = -6\, \mathrm{sech}^2 x$. The difference between this case and the case discussed earlier appears to be minor. However, as we will see, the difference in the solutions is significant. The selected initial condition corresponds to a Pöschel–Teller potential with a depth of $N = 2$. The discussion of the eigenvalue problem in Section 5.5 shows that the eigenvalues are given by $\kappa_1 = 1$ and $\kappa_2 = 2$. The corresponding eigenfunctions are

$$\psi_2^1 = \sqrt{\tfrac{3}{2}} \tanh x \operatorname{sech} x \tag{3.3.42}$$

$$\psi_2^2 = \tfrac{\sqrt{3}}{2} \operatorname{sech}^2 x. \tag{3.3.43}$$

The normalization constants c_1 and c_2 for this case are given by

$$c_1 = \sqrt{6} \qquad \text{and} \qquad c_2 = 2\sqrt{3}. \tag{3.3.44}$$

The time evolution of c is determined by

$$c_1(t) = \sqrt{6}\, e^{4t}, \tag{3.3.45}$$

$$c_2(t) = 2\sqrt{3}\, e^{32t}. \tag{3.3.46}$$

In close analogy to $N = 1$, we get the M terms of the Marchenko equation by using relation (3.3.31) in the form

$$M(x; t) = 6\, e^{8t-x} + 12\, e^{64t-2x}. \tag{3.3.47}$$

The Marchenko equation itself is given by

$$K(x, z; t) + 6\, e^{8t-(x+z)} + 12\, e^{64t-2(x+z)} +$$
$$\int_x^\infty K(x, y; t)\, (6\, e^{8t-(y+z)} + 12\, e^{64t-2(y+z)})\, dy = 0. \tag{3.3.48}$$

We obtain the solution of Equation (3.3.48) in the form

$$K(x, z; t) = K_1(x; t)\, e^{-z} + K_2(x; t)\, e^{-2z} \tag{3.3.49}$$

by again using a separation ansatz for K. In the general case of N eigenvalues, we can use the ansatz

$$K(x, z; t) = \sum_{n=1}^{N} K_n(x; t)\, e^{-nz} \tag{3.3.50}$$

to reduce the integral equation to an algebraic relation. Since e^{-z} and e^{-2z} are linearly independent functions, we can derive from Equation (3.3.48) the following system of equations:

$$K_1 + 6\, e^{8t-x} + 6\, e^{8t}\, (K_1 \int_x^\infty e^{-2y}\, dy + K_2 \int_x^\infty e^{-3y}\, dy) = 0, \tag{3.3.51}$$

$$K_2 + 12\, e^{64t-2x} +$$
$$12\, e^{64t}\, (K_1 \int_x^\infty e^{-3y}\, dy + K_2 \int_x^\infty e^{-4y}\, dy) = 0. \tag{3.3.52}$$

Integrating Equations (3.3.51) and (3.3.52), we get a linear system of equations for the unknowns K_i:

$$\begin{pmatrix} 1 + 3\, e^{8t-2x} & 2\, e^{8t-3x} \\ 4\, e^{64t-3x} & 1 + 3\, e^{64t-4x} \end{pmatrix} \begin{pmatrix} K_1 \\ K_2 \end{pmatrix} = \begin{pmatrix} -6\, e^{8t-x} \\ -12\, e^{64t-2x} \end{pmatrix}. \tag{3.3.53}$$

For cases with $N > 2$, we get a general system of equations:

$$A.K = B,$$ (3.3.54)

where

$$A_{n,m} = \delta_{n,m} + \frac{c_m^2(0)}{m+n} e^{8\,m^3\,t - (m+n)\,x}$$ (3.3.55)

and

$$B_n = -c_n^2(0)\, e^{8\,n^3\,t - n\,x}.$$ (3.3.56)

The final solution reads

$$u(x,\,t) = -2\,\frac{\partial^2}{\partial x^2}\,\log\big|A\big|.$$ (3.3.57)

Equation (3.3.57) is the general representation of the solution for the KdV equation. For the specific case with $N = 2$, we get

$$K_1(x;\,t) = \frac{6\,(e^{72\,t - 5\,x} - e^{8\,t - x})}{D(x,t)},$$ (3.3.58)

$$K_2(x;\,t) = -\frac{12\,(e^{64\,t - 2\,x} - e^{72\,t - 4\,x})}{D(x,t)}.$$ (3.3.59)

The determinant $D(x,\,t) = \det A = |A|$ of Equation (3.3.53) is

$$D(x,\,t) = 1 + 3\,e^{8\,t - 2\,x} + 3\,e^{64\,t - 4\,x} + e^{72\,t - 6\,x}.$$ (3.3.60)

The solution of the KdV equation then reads

$$u(x,\,t) = -2\,\frac{\partial}{\partial x}\,(K_1\,e^{-x} + K_2\,e^{-2\,x})$$
$$= -12\,\frac{3 + 4\cosh(2\,x - 8\,t) + \cosh(4\,x - 64\,t)}{(3\cosh(x - 28\,t) + \cosh(3\,x - 36\,t))^2}.$$ (3.3.61)

This type of solution is called a bisoliton solution in the theory of inverse scattering. To make the term soliton more understandable, we examine the behavior of solution (3.3.61) in a certain time domain. Since the KdV equation is invariant with respect to a Galilean transformation, we can use $t < 0$ in our calculations. A sequence of time steps illustrating Equation (3.3.61) is presented in Figure 3.3.2-3.3.4. In order to give the impression of a wave packet, we have plotted the negative amplitude of the solution u in this figure. Initially, there are two separated peaks. As time passes, the two humps overlap and form a single peak at time $t = 0$, which represents the initial solution $u_0(x) = -6\,\mathrm{sech}^2 x$. For times $t > 0$, we observe that the single peak located at $x = 0$ splits into two peaks with differing amplitudes. We observe that wave packets with larger amplitudes split from those with smaller amplitudes. Larger wave packets travel faster than smaller ones. If we compare the soliton movement before and after the collision of pulses, we observe during the scattering process that neither the shapes nor the velocities of the pulses change. The term soliton originates from its insensitivity to any variance in the scattering process. This phenomenon was first observed by Zabusky and Kruskal [3.5]. Another characteristic of solitons is that larger pulses travel faster whereas smaller pulses move more slowly. This means that larger pulses will overtake smaller ones during the evolution of motion. We can understand this evolution by examining the propagation velocity with respect to the amplitude of the solitons.

From Figure 3.3.2, we note that for times $|t| \to \infty$ the shape of the solitons remains stable. As already mentioned, the shape of the pulses is recovered in a scattering process. However, the phase of the pulses does not stay continuous. It smoothly changes at the interaction of the solitons. A two-soliton scattering is pictured in Figure 3.3.3, created with **ContourPlot[]**. We observe in this plot that smaller packets retard whereas larger ones advance.

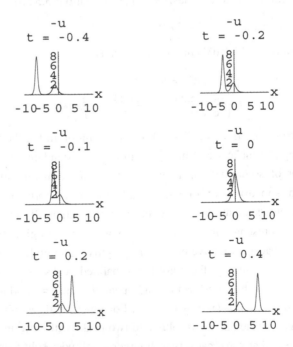

Figure 3.3.2.   Soliton solution of the KdV equation. The initial condition is $u(x, 0) = -6 \, \text{sech}^2 x$.

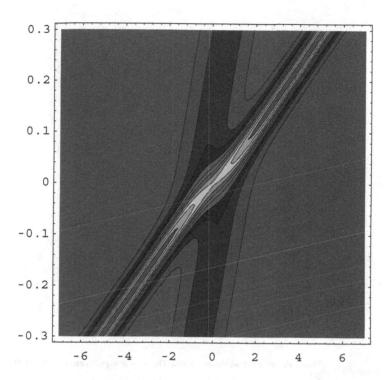

Figure 3.3.3.   Contour plot of the bisoliton solution. The space coordinate x is plotted horizontally and time t is plotted vertically. We can clearly detect the discontinuity of the phase in the contour plot at $t=0$. The gap occurs in the spatial direction.

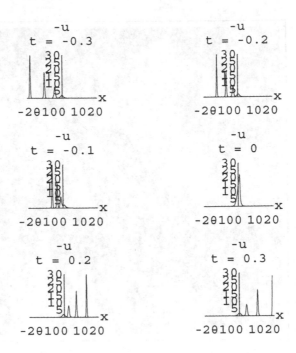

Figure 3.3.4. Time series for a quartic soliton solution. The given time points are $t = -0.5, 0.00001$, and 0.3.

The *Mathematica* functions needed to create the figures for the soliton movement are collected in the package **KDVAnalytic`**. The function needed to plot the solitons is **Soliton[]** and a graphical representation of an N-soliton solution is obtained by using the function **PlotKDV[]**. An example of a quartic soliton solution is given in Figure 3.3.4, created by calling **PlotKdV[-0.5,0.5,0.02,4]**. The four pictures created in the time domain ranging from $t = -0.5$ up to $t = 0.5$ in steps of $\Delta t = 0.02$ are collected in one picture by using **Show[]** in connection with **GraphicsArray[]**.

To demonstrate the application of functions from **KDVAnalytic`**, we first calculate a one-soliton solution by

```
Soliton[x, t, 1]
```

$$-\frac{8\,e^{8t+2x}}{(e^{8t}+e^{2x})^2}$$

Next, we generate a flip chart movie for a three-soliton collision by

```
PlotKdV[-1, 1, 0.1, 3]
```

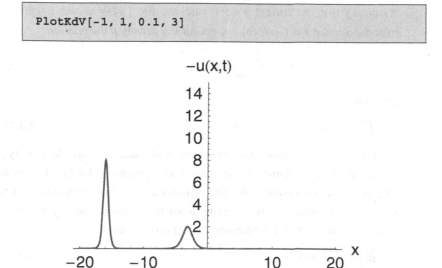

3.4 Conservation Laws of the Korteweg–de Vries Equation

Conservation laws such as the conservation of energy are central quantities in physics. The conservation of angular momentum is equally important to quantum mechanics as it is to classical mechanics. Conservation laws imply the existence of invariant quantities (e.g., when applied to the scattering of molecules). The Boltzmann equation is an example, as the particle density remains constant, since particles are neither created nor destroyed.

3.4.1 Definition of Conservation Laws

Denoting the macroscopic particle density with $\rho(x, t)$ and the streaming velocity with $v(x, t)$, we can express the conservation law in the differential form of a continuity equation:

$$\partial_t \rho(x, t) + \partial_x(\rho v) = 0. \tag{3.4.1}$$

Assuming that the current $j = \rho v$ vanishes for $|x| \to \infty$ and integrating over the domain $x \in (-\infty, \infty)$, we get for the density ρ the relation

$$\frac{d}{dt}\left(\int_{-\infty}^{\infty} \rho\, dx\right) = -\rho v \Big|_{-\infty}^{\infty} = 0, \tag{3.4.2}$$

and thus

$$\int_{-\infty}^{\infty} \rho\, dx = \text{const.} \tag{3.4.3}$$

Equation (3.4.3) expresses the conservation of mass although the density ρ follows the time evolution in accordance with Equation (3.4.1). The simple idea of mass conservation in fluid dynamics can also be transformed to more general situations. If we write down for a general density T and its corresponding current J a continuity equation such as

$$\partial_t T + \partial_x J = 0, \tag{3.4.4}$$

we find the related conservation law. To extend the formulation of the general continuity equation to nonlinear partial differential equations, we assume that T and J depend on t, x, u, u_x, u_{xx}, and so forth, but not on u_t. If we retain the assumption that $J(x \to \pm\infty) \to 0$, then Equation (3.4.4) can be integrated over all space as was done for Equation (3.4.1), getting

$$\frac{d}{dt}\int_{-\infty}^{\infty} T\, dx = 0 \tag{3.4.5}$$

or

$$\int_{-\infty}^{\infty} T\, dx = \text{const.} \tag{3.4.6}$$

The quantity defined by Equation (3.4.6) is an integral of motion in the theory of nonlinear PDEs.

As an example, we consider the KdV equation

$$u_t - 6u\, u_x + u_{xxx} = 0. \tag{3.4.7}$$

The KdV equation already takes the form of a continuity equation. $T_1 = u$ is the density and $J = u_{xx} - 3\,u^2$ is the current. If the density T is integrable and $\partial_x J$ vanishes at the points $x = \pm\infty$, we can write

$$\int_{-\infty}^{\infty} u(x,\,t)\,dx = \text{const.} \tag{3.4.8}$$

Equation (3.4.8) must be satisfied for all solutions of the KdV equation satisfying the conditions listed earlier. However, not all solutions of the KdV equation satisfy the asymptotic relations. For example, the conservation laws do not apply to periodic solutions of the KdV equation.

Another conserved quantity can be obtained if Equation (3.4.7) is multiplied by u. In this case,

$$\partial_t(\tfrac{1}{2}\,u^2) + \partial_x(u\,u_{xx} - \tfrac{1}{2}\,u_x^2 - 2\,u^3) = 0. \tag{3.4.9}$$

The second conserved quantity is given by $T_2 = u^2$, which directly integrates into

$$\int_{-\infty}^{\infty} u^2\,dx = \text{const.} \tag{3.4.10}$$

This notation holds for solutions vanishing sufficiently rapidly at $|x| \to \infty$. The physical interpretation of these equations is that relation (3.4.8) represents conservation of mass and that Equation (3.4.10) represents conservation of momentum (compare also Section 3.2). We have thus derived two conserved quantities by simple manipulations of the KdV equation. The question now is whether we can derive other conserved quantities from the KdV and how these quantities are related to each other. This question was first discussed by Miura et al. [3.3]. They observed that there are a large number of conserved quantities for the KdV equation. They discovered that, in fact, there exists an infinite number of conserved quantities for the KdV equation. For example,

$$T_3 = u^3 + \tfrac{1}{2}\,u_x^2, \tag{3.4.11}$$
$$T_4 = 5\,u^4 + 10\,u\,u_x^2 + u_{xx}^2. \tag{3.4.12}$$

T_3 can be identified as the energy density. The higher densities T_n for $n > 3$ have no physical interpretation in terms of energy, momentum and so forth. Other conserved quantities are obtained algorithmically. In the following, we show how Miura et al. constructed the infinite hierarchy of constants of motion.

3.4.2 Derivation of Conservation Laws

Miura et al. [3.3] made an important step in understanding the phenomenon of invariants in nonlinear PDEs. The tool they invented is a transformation vehicle which linearizes the nonlinear PDE. Today, this tool is known as the Miura transformation of the KdV equation to the modified KdV equation (mKdV):

$$v_t - 6 v^2 v_x + v_{xxx} = 0. \tag{3.4.13}$$

By transforming the field v to the field u according to

$$u(x, t) = v^2(x, t) + v_x(x, t), \tag{3.4.14}$$

solutions of Equation (3.4.13) are also solutions of the KdV equation. The Miura transformation $v = \psi(x, t)/\psi_x(x, t)$ connects the KdV equation with its related Sturm–Liouville problem. The Miura transformation (3.4.14) is primarily used for the construction of conservation laws. If, for example, we replace field v in Equation (3.4.14) by

$$v = \frac{1}{2\varepsilon} + \varepsilon w, \tag{3.4.15}$$

where ε is an arbitrary parameter, we get the Miura transformation for w in the form

$$u = \frac{1}{4\varepsilon^2} + w + \varepsilon^2 w^2 + \varepsilon w_x. \tag{3.4.16}$$

If we additionally assume the Galilean invariance for u to be ($\tilde{u} = u + \lambda$), we can simplify relation (3.4.16) to

$$u = w + \varepsilon w_x + \varepsilon^2 w^2. \tag{3.4.17}$$

This transformation connecting w with u is called a Gardner transformation. Substituting the transformation (3.4.17) into the KdV equation (3.4.7) gives us

$$\begin{aligned}
u_t - 6\, u\, u_x + u_{xxx} = {}& \\
w_t + \varepsilon\, w_{xt} + 2\,\varepsilon^2\, w\, w_t - {}& \\
6\,(w + \varepsilon\, w_x + \varepsilon^2\, w^2)\,(w_x + \varepsilon\, w_{xx} + 2\,\varepsilon^2\, w\, w_x) + {}& \\
w_{xxx} + \varepsilon\, w_{xxxx} + 2\,\varepsilon^2\,(w\, w_x)_{xx} = {}& \\
(1 + \varepsilon\, \tfrac{\partial}{\partial x} + 2\,\varepsilon^2\, w)\,(w_t - 6\,(w + \varepsilon^2\, w^2)\, w_x + w_{xxx}). &
\end{aligned} \tag{3.4.18}$$

As is the case for the Miura transformation, u is a solution of the KdV equation and thus w is also a solution of the KdV equation:

$$w_t - 6 (w + \varepsilon^2 w^2) w_x + w_{xxx} = 0. \tag{3.4.19}$$

If we set the parameter to be $\varepsilon = 0$, Equation (3.4.19) reduces to the KdV equation. For this case, the Gardner transformation yields the identity transformation $u = w$. The Gardner transformation is closely related to a continuity equation of the form

$$\partial_t w + \partial_x (w_{xx} - 3 w^2 - 2 \varepsilon^2 w^3) = 0. \tag{3.4.20}$$

Thus, we get

$$\int_{-\infty}^{\infty} w \, dx = \text{const.} \tag{3.4.21}$$

(i.e., another conserved quantity). To construct the conservation laws of the KdV equation by an algorithm, we use the parameter ε. The important aspect of this operation is that for $\varepsilon \to 0$, w converges to u. For this reason, we expand field w as a power series in ε:

$$w(x, t; \varepsilon) = \sum_{n=0}^{\infty} \varepsilon^n w_n(x, t). \tag{3.4.22}$$

From Equation (3.4.21) it follows

$$\int_{-\infty}^{\infty} w \, dx = \sum_{n=0}^{\infty} \varepsilon^n \int_{-\infty}^{\infty} w_n(x, t) \, dx = \text{const.}, \tag{3.4.23}$$

or

$$\int_{-\infty}^{\infty} w_n \, dx = \text{const.} \quad \text{for} \quad n = 0, 1, 2, \dots. \tag{3.4.24}$$

The expansion of the Gardner transformation (3.4.17) yields

$$\sum_{n=0}^{\infty} \varepsilon^n w_n = u - \varepsilon \sum_{n=0}^{\infty} \varepsilon^n w_{nx} - \varepsilon^2 \left(\sum_{n=0}^{\infty} \varepsilon^n w_n\right)^2. \tag{3.4.25}$$

The conserved quantities resulting from the first terms of this expansion are

$$w_0 = u, \tag{3.4.26}$$
$$w_1 = -w_{0x} = -u_x, \tag{3.4.27}$$
$$w_2 = -w_{1x} - w_0^2 = u_{xx} - u^2, \tag{3.4.28}$$
$$w_3 = -w_{2x} - 2 w_0 w_1 = -(u_{xx} - u^2)_x + 2 u u_x. \tag{3.4.29}$$

The quantities w_1 and w_3 are given by total differentials and thus provide new information on the conservation laws.

Since the construction of the invariants of motion follows from a completely algorithmic procedure, *Mathematica* can be used to derive the higher densities of the conservation laws. Indeed, a calculation by hand immediately shows us that a manual approach is very cumbersome. However, *Mathematica* can do all the calculations for us.

The algorithm to derive the conserved densities starts out from a power series expansion of the field w. The comparison of equal powers of ε in Equation (3.4.25) gives us the expressions for the w_n's. If we replace the w_n's by the w_{n-1}'s, we get a representation of function u. The steps used to carry out the calculation are summarized in the package **KdVIntegrals`**. The **Gardner[]** function activates our calculation of conserved quantities. Given an integer as an argument, **Gardner[]** creates the first n conserved densities. These densities are collected in a list. Applying **Integrate[]** to the result of **Gardner[]**, all even densities result in an integral of motion. Results of a calculation with $n = 6$ are as follows:

```
g6=Gardner[u,x,t,5]
```

$$\{u(x, t), -u^{(1,0)}(x, t), u^{(2,0)}(x, t) - u(x, t)^2, 4u(x, t)u^{(1,0)}(x, t) - u^{(3,0)}(x, t),$$
$$-5u^{(1,0)}(x, t)^2 - 4u(x, t)u^{(2,0)}(x, t) - 2u(x, t)(u^{(2,0)}(x, t) - u(x, t)^2) +$$
$$u^{(4,0)}(x, t), 14u^{(1,0)}(x, t)u^{(2,0)}(x, t) + 4u^{(1,0)}(x, t)(u^{(2,0)}(x, t) - u(x, t)^2) -$$
$$2u(x, t)(4u(x, t)u^{(1,0)}(x, t) - u^{(3,0)}(x, t)) + 4u(x, t)u^{(3,0)}(x, t) +$$
$$2u(x, t)(u^{(3,0)}(x, t) - 2u(x, t)u^{(1,0)}(x, t)) - u^{(5,0)}(x, t)\}$$

After integrating the list, we obtain

```
Integrate[g6, x]
```

$$\left\{\int u(x, t)\,dx, -u(x, t), \int(u^{(2,0)}(x, t) - u(x, t)^2)\,dx,\right.$$
$$2u(x, t)^2 - u^{(2,0)}(x, t), \int\left(-5u^{(1,0)}(x, t)^2 - 4u(x, t)u^{(2,0)}(x, t) -\right.$$
$$2u(x, t)(u^{(2,0)}(x, t) - u(x, t)^2) + u^{(4,0)}(x, t)\right)dx,$$
$$\left.-\frac{16}{3}u(x, t)^3 + 8u^{(2,0)}(x, t)u(x, t) + 5u^{(1,0)}(x, t)^2 - u^{(4,0)}(x, t)\right\}$$

3.5 Numerical Solution of the Korteweg–de Vries Equation

Our considerations of the solutions of the KdV equations have so far been restricted to reflectionless potentials and thus we have used a special type of potential (Pöschel–Teller potential) in the analytic calculations. In this section, we examine solutions of the KdV equation for arbitrary potentials $u(x, 0)$. For an arbitrary potential $u(x, 0)$, we cannot expect the reflection coefficient to be $b(k) = 0$. For a reflectionless potential, we solve the Marchenko equation by a separation ansatz. For $b(k) \neq 0$, however, there is no analytic procedure available to solve the Marchenko equation. In this case, the KdV equation can be solved numerically. There are several procedures for finding numerical solutions of the KdV equation. An overview of the various integrating methods is given by Taha and Ablowitz [3.4].

Nonlinear evolution equations are solvable by a pseudospectral method or by difference methods. With respect to the difference methods, there are several versions of the standard Euler method known as leap-frog and Crank–Nicolson procedures. For our numerical solution of the KdV equation, we use the leap-frog procedure as developed by Zabusky and Kruskal [3.5].

All of the difference methods represent the continuous solution $u(x, t)$ for discrete points in space and time. In the process of discretization, the space and time coordinates are replaced by $x = m h$ and $t = n k$. $m = 0, 1, ..., M$, $n = 0, 1, 2, ...$, h, and k determine the step lengths in the spatial and temporal directions. Since the x domain of integration is restricted to an interval of finite length, we choose $h = 2\pi/M$ for the step length in the x-direction. The continuous solution $u(x, t)$ is approximated for each integration step by $u(x, t) = u_m^n$; that is, steps h and k have to be chosen properly to find convergent solutions as follows.

All discretization procedures differ in the representation of their derivatives. The main challenge of the discretization procedure is to find the proper representation of the needed derivatives. Errors are inevitable in this step and we have to settle for an approximate solution. Various representations of the derivatives give us a varying degree of accuracy for the representation of the solution. The leap-frog method of

$$u_t - 6 u u_x + u_{xxx} = 0 \tag{3.5.1}$$

by the formula

$$u_m^{n+1} = u_m^{n-1} + \frac{6k}{3k} (u_{m+1}^n + u_m^n + u_{m-1}^n) (u_{m+1}^n - u_{m-1}^n) -.$$
$$\frac{k}{h^3} (u_{m+2}^n - 2 u_{m+1}^n + 2 u_{m-1}^n - u_{m-2}^n). \tag{3.5.2}$$

The first term on the right-hand side of Equaton (3.5.2) represents the first derivative with respect to time. The second term gives a representation of the nonlinearity in the KdV equation. The last term in the sum of the right-hand side describes the dispersion term of third order in the KdV. The main advantage of the Zabusky and Kruskal procedure is the conservation of mass in the integration process $\sum_{m=0}^{M-1} u_m^n$. Another aspect of this discretization procedure is the representation of nonlinearity by $\frac{1}{3} (u_{m+1}^n + u_m^n + u_{m-1}^n)$. In this representation, the energy is conserved up to second order:

$$\frac{1}{2} \sum_{m=0}^{M-1} (u_m^n)^2 - \frac{1}{2} \sum_{m=0}^{M-1} (u_m^{n-1})^2 = O(k^3) \quad \text{for} \quad k \to 0 \tag{3.5.3}$$

if u is periodic or vanishes sufficiently rapidly at the integration end points. Since the Zabusky and Kruskal procedure is a second-order method in the time domain, we face the problem of specifying the initial conditions for the terms u_m^n and u_m^{n-1}. This problem can be solved if we use as a first step of integration an Euler procedure given by

$$u_m^{n+1} = u_m^n + \frac{6k}{3k} (u_{m+1}^n + u_m^n + u_{m-1}^n) (u_{m+1}^n - u_{m-1}^n) -$$
$$\frac{k}{h^3} (u_{m+2}^n - 2 u_{m+1}^n + 2 u_{m-1}^n - u_{m-2}^n). \tag{3.5.4}$$

To find stable solutions for this integration process, we have to choose the time and space steps appropriately. If we assume linear stability of the solution procedure, we have to take the following relation into account:

$$k \le \frac{h^3}{4 + h^2 |u|}, \tag{3.5.5}$$

where $|u|$ denotes the maximum magnitude of u. The process of integration includes the following steps:

1. Create the initial conditions.

2. Execute the first step of the integration by applying the simple Euler procedure using relations (3.5.4).

3. Iterate the following steps by using Equation (3.5.2).

4. Create a graphical representation of the results for equal time intervals.

The above four steps for integrating the KdV equation are contained in the package **KdVNumeric`**. **KdVNIntegrate[]** activates the integration process. **KdVNIntegrate[]** needs steps h and k, the number of points used in the x domain, and the initial solution for $t = 0$ as input parameters. Results of an integration with the initial condition $u(x, 0) = -6 \operatorname{sech}^2 x$ are given in Figure 3.5.1. As we know from our analytical considerations in the previous section, we expect a bisoliton solution. Choosing a larger amplitude in the initial condition $u(x, 0) = -10 \operatorname{sech}^2 x$, we get two solution components. In addition to the soliton properties, we observe a radiation solution in Figure 3.5.2. The radiation part of the solution moves in the opposite direction to that of the soliton and decreases in time.

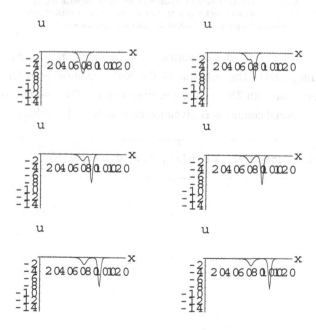

Figure 3.5.1.   Numerical solution of the KdV equation for the initial condition $u(x, 0) = -6 \operatorname{sech} x$. The time points shown from left to right and top to bottom are $t = \{0, 0.16, 0.32, 0.64\}$. The calculation is based on 128 points in the x domain corresponding to a step size of $h = 0.2$. The steps in the time domain are $k = 0.002$.

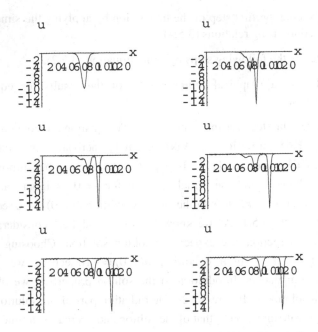

Figure 3.5.2. Numerical solution of the KdV equation for the initial condition $u(x, 0) = -10 \operatorname{sech} x$. The time points shown from left to right and top to bottom are $t = \{0, 0.16, 0.32\}$. The calculation is based on 128 points in the x-domain with a step size of $h = 0.2$.

The following cell demonstarates the application of the function **KdVNIntegrate[]**. The solution of the KdV equation is generated on a spatial grid line with 256 points. The time step is 0.001 and the spatial step is 0.2. The initial condition is given by the function $-12 \operatorname{sech}(x)$.

```
KdVNIntegrate[-12 Sech[x], 0.2, 0.001, 256]
```

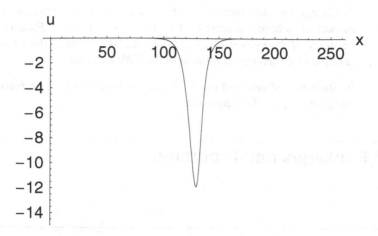

We observe fom the results that a four soliton plus and radiation is generated. The three solitons move to the right where the radiation moves to the left.

3.6 Exercises

1. Using the package **KdVEquation`**, find the type of differential equation for approximating orders $n \geq 3$. Does this approximation change the nonlinearity of the equation? What kinds of effect occur in higher approximations?

2. Change the package **KdVEQuation`** so that you can treat arbitrary dispersion relations. *Caution*: Make a copy of the original package first!

3. Examine the motion of the four solitons of the KdV equation. Study the phase gap in the contour plot of the four solitons.

4. Demonstrate that the odd densities of the conservation laws of the KdV equation w_{2n+1} ($n = 0, 1, 2, \ldots$) are total differentials of the w_{2n}'s.

5. Reexamine the determination of eigenvalues for the anharmonic oscillator. Discuss the link between the eigenvalue problem and the KdV equation.

6. Derive a single soliton solution by using the inverse scattering method for the KdV equation.

7. Examine the numerical solution of the KdV equation for initial conditions which do not satisfy $b(k) = 0$.

8. Change the step intervals in the space and time parameters of the numerical solution procedure for the KdV equation. Examine the accuracy of the numerical integration process. Compare the numerical solution to the analytical solution of the KdV equation.

9. Study the influence of the discretization number M in the numerical integration of the KdV equation.

3.7 Packages and Programs

3.7.1 Solution of the KdV Equation

The following package implements the solution steps for the KdV equation discussed in Section 3.3:

```
BeginPackage["KdVAnalytic`"];

Clear[PlotKdV,c2,Soliton];

Soliton::usage = "Soliton[x_,t_,N_] creates the N
soliton solution of the KdV
equation.";

PlotKdV::usage = "PlotKdV[tmin_,tmax_,dt_,N_]
calculates a sequence of
pictures for the N soliton solution of the KdV
equation. The time interval
of the representation is [tmin,tmax]. The variable
dt measures the length
of the time step.";

Begin["`Private`"];

(* --- squares of the normalization constants c_n
--- *)

c2[n_, N_] := Block[{h1,x},
    h1 = LegendreP[N, n, x]^2/(1-x^2);
    h1 = Integrate[h1, {x, -1, 1}]]

(* --- N soliton solution --- *)
```

```
Soliton[x_,t_,N_] :=
  Block[{cn,A,x,t,deltanm,u},
(* --- calculate normalization constants --- *)
    cn = Table[c2[i, N], {i, 1, N}];
(* --- create the coefficient matrix A --- *)
    A = Table[
          If[n==m, deltanm = 1, deltanm=0];
          deltanm + (cn[[m]] Exp[8 m^3 t - (m + n)
x])/(m + n),
          {m, 1, N}, {n, 1, N}];
(* --- determine the solution --- *)
    u = -2 D[Log[Det[A]],{x,2}];
    u = Expand[u];
    u = Factor[u]]


(* --- time series of the N soliton solution --- *)

PlotKdV[tmin_,tmax_,dt_,N_]:=Block[{p1,color,u},
(* --- create the N soliton --- *)
    u = Soliton[x,t,N];
(* --- plot the N soliton --- *)
    Do[

If[t>0,color=RGBColor[0,0,1],color=RGBColor[1,0,0]];
    Plot[-u,{x,-20,20},PlotRange->{0,15},
          AxesLabel->{"x","-u(x,t)"},
          DefaultColor->Automatic,
          PlotStyle->{{Thickness[1/170],color}}],
    {t,tmin,tmax,dt}]]
End[];
EndPackage[];
```

3.7.2 Conservation Laws for the KdV Equation

The following package is an implementation of the determination of conservation laws for the KdV equation discussed in Section 3.4:

```
BeginPackage["KdVIntegrals`"];

Clear[Gardner];

Gardner::usage = "Gardner[u_,x_,t_,N_] calculates
the densities of the integrals of
motion for the KdV equation using Gardner's method.
```

```
The integrals are
determined up to the order N. u, x, t are the
symbols for dependent and independet variables,
respectively.";


Begin["`Private`"];


Gardner[u_,x_,t_,N_] :=
        Block[{expansion,eps,x,t,sublist
={},list1={},list2},
        list2=Table[1, {i,1,N+1}];

(* --- representation of a Gardner expansion  --- *)
        expansion = Expand[
            Sum[eps^n w[x,t,n] - eps^(n+1)
D[w[x,t,n],x],
            {n,0,N}] -
            eps^2 (Sum[eps^n w[x,t,n], {n,0,N}])^2 -
u[x,t]
            ];
(* --- compare coefficients --- *)
        Do[AppendTo[list1,

Expand[Coefficient[expansion,eps,i]-w[x,t,i]]],
        {i,0,N}];
        list2[[1]] = -list1[[1]];
(* --- define replacements and application of the
replacements --- *)
        Do[sublist={};

Do[AppendTo[sublist,w[x,t,i]->list2[[i+1]] ],
            {i,0,N}];

AppendTo[sublist,D[w[x,t,n],x]->D[list2[[n+1]],x]];
        list2[[n+2]] = list1[[n+2]] /. sublist,
        {n,0,N-1}];
        list2
        ];
End[];
EndPackage[];
```

3.7.3 Numerical Solution of the KdV Equation

The following package provides functions for the numerical solution of the KdV equation discussed in Section 3.5:

```
BeginPackage["KdVNumeric`"];

Clear[KdVNIntegrate];

KdVNIntegrate::usage =
"KdVNIntegrate[initial_,dx_,dt_,M_] carries out a
numerical
integration of the KdV equation using the procedure
of Zabusky & Kruskal.
The input parameter initially determines the initial
solution in the procedure;
e.g. -6 Sech^2[x]. The infinitesimals dx and dt are
the steps with respect
to the spatial and temporal directions. M fixes the
number of steps along
the x-axis.";

Begin["`Private`"];

KdVNIntegrate[initial_,dx_,dt_,M_]:=Block[
            {uPresent, uPast, uFuture, initialh, m,
n},
(* --- transform the initial conditions on the grid
--- *)
      initialh = initial /. f_[x_] -> f[(m-M/2) dx];
      h = dx;
      k = dt;
(* --- calculate the initial solutions on the grid
--- *)
      uPast = Table[initialh, {m,1,M}];
(* --- initialization of the lists containing the
grid points
            uPresent = present   (m)
            uFuture = future     (m+1)
            uPast = past         (m-1)           --- *)
      uPresent = uPast;
      uFuture = uPresent;
      ik = 0;
(* --- iteration for the first step --- *)
      Do[
      uPresent[[m]] = uPast[[m]] + 6 k (uPast[[m+1]] +
            uPast[[m]] + uPast[[m-1]])
            (uPast[[m+1]] - uPast[[m-1]])/(3 h) -
```

```
                k (uPast[[m+2]] - 2 uPast[[m+1]] + 2
uPast[[m-1]] -
                uPast[[m-2]])/h^3,
        {m,3,M-2}];
(* --- iterate the time --- *)
        Do[
(* --- iterate the space points --- *)
          Do[
        uFuture[[m]] = uPast[[m]] + 6 k
(uPresent[[m+1]] +
                uPresent[[m]] + uPresent[[m-1]])
                (uPresent[[m+1]] - uPresent[[m-1]])/(3
h) -
                k (uPresent[[m+2]] - 2 uPresent[[m+1]]
+
                2 uPresent[[m-1]] -
                uPresent[[m-2]])/h^3,
        {m,3,M-2}];
(* --- exchange lists --- *)
        uPast = uPresent;
        uPresent = uFuture;
(* --- plot a time step --- *)
        If[Mod[n,40] == 0,
          ik = ik + 1;
(*--- plots are stored in a[1], a[2], ... a[6] ---*)
          a[ik] = ListPlot[uFuture,
                    AxesLabel->{"x","u"},
                    Prolog->Thickness[0.001],
                    PlotJoined->True,
                    PlotRange->{-15,0.1}]],
        {n,0,500}]
        ];
End[];
EndPackage[];
```

References

Volume I

Chapter 1

[1.1] S. Wolfram, The *Mathematica* book, 5th ed. Wolfram Media/Cambridge University Press, Cambridge, 2003.

[1.2] M. Abramowitz and I.A. Stegun, Handbook of Mathematical Functions. Dover Publications, New York, 1968.

[1.3] N. Blachman, *Mathematica*: A Practical Approach. Prentice-Hall, Englewood Cliffs, 1992.

[1.4] Ph. Boyland, A. Chandra, J. Keiper, E. Martin, J. Novak, M. Petkovsek, S. Skiena, I. Vardi, A. Wenzlow, T. Wickham-Jones, D. Withoff, and others, Technical Report: Guide to Standard *Mathematica* Packages, Wolfram Research, Champaign, 1993.

Chapter 2

[2.1]   R. Maeder, Programming in *Mathematica*. Addison-Wesley, Redwood City, CA,1991.

[2.2]   L.D. Landau and E.M. Lifshitz, Mechanics. Addison-Wesley, Reading, MA, 1960.

[2.3]   J. B. Marion, Classical Dynamics of Particles and Systems. Academic Press, New York, 1970.

[2.4]   R. Courant and D. Hilbert, Methods of Mathematical Physics, Vols. 1 and 2. Wiley–Interscience, New York, 1953.

[2.5]   R.H. Dicke, Science **124**, 621 (1959).

[2.6]   R.V. Eötvös, Ann. Phys. **59**, 354 (1896).

[2.7]   L. Southerns, Proc. Roy. Soc. London, **A84**, 325 (1910).

[2.8]   P. Zeeman, Proc. Amst., **20**, 542 (1917).

[2.9]   G. Baumann, Symmetry Analysis of Differential Equations Using *Mathematica*. Springer-Verlag, New York, 2000.

[2.10] H. Geiger and E. Marsden, The laws of deflexion of α particles through large angles. Phil. Mag., **25**, 605 (1913).

[2.11] Ph. Blanchard and E. Brüning, Variational Methods in Mathematical Physics. Springer-Verlag, Wien, 1982.

Chapter 3

[3.1]   F. Calogero and A. Degasperis, Spectral Transform and Solitons: Tools to Solve and Investigate Nonlinear Evolution Equations. North-Holland, Amsterdam, 1982.

[3.2]   V.A. Marchenko, On the reconstruction of the potential energy from phases of the scattered waves. Dokl. Akad. Nauk SSSR, **104**, 695 (1955).

[3.3] R.M. Miura, C. Gardner, and M.D. Kruskal. Korteweg–de Vries equation and generalizations. II. Existence of conservation laws and constants of motion. J. Math. Phys., **9**, 1204 (1968).

[3.4] T.R. Taha and M.J. Ablowitz, Analytical and numerical solutions of certain nonlinear evolution equations. I. Analytical. J. Comput. Phys., **55**, 192 (1984).

[3.5] N.J. Zabusky and M.D. Kruskal, Interactions of 'solitons' in a collisionless plasma and the recurrence of initial states. Phys. Rev. Lett. **15**, 240 (1965).

Volume II

Chapter 4

[4.1] G. Arfken, Mathematical Methods for Physicists. Academic Press, New York, 1966.

[4.2] P.M. Morse and H. Feshbach, Methods of Theoretical Physics. McGraw-Hill, New York, 1953.

[4.3] W. Paul, O. Osberghaus, and E. Fischer, Ein Ionenkäfig. Forschungsbericht des Wissenschafts- und Verkehrsministeriums Nordrhein-Westfalen, **415**, 1 (1958). H. G. Dehmelt, Radiofrequency Spectroscopy of stored ions I: Storage. Adv. Atomic Mol. Phys., **3**, 53 (1967). D. J. Wineland, W.M. Itano and R.S. van Dyck Jr., High-resolution spectroscopy of stored ions, Adv. Atomic Mol. Phys., **19**, 135 (1983).

[4.4] F.M. Penning, Die Glimmentladung bei niedrigem Druck zwischen koaxialen Zylindern in einem axialen Magnetfeld. Physica **3**, 873 (1936). D. Wineland, P. Ekstrom, and H. Dehmelt, Monoelectron oscillator, Phys. Rev. Lett., **31**,1279 (1973).

[4.5]   G. Baumann, The Paul trap: a completely integrable model? Phys. Lett. **A 162**, 464 (1992).

Chapter 5

[5.1]   E. Schrödinger, Quantisierung als Eigenwertproblem. Ann. Phys., **79**, 361 (1926).

[5.2]   N. Rosen and P.M. Morse, On the vibrations of polyatomic molecules. Phys. Rev., **42**, 210 (1932).

[5.3]   G. Pöschel and E. Teller, Bemerkungen zur Quantenmechanik des anharmonischen Oszillators. Z. Phys., **83**, 143 (1933).

[5.4]   W. Lotmar, Zur Darstellung des Potentialverlaufs bei zweiatomigen Molekülen. Z. Phys., **93**, 518 (1935).

[5.5]   S. Flügge, Practical Quantum Mechanics I and II. Springer-Verlag, Berlin, 1971.

[5.6]   C. Cohen-Tannoudji, B. Diu, and F. Laloë, Quantum Mechanics I and II. John Wiley & Sons, New York, 1977.

[5.7]   J.S. Rowlinson, Mol. Phys., **6**, 75 (1963).

[5.8]   J.E. Lennard-Jones, Proc. Roy. Soc., **A106**, 463 (1924).

[5.9]   F. London, Z. Phys., **63**, 245 (1930).

[5.10] J.O. Hirschfelder, R.F. Curtiss, and R.B. Bird, Molecular Theory of Gases and Liquids. Wiley & Sons, New York, 1954.

[5.11] E.A. Mason and T.H. Spurling, The Virial Equation of State. Pergamon Press, Oxford, 1969.

[5.12] D.A. McQuarrie, Statistical Thermodynamics. Harper and Row, New York 1973, p. 307.

[5.13] O. Sinanoglu and K.S. Pitzer, J. Chem. Phys., **31**, 960 (1959).

[5.14]   D.G. Friend, J. Chem. Phys., **82**, 967 (1985).

[5.15]   T. Kihara, Suppl. Progs. Theor. Phys., **40**, 177 (1967).

[5.16]   D.E. Stogryn and J.O. Hirschfelder, J. Chem. Phys., **31**, 1531 (1959).

[5.17]   R. Phair, L. Biolsi, and P.M. Holland, Int. J. Thermophys., **11**, 201 (1990).

[5.18]   F.H. Mies and P.S. Julienne, J. Chem. Phys., **77**, 6162 (1982).

Chapter 6

[6.1]   W. Rindler, Essential Relativity. Springer-Verlag, New York, 1977.

[6.2]   C.W. Misner, K.S. Thorne, and J.A. Wheeler, Gravitation. Freeman, San Francisco, 1973.

[6.3]   H. Stephani, General Relativity: An Introduction to the Gravitational Field. Cambridge University Press, Cambridge, 1982.

[6.4]   M. Berry, Principles of Cosmology and Gravitation. Cambridge University Press, Cambridge, 1976.

Chapter 7

[7.1]   T.W. Gray and J. Glynn, Exploring Mathematics with *Mathematica*. Addison-Wesley, Redwood City, CA, 1991.

[7.2]   T.F. Nonnenmacher, G. Baumann, and G. Losa, Self organization and fractal scaling patterns in biological systems. In: Trends in Biological Cybernetics, World Scientific, Singapore, Vol. 1, 1990, p. 65.

[7.3]   A. Barth, G. Baumann, and T.F. Nonnenmacher, Measuring Rényi-dimensions by a modified box algorithm. J. Phys. A: Math. Gen., **25**, 381 (1992).

[7.4]   B. Mandelbrot, The Fractal Geometry of Nature. W.H. Freeman, New York, 1983.

[7.5]   A. Aharony, Percolation. In: Directions in Condensed Matter Physics (Eds. G. Grinstein and G. Mazenko). World Scientific, Singapore, 1986.

[7.6]   T. Grossman and A. Aharony, Structure and perimeters of percolation clusters. J. Phys. A: Math. Gen., **19**, L745 (1986).

[7.7]   P.G. Gennes, Percolation – a new unifying concept. Recherche, **7**, 919 (1980).

[7.8]   S.F. Lacroix, Traité du Calcul Différentiel et du Calcul Intégral. 2nd ed., Courcier, Paris, 1819, Vol. 3, pp. 409–410.

[7.9]   L. Euler, De progressionibvs transcendentibvs, sev qvarvm termini generales algebraice dari negvevnt. Comment Acad. Sci. Imperialis Petropolitanae, **5**, 36, (1738).

[7.10]   K.B. Oldham and J. Spanier, The Fractional Calculus. Academic Press, New York, (1974).

[7.11]   K.S. Miller and B. Ross, An Introduction to the Fractional Calculus and Fractional Differential Equations. John Wiley & Sons, New York, 1993.

[7.12]   G.F.B. Riemann, Gesammelte Werke. Teubner, Leipzig, 1892, pp.353–366,.

[7.13]   J. Liouville, Mémoiresur le calcul des différentielles à indices quelconques. J. École Polytech., **13**, 71 (1832).

[7.14]   H. Weyl, Bemerkungen zum Begriff des Differentialquotienten gebrochener Ordnung. Vierteljahresschr. Naturforsch. Ges. Zürich, **62**, 296 (1917).

[7.15] H.T. Davis, The Theory of Linear Operators. Principia Press, Bloomington, 1936.

[7.16] B. Riemann, Über die Anzahl der Primzahlen unter einer gegebenen Größe. Gesammelte Math. Werke, 136-144, (1876).

[7.17] E. Cahen, Sur la fonction $\zeta(s)$ de Riemann et sur des fonctions analoges. Ann. Ecole Normale, **11**, 75 (1894).

[7.18] H. Mellin, Über die fundamentale Wichtigkeit des Satzes von Cauchy für die Theorie der Gamma- und der hypergeometrischen Funktion. Acta Soc. Fennicae, **21**, 1 (1896).

[7.19] H. Mellin, Über den Zusammenhang zwischen den linearen Differential- und Differenzengleichungen. Acta Math., **25**, 139 (1902).

[7.20] F. Oberhettinger, Mellin Transforms. Springer-Verlag, Berlin, 1974.

[7.21] G. Baumann, Symmetry Analysis of Differential Equations using *Mathematica*. Springer-Verlag, New York, 2000.

[7.22] J.B. Bates and Y.T. Chu, Surface topography and electrical response of metal-electrolyte interfaces. Solid State Ionics, **28-30**, 1388 (1988).

[7.23] H. Scher and E.W. Montroll, Anomalous transit-time dispersion in amorphous solids. Phys. Rev. B, **12**, 2455 (1975).

[7.24] K.S. Cole and R.H. Cole, Dispersion and absorption in dielectrics. J. Chem. Phys., **9**, 341 (1941).

[7.25] W.G. Glöckle, Anwendungen des fraktalen Differentialkalküls auf Relaxationen. PhD Thesis, Ulm, 1993.

[7.26] R. Metzler, Modellierung spezieller dynamischer Probleme in komplexen Materialien. PhD Thesis, Ulm, 1996.

[7.27]   H. Schiessel and A. Blumen, Mesoscopic pictures of the sol-gel transition: Ladder models and fractal networks. Macromolecules, **28**, 4013 (1995).

[7.28]   T.F. Nonnenmacher, On the Riemann-Liouville fractional calculus and some recent applications. Fractals, **3**, 557 (1995).

[7.29]   B.J. West and W. Deering, Fractal physiology for physicists: Lévy statistics. Phys. Rep. **246**, 1 (1994).

[7.30]   W. Wyss, The fractional diffusion equation. J. Math. Phys., **27**, 2782 (1986).

[7.31]   B. O'Shaugnessy and I. Procaccia, Analytical solutions for diffusion on fractal objects. Phys. Rev. Lett., **54**, 455 (1985).

[7.32]   W.R. Schneider and W. Wyss, Fractional diffusion and wave equations. J. Math. Phys., **30**, 134 (1989).

[7.33]   R. Metzler, W.G. Glöckle, and T.F. Nonnenmacher, Fractional model equation for anomalous diffusion. Physica, **211A**, 13 (1994).

[7.34]   A. Compte, Stochastic foundations of fractional dynamics. Phys. Rev. E, **53**, 4191 (1996).

[7.35]   B.J. West, P. Grigolini, R. Metzler, and T.F. Nonnenmacher, Fractional diffusion and Lévy stable processes. Phys. Rev. E, **55**, 99 (1997).

Index